Plant Nematodes
Methodology, Morphology, Systematics, Biology and Ecology

Plant Nematodes
Methodology, Morphology, Systematics, Biology and Ecology

Mujeebur Rahman Khan
Department of Plant Protection
Faculty of Agricultural Sciences
Aligarh Muslim University
Aligarh, India

Enfield (NH)　　　Jersey　　　Plymouth

SB
998
.N4
K43
2008

Science Publishers *www.scipub.net*
234 May Street
Post Office Box 699
Enfield, New Hampshire 03748
United States of America

General enquiries : *info@scipub.net*
Editorial enquiries : *editor@scipub.net*
Sales enquiries   : *sales@scipub.net*

Published by Science Publishers, Enfield, NH, USA
An imprint of Edenbridge Ltd., British Channel Islands
Printed in India

© 2008 reserved

ISBN 978-1-57808-533-0

Library of Congress Cataloging-in-Publication Data

Khan, Mujeebur Rahman.
  Plant nematodes: methodology, morphology, systematics, biology and ecology/Mujeebur Rahman Khan.
     p.cm.
  Includes bibliographical references and index.
  ISBN 978-1-57808-533-0 (hardcover)
  1. Plant nematodes. I. Title.
   SB998.N4K43 2008
   632'.6257--dc22

2008008868

All rights reserved. No part of this publication may be reproduced,
stored in a retrieval system, or transmitted in any form or by any
means, electronic, mechanical, photocopying or otherwise, without
the prior permission of the publisher, in writing. The exception to
this is when a reasonable part of the text is quoted for purpose of
book review, abstracting etc.

This book is sold subject to the condition that it shall not, by way
of trade or otherwise be lent, re-sold, hired out, or otherwise circulated
without the publisher's prior consent in any form of binding or cover
other than that in which it is published and without a similar condition
including this condition being imposed on the subsequent purchaser.

Affectionately and Respectfully to
Abba and Ammi

Preface

The current trend is to publish edited books with chapters contributed by different authors, each providing detailed information on a specific topic, especially of an applied nature. Such contributions are useful to researchers in a particular area, but less helpful to a vast majority that includes largely undergraduate and graduate students, general researchers and teachers who face a lot of difficulty in acquiring basic and fundamental knowledge of the subject. The task of collecting relevant information becomes cumbersome, as one has to search, consult and/or surf through a large number of books/journals/websites, and sometimes have to rely on obsolete and ageing material. In the current age, authored books are seldom published, as this contribution necessarily needs an author who has broad and in-depth knowledge, coupled with a strong temperament of dedication and hard work.

Plant Nematodes... is an authored book, which is written to fulfill the needs of students, general researchers and teachers pursuing study in Plant Nematology. The author holds Ph.D. in Nematology and has been teaching various aspects of Plant Nematology at PG level for the last 20 years. This book is prepared with the objective to alleviate the difficulties teachers or students face while preparing a lecture or notes. It explicates the basic aspects of Nematology, viz., Methodology, Morphology, Systematics and Classification, Biology, Physiology and Ecology. It is designed and written in an explicit manner so that UG students who do not have a nematology background can easily perceive the basics of the discipline, and at the same time students of higher classes as well as teachers, scientists and researchers could get in-depth and up-to-date information on various fundamental and advanced aspects of nematology. Using all means of information, old and recent literature has been searched to compile the chapters. To make a chapter understandable and palpable, the descriptions are supported by simple and self-explanatory figures and diagrams. The book contains over 80 schematic and natural diagrams. The first chapter, Introduction gives general information and historic development of plant nematology with enumeration of nematological societies, research journals and important books. This chapter also elaborates the agricultural importance of plant nematodes. The second chapter is on Methodology and Techniques that describes various methods and procedures of isolation, fixing, staining and mounting of free-living, plant parasitic and entomopathogenic nematodes supported by numerous diagrams; preparation of nematodes for electron microscopy and PCR based tests for molecular studies of nematodes. All kinds of minor to major morphological structures of nematodes with the help of diagrams are described explicitly in the third chapter, Morphology of Nematodes. The next chapter, Systematics and Classification provides basic information on the subject, its history and a brief description of the International Code of Nomenclature followed by the classification from the phylum to family along with the key of identification. Thereafter, the chapter, Biology of Nematodes deals with the nervous, digestive, excretory and reproductive systems and physiology of nematodes. This chapter also covers development of nematodes. In the chapter Feeding and Trophic Relationship, feeding behaviour of nematodes, mode of parasitism and host parasite relationship with regard to morphological and physiological modifications in the host have been discussed in detail. The last chapter in the book is on the Ecology of Nematodes that describes the survival, reproduction and population buildup of nematodes. The chapter contains detailed information on population dynamics of nematodes, and their relationship with biotic and abiotic components of the ecosystem, and the interaction with plant bacteria, fungi, viruses, insects, rhizobium, mycorrhizal fungi and cohabiting phytonematodes and entomopathogenic nematodes. To ease in the interpretation and to understand various abbreviations, terms, etc. used in the text, an exhaustive list of elaborated abbreviations and symbols, and Glossary of relevant terms have been given. A list of references cited and Subject Index have been presented at the end.

15 August, 2008   *Mujeebur Rahman Khan*

Acknowledgements

First of all, I beg to express profound gratitude and indebtedness to the Almighty Allah; without his benevolence and consecration that guided me through all channels to work in cohesion and coordination, this publication would never have been possible.

I wish to start the acknowledgement section by thanking the scientists of nematology and allied disciplines for their immensely important researches which served as an invaluable base line material for the preparation of manuscript of the present book, **Plant Nematodes...**, and for granting kind permission to use the information especially the photographs. I have tried to acknowledge the authors specifically in the figure caption, but at a few instances I had failed because of non availability of the source, hence the anonymous authors are acknowledged with sincere thanks for their contribution used in the book. I wish to express my appreciation for the valuable assistance provided by my students, Fayaz and Shahana during preparation of this document, and also by Reshu, Mahmud, Arshad, Uzma and Abdussalam. The moral support extended by my Ph.D. guide, Prof. M. Wajid Khan (Late), Founder Nematologist Prof. Abrar Mustafa Khan, Founder Director Prof. M. S. Jairajpuri, Ex-Director Prof. G. M. Khan, Dr. M. R. Siddiqi (CABI), Prof. J.N. Sasser, Prof. R.A. Reinert (NCSU, USA), Dr. John J. Chitambar (CDFA, USA) and colleagues/friends, Dr. Akhtar Haseeb, Dr. M. Shafiq Ansari, Dr. Shabbir Ashraf, Dr. Abrar Khan, Dr. Badsha Khan, Dr. Qayyum Husain, Dr. Saghir Khan, Dr. Akram A. Khan, Dr. A. Malik, Dr. Iqbal Ahmad, Dr. P.Q. Rizvi, Dr. R.U. Khan and Dr. M. Haseeb, has been a strength in sustaining my academic activities, especially in adversities. I wish to put on records my sincere thanks to all the teachers especially Prof. Ziauddin A. Siddiqi, Prof. A.K.M. Ghouse, Prof. Wazahat Husain, Prof. Saeed A. Siddiqi, Prof. S. K. Saxena, Prof. Khalid Mahmood, Prof. S. Israr Husain, Prof. M. Akram, Prof. Mashkoor Alam, Prof. M.M.R.K. Afridi, Prof. Ziauddin Ahmad, Prof. M. Iqbal (HU), Prof. S. Q. A. Naqvi, Prof. A. K. M. Ghose, Prof. Samiullah, Prof. Ainul Haq Khan, Prof. Arif Inam, Prof. Aqil Ahmad, Prof. Saleemuddin, Prof. Shujauddin, Prof. Bahar A.

Siddiqi, Dr. Aslam Pervez, Dr. Adam Shafi, Dr. Shakeel A. Khan and others who taught me during graduate and post-graduate studies and contributed a lot to make me finally a nematologist. A partial grant sanctioned by the Aligarh Muslim University, to prepare the manuscript of the present book for publication is gratefully acknowledged. The help offered by M/s Amir, Mr. Raghib, Aman Sb, Akbar, Iftitah, Sabir, Adil, Shariq and Ambar is acknowledged with appreciation.

Above all, I am highly indebted to my beloved Ammi and Abba for their savvy tutelage and benedictions throughout my life that I do miss today. They inspired and encouraged me for dedication in research, and felt a pride whatever little I achieved ever. I am grateful to my brothers and sisters, Bhaijan, Bajibi, Chotebhaijan and Chotibaji, without their care, help and love I could not have been whatever I am today. I immensely thank Asrar Bhai, Khalid Bhai and Afaq Bhai, my Bhabies, Afsar, Afzal, Khalas and Mamoos for the affection, appreciation and concern shown for me. The love for me and interest in my work expressed by my loving youngers, Fahad, Taimoor, Sheena, Saif, Saniya, Juveria, Imad, Saad and Jumana have been invaluable assets. I am extremely thankful to my wife Huma and children Arham, Ayesha and Ahzam for the support, care and affection showered upon me, and apologize for the neglect and loss they may have suffered due to the extra time that I had to devote for writing of this book.

Mujeebur Rahman Khan

Contents

Preface	*vii*
Acknowledgements	*ix*
1. INTRODUCTION	**1**
Introduction to Plant Nematology	1
History of Plant Nematology	3
Agricultural Importance of Plant Nematodes	8
2. METHODOLOGY	**14**
Sampling for Nematode Assay	14
Soil Sampling Technique	14
Sampling Pattern	14
Storage of Samples	16
Soil Samples	16
Plant Material	16
Extraction of Nematodes from Soil	17
Extraction of Vermiform Nematodes	17
Elutriation Techniques	18
Extraction of Cysts of *Heterodera* and *Globodera*	24
Extraction of Nematodes from Plant Tissue	27
Direct Examination of Plant Material	27
Baermann Funnel Technique	27
Maceration Filtration Technique	28
Mistifier Extraction Technique	28
Root Incubation Technique	29
Shoot Incubation Technique	29
Matchstick Extraction Technique	29
Extraction of Cysts and Egg Masses	30
Extraction of Eggs of Root-Knot Nematodes	30

Extraction of Entomopathogenic Nematodes from Soil	30
Collection of Soil Samples for EPN	30
Isolation of EPN	31
Rearing of EPN	31
Bioassay of EPN	32
Processing of Nematodes	33
Picking of Vermiform Nematodes	33
Anaesthetizing and Killing of Nematodes	33
Fixing of Nematodes	34
Processing of Nematodes for Mounting	36
Mounting of Nematodes	38
Preparation of Enface View, Perineal Pattern and Vulval Cone	40
Enface View	40
Perineal Pattern	40
Vulval Cone or Cone Top	41
Preservation and Staining of Nematodes in Plant Material	41
Fixation and Preservation of Plant Material	41
Staining of Nematodes in Plant Tissue	42
Electron Microscopy of Nematodes	43
Transmission Electron Microscopy of Nematodes	45
Scanning Electron Microscopy of Nematodes	48
Biochemical and Molecular Techniques	52
Biochemical Techniques	52
Molecular Techniques (DNA Study)	54
Isolation of DNA from Nematodes	55
Amplification of DNA Isolated from Nematodes	56
Gel Electrophoresis	59
Agarose Gel Electrophoresis	59
Running of Electrophoresis	63
Polyacrylamide Gel Electrophoresis (PAGE)	64
Running of Electrophoresis	67
3. MORPHOLOGY OF NEMATODES	**70**
General Structure of Nematodes	70
Body Shape	71
Body Wall	72
Cuticle	73

Hypodermis	76
Somatic musculature	77
Cuticular Markings	78
Transverse Striations and Annulations	78
Longitudinal Striations	80
Longitudinal Lines	81
Head	82
Tail	84
4. SYSTEMATICS AND CLASSIFICATION OF NEMATODES	**87**
Proposed Classification of Nematodes	91
Phylum Nemata	91
Class 1: Adenophorea	100
Subclass A: Enoplia	100
Superorder 1: Marenoplica	101
Order 1: Enoplida	101
Superfamily 1: Oxystominoidea	101
Superfamily 2: Enoploidea	101
Order 2: Oncholaimida	102
Order 3: Tripylida	102
Suborder 1: Tripylina	102
Suborder 2: *Ironina inquirenda*	103
Superorder 2: Terrenoplica	103
Order 1: *Isolaimida inquirenda*	103
Order 2: Mononchida	103
Superfamily 1: Mononchoidea	104
Superfamily 2: Bathyodontoidea	104
Order 3: Dorylaimida	104
Suborder 1: Dorylaimina	105
Superfamily 1: Dorylaimoidea	105
Family: Longidoridae	106
Key to Subfamilies of longidoridae	106
Superfamily 2: Actinolaimoidea	107
Superfamily 3: Belondiroidea	107
Suborder 2: Diphtherophorina	107
Family: Trichodoridae	108
Suborder 3: Campydorina	108
Suborder 4: Nygolaimina	108

Order 4: Stichosomida	108
Superfamily 1: Trichocephaloidea	109
Superfamily 2: Mermithoidea	109
Superfamily 3: Echinomermelloidea	109
Subclass B: Chromadoria	109
Order 1: Chromadorida	110
Superfamily 1: Chromadoroidea	110
Superfamily 2: Choanolaimoidea	111
Order 2: Desmoscolecida	111
Order 3: Desmodorida	111
Superfamily 1: Desmodoroidea	111
Superfamily 2: Draconematoidea	112
Order 4: Monhysterida	112
Order 5: Araeolaimida	112
Suborder 1: Araeolaimina	113
Superfamily 1: Araeolaimoidea	113
Superfamily 2: Plectoidea	113
Class 2: Secernentea	114
Subclass A: Rhabditia	114
Order 1: Rhabditida	115
Suborder 1: Rhabditina	115
Superfamily 1: Rhabditoidea	115
Superfamily 2: Bunonematoidea	115
Superfamily 3: Cosmocercoidea	116
Superfamily 4: Oxyuroidea	116
Superfamily 5: Heterakoidea	116
Suborder 2: Cephalobina	116
Superfamily 1: Panagrolaimoidea	117
Order 2: Strongylida	117
Superfamily 1: Strongyloidea	117
Superfamily 2: Ancylostomatoidea	118
Superfamily 3: Trichostrongyloidea	118
Subclass B: Spiruria	118
Order 1: Ascaridida	119
Superfamily 1: Ascaridoidea	119
Superfamily 2: Seuratoidea	119
Superfamily 3: Camallanoidea	120

Superfamily 4: Dioctophymatoidea	120
Superfamily 5: Muspiceoidea *incertae sedis*	120
Order 2: Spirurida	121
Superfamily 1: Spiruroidea	121
Superfamily 2: Drilonematoidea	121
Superfamily 3: Physalopteroidea	122
Superfamily 4: Dracunculoidea	122
Superfamily 5: Diplotriaenoidea	122
Superfamily 6: Filarioidea	123
Subclass C: Diplogasteria	123
Order 1: Diplogasterida	124
Order 2: Tylenchida	124
Suborder 1: Tylenchina	125
Key to Infraorders of Tylenchina	125
Infraorder 1: Tylenchata	126
Key to Families of Tylenchoidea	126
Family: Tylenchidae	127
Key to Subfamilies of Tylenchidae	127
Family: Ecphyadophoridae	128
Key to Subfamilies of Ecphyadophoridae	128
Family: Atylenchidae	128
Key to Subfamilies of Atylenchidae	129
Family: Tylodoridae	129
Key to Subfamilies of Tylodoridae	129
Infraorder 2: Anguinata (Infraord. n.)	130
Key to Families of Anguinoidea	131
Key to Subfamilies of Anguinidae	131
Family: Sychnotylenchidae Neoditylenchidae	132
Suborder 2: Hoplolaimina	132
Key to Superfamilies of Hoplolaimina	133
Key to Families of Hoplolaimoidea	134
Key to Subfamilies of Hoplolaimidae	136
Key to Subfamilies of Rotylenchulidae	136
Family: Pratylenchidae	137
Key to Subfamilies of Pratylenchidae	138
Family: Meloidogynidae	138
Key to Subfamilies of Meloidogynidae	139
Family: Heteroderidae	139

Key to Subfamilies of Heteroderidae	140
Superfamily 2: Dolichodoroidea	140
Key to Families of Dolichodoroidea	142
Family: Dolichodoridae	142
Key to Subfamilies of Dolichodoridae	143
Family: Belonolaimidae	143
Family: Telotylenchidae	143
Key to Subfamilies of Telotylenchidae	144
Family: Psilenchidae	144
Key to Subfamilies of Psilenchinae	145
Suborder 3: Criconematina	145
Key to Superfamilies of Criconematina	146
Superfamily 1: Criconematoidea	146
Key to Subfamilies of Criconematidae	147
Superfamily 2: Hemicycliophoroidea	148
Key to Families of Hemicycliophoroidea	148
Family: Hemicycliophoridae	148
Family: Caloosiidae	149
Superfamily 3: Tylenchuloidea	149
Key to Families of Tylenchuloidea	150
Family: Tylenchulidae	150
Family: Sphaeronematidae	151
Key to Subfamilies of Sphaeronematidae	151
Family: Paratylenchidae Tylenchocriconematidae	151
Key to Subfamilies of Paratylenchidae	152
Suborder 4: Hexatylina Sphaerulariina	152
Key to Superfamilies of Hexatylina	154
Superfamily 1: Sphaerularioidea	154
Key to Families of Sphaerularioidea	155
Family: Neotylenchidae	155
Key to Subfamilies of Neotylenchidae	156
Family: Sphaerulariidae	156
Family: Paurodontidae	157
Family: Allantonematidae	157
Key to Subfamilies of Allantonematidae	158
Superfamily 2: Iotonchioidea	158
Key to Families of Iotonchioidea	159
Family: Iotonchiidae	159
Family: Parasitylenchidae	159

Key to Subfamilies of Parasitylenchidae	160
Order 3: Aphelenchida	160
Suborder 1: Aphelenchina	161
Key to Superfamilies of Aphelenchina	161
Superfamily 1: Aphelenchoidea	162
Key to Families of Aphelenchoidea	162
Family: Aphelenchidae	162
Family: Paraphelenchidae	162
Superfamily 2: Aphelenchoidoidea	163
Key to Families of Aphelenchoidoidea	163
Family: Aphelenchoididae	164
Key to Subfamilies of Aphelenchoididae	164
Family: Seinuridae	164
Family: Ektaphelenchidae	164
Family: Acugutturidae	165
Key to Subfamilies of Acugutturidae	165
Family: Parasitaphelenchidae	165
Key to Subfamilies of Parasitaphelenchidae	165
Family: Entaphelenchidae	166
5. BIOLOGY OF NEMATODES	**167**
Egg Development	167
Embryonic Development	167
Post-embryonic Development	169
Reproduction	170
Female Reproductive System	170
Male Reproductive System	174
Digestive System	178
Excretory System	187
Nervous System and Sensory Structures	189
Anterior Nervous System	190
Posterior Nervous System	191
Sensory Structures	192
Nematode Physiology	200
Respiration	200
Metabolism	201
Carbohydrate Metabolism	201
Lipid and Sterol Metabolism	204

Protein Metabolism	207
Physiology of Reproduction	207
Physiology of Hatching	208
Dormancy and Quiescence	208

6. FEEDING AND TROPHIC RELATIONSHIP OF NEMATODES — 211

Feeding and Feeding Behaviour	211
Mode of Parasitism	216
Parasitism by the Order Dorylaimida	216
Parasitism by the Order Tylenchida	218
Parasites of Aboveground Parts	223
Host Parasite Relationship	224
Morphological Modifications	225
Physiological Modifications	234

7. ECOLOGY OF NEMATODES — 237

Population Dynamics of Plant Nematodes	237
Nematode Characteristics	238
Host Characteristics	240
Biotic Factors	243
Abiotic Factors	245
Interaction of Nematodes with other Organisms	249
Disease Etiology	250
Nematode-Fungus/Bacteria Interaction	251
Nematode-Virus Interaction	257
Nematode-Rhizobium Interaction	260
Nematode-Mycorrhizal Fungi Interaction	263
Nematode-Insect Interaction	268
Nematode-Nematode Interaction	271
References	**279**
Abbreviations, Symbols and Glossary	**298**
Subject Index	**333**

1

Introduction

INTRODUCTION

Nematodes constitute the largest and most ubiquitous groups of the animal kingdom. In fact, by number they are 80-90% of all the multicellular animals. Basically, the nematodes are aquatic animals thriving best in water but they have adapted terrestrial habitats, and are mostly found in soil in all geographical regions and habitats, from snowy mountains to deserts. In soil, they are the most common of all the soil fauna numbering 1.8-120 millions per square meter of soil. Fortunately, only a fraction of this number possesses the ability of parasitizing plants, and the rest are free-living surviving on various substrates.

Nematodes are vermiform or thread like (derived from the Greek words, *nema-* for thread and *oides* for form or like), unsegmented, usually circular in cross-section and bilaterally symmetrical (Fig. 1). The body is flexible, usually elongate, filiform, cylindrical to fusiform and tapering towards the ends. Adults of plant nematodes may be 0.3 - 0.5 mm (*Paratylenchus pratensis*) to 11 mm long (*Paralongidorus maximus*). Animal nematodes are, however, much bigger. *Ascaris lumbricoides* reaches 40 cm length and the female *Placentonema gigantissimum* from whale fish may be 8.5 m long. Nematodes move in soil through pore space between the soil particles and they need some moisture for survival. The food of nematodes is invariably some source of protoplasm, such as plants, fungal hyphae, algae, bacteria, protozoa and even nematodes. On the basis of diet and habit, the nematodes may be divided under four categories.

Microbial Feeders or Microbivorous Nematodes

Dead organic matter which reveals high population of nematodes, does not directly form their diet. It serves as a substrate for the better activity and multiplication of microorganisms on which the nematodes feed.

Fig 1 Different plant nematodes on a glass slide under low power of magnification (4x)

Many genera of the order Aphelenchida, e.g., *Aphelenchus, Aphelenchoides, Rhadinaphelenchus, Bursaphelenchus* etc., and also a few of Tylenchida (*Ditylenchus* spp.) may feed on fungi, bacteria, diatoms, etc. *Aphelenchoides hamatus, A. composticola* and *D. myceliophagus* feed on the spores of pathogenic and mushroom fungi. *Panagrolaimus* and *Poikilolaimus* species feed on numerous bacteria (Muschiol and Traunspurger, 2007).

Predaceous Nematodes

Certain members of Mononchida, Dorylaimida and Aphelenchida feed on protozoa, nematodes, insects, etc. The mouth of *Clarkus papillatus* (Mononchida) is equipped with teeth, whereas *Panagrellus redivivus* (Dorylaimida) has a piercing and sucking type of feeding apparatus which is used during predation upon nematodes. *Seinura* spp. (Aphelenchida) are much smaller than their prey nematodes. For predation, the nematode first injects saliva into the body to immobilize the prey followed by ingestion of partially digested body contents of the prey nematode. Another important category of predatory nematodes is entomophagous nematodes, which are of economic importance with regard to insect biocontrol. These nematodes feed upon certain insects, e.g., *Steinernema* (= *Neoaplectana*) *carpocapsae* enter into the insect body (codling moth, *Cydia pomonella*) through natural openings or cuticle, and reach the haemocoel where they are able to feed due to symbiotic association with a bacterium, *Achromobacter nematophilus*. Similarly, *Steinernema kraussei* with symbiosis of *Flavibacterium cytophaga* feeds upon sawfly larvae (*Cephalcia abietis*). *Hexamermis* spp. possess ability to feed without the aid of symbionts on shoot borers, cutworms, etc.

Plant Feeders or Phytonematodes

This group of nematodes is very important with regard to pathogenic effects on crops and their productivity. Among the pathogens, largest plant diseases with specific and visible symptoms are caused by fungi, bacteria and viruses. The diseases caused by nematodes come next. Nematodes cause damage to plants by injuring and feeding on the root hairs, epidermal cells, cortical and/or stealer cells. A large number of nematodes are ectoparasites which feed on root surface, e.g., *Tylenchus, Rotylenchus, Tylenchorhynchus, Belonolaimus, Hoplolaimus, Trichodorus, Longidorus*, etc. Root-knot nematodes (*Meloidogyne* spp.), cyst forming nematodes (*Heterodera* spp.) and root-lesion nematode (*Pratylenchus* spp.) are endoparasites and enter fully inside the host root. Citrus nematode (*Tylenchulus semipenetrans*) and reniform nematode (*Rotylenchulus reniformis*) are semi-endoparasites, the posterior half of the body remains outside the host in soil. The worst effect of nematode parasitism is debilitation of the plant even without appearance of any symptom. In addition to direct damage, nematodes often aid or aggravate the diseases caused by fungi, bacteria and viruses, or the resistance of cultivars is broken. Hairy root of roses caused by *Agrobacterium rhizogenes* is of minor importance, but in the presence of *Pratylenchus vulnus* the disease becomes severe. The fusarium wilt resistant cultivars of cotton become susceptible in the presence of root-knot nematodes. Plant nematodes may also act as vectors for bacteria, fungi and viruses, e.g., *Anguina tritici* carries *Clavibacter tritici* or *Dilophospora alopecuri* to shoot meristem of wheat. Ringspot viruses (NEPO viruses), e.g., tobacco ring spot virus, are transmitted by *Xiphinema* and *Longidorus* species. *Trichodorus* and *Paratrichodorus* species act as vectors for certain tobra viruses such as tobacco rattle and pea early browning viruses.

Miscellaneous Nematodes

This category includes those nematodes whose feeding habits are nonspecific, and are referred to as saprophytic nematodes or free-living nematodes. Food requirement of such nematodes is limited and any organic source of plant or animal origin may sustain them.

HISTORY OF PLANT NEMATOLOGY

Knowledge of nematode parasites of animals is almost as ancient as the history of human civilization itself. The roundworm, *Ascaris lumbricoides* and guinea worm, *Dracunculus medinensis* were known to Egyptians as early as 1550-1530 BC. Knowledge of nematode parasites of plants in comparison to that of animals is relatively of much recent origin. The first phytonematode, seed-gall nematode, *Anguina tritici* which causes ear-cockle of wheat was first observed by a catholic priest, John Turbevill Needham in the year 1743 in England. Needham saw the nematode under a crude microscope available at that time, and communicated this observation to the President of the Royal Society, London, stating that small black grains of smutty wheat had soft fibrous substances, which upon soaking in water took life and yielded a large number of motile worms (Needham, 1744). Scopoli (1777) studied the nematode and proposed the genus *Anguina* as *Angvina,* but it was not accepted. In 1799 J. G. Steinbuch named the nematode as *Vibrio tritici* and also described another similar nematode as *Vibrio agrostis*. The present genus *Anguina* was named by a Russian nematologist, I. N. Filipjev in 1936.

After *Anguina*, the next, root-knot nematode (*Meloidogyne* spp.) was discovered in 1855 by M. J. Berkeley, who noticed galls or knots on cucumber roots growing in a glasshouse in England (Berkeley, 1855). He observed white larvae and eggs of the nematode which he named *Vibrio*. M. Cornu in 1879 made a scientific description of the nematode and called it *Anguillula marioni*. Goeldi (1887) proposed the present genus *Meloidogyne* but this nomenculature was not recognized at that time. Untill 1932 root-knot nematodes had been known as *Heterodera radicicola*. Chitwood (1949) separated root-knot nematodes from *Heterodera* and named the present genus, *Meloidogyne* which was originally proposed by Goeldi.

By the middle of the 19th century (1850s), nematology had been established in Germany, where Julius Kuhn in 1857 described a stem and bulb nematode, *Anguillula dipsaci* (now, *Ditylenchus dipsaci*) from teasel (Kuhn, 1857). In 1859, H. Schacht in Germany reported the cyst forming nematode on sugar beet. This was the first nematode that drew farmers' concern towards significance of nematodes in crop production. The nematode inflicted serious crop damage to beet, which was a very important crop for the economy of Central Europe during that period. In 1871 Schmidt described the sugar beet nematode and in recognition of contribution of H. Schacht named the nematode after his name as *Heterodera schachtii*. Cyst forming nematode also enjoys the recognition of being the first plant parasitic nematode to be treated with a nematicide, carbon disulphide (CS_2) by Kuhn in Germany in 1871 (Kuhn, 1881). Kuhn also conducted studies on trap crops and crop rotation to interrupt the life cycle of the nematode, his recommendations still continue as a major approach to nematode control.

The first type species of the order Aphelenchida, *Aphelenchus avenae* was described by H. Charlton Bastian in 1865. Bastian published an excellent monograph, describing 100 new species. He realized that information on nematodes by that time was insufficient for a philosophical classification; for this reason he made tables to assist in the classification (Bastian, 1865). I. G. de Man produced a series of excellent monographs in 1876, 1880, 1884 and 1924. In the 1884 monograph, he used the formulae of α, β, γ for nematode measurements, which is still used in morphometrics as a, b, c, (de Man, 1884). Ritzema Bos (1891) in England described *Aphelenchoides fragariae* on strawberry. *A. ritzemabosi* and *A. besseyi* are major destructive pests of chrysanthemum and rice which were described by Schwartz (1911) and Christie (1942), respectively. In 1894, M. Fischer created the genus, *Aphelenchoides* and transferred a number of species of *Aphelenchus* to this genus (Fischer, 1894). Other plant parasitic nematodes, such as *Radopholus, Pratylenchus, Paratylenchus, Rotylenchulus, Hoplolaimus, Longidorus, Xiphinema, Trichodorus,* etc., were discovered during the end of the 19th century and in the begining of the 20th century (Siddiqi, 1986, Hunt, 1993).

During the early years of the last century, nematology made rapid progress. Major developments, however, occurred in the description of nematodes, their classification, laboratory techniques and visible effects on plants. Dr. Nathan Angustus Cobb coined the term **Nematology** for the science dealing with the study of nematodes and is referred to as the **Father of Plant Nematology**. His monograph in 1893 on **Nematodes, mostly Australian and Fijiian** contained several tylenchid nematodes (Cobb, 1893). He developed several techniques to study nematodes which are still used (Cobb, 1918). Other pioneer researchers in nematology are Tom Goodey of the U.K. and I. N. Filipjev of USSR. They contributed books during the periods of scarcity of relevant printed information on nematodes. Goodey published two exhaustive books on nematodes, **Plant Parasitic Nematodes and the Diseases They Cause** in 1933 and **Soil and Fresh Water Nematodes** in 1951. The latter was rewritten and elaborated by his son, J. Basil Goodey in 1963. Filipjev wrote a book in Russian in 1934 on **Harmful and Useful Nematodes in Rural Economy,** which was translated in English in 1941 under the title, **A Manual of Agricultural Helminthology**. After Cobb's death in 1932, Gotthold Steiner became the leading nematologist in the U.S.A. Steiner in 1949 introduced nematodes to farmers and growers through a publication, **Plant Parasitic Nematodes the Growers should Know**.

In 1955, when Dr. M. R. Siddiqi joined post-graduate research (Ph.D.) on phytonematodes at the Aligarh Muslim University, not even his guide, Dr. M. A. Basir could have visualized that within a decade this budding nematologist would contribute so enormously that nematology would rely on his research. Dr. Siddiqi has described hundreds of families, genera and species of terrestrial, plant parasitic and insect nematodes. His book, **Tylenchida - Parasites of Plants and Insects,** has not only validated existing genera and species of nematodes, but has also provided a standard text for global studies in systematics and taxonomy of Tylenchida (Siddiqi, 2000). Abrar M. Khan, A. Coomans, A. A. Paramonov, A. C. Tarjan, A. D. Grisse, A. M. Golden, A. R. Maggenti, A. R. Seshadri, B. G. Chitwood, B. M. Zuckerman, B. R. Kerry, D. J. Eisenback, D. J. Raski, D. J. Hooper, E. Geraert, E. Khan, E. L. Krall, E. S. Kirjanova, F. Lamberti, G. R. Stirling, G. Thorne, G. Steiner, H. Micoletzky, H. R. Wallace, I. Andrassy, J. A. Otto Butschli, J. F. Southey, J. M. Webster, J. N. Sasser, J. R. Christie, K. Evans, L. Taylor, M. Luc, M. G. K. Jones, M. P. Starr, M. R. Siddiqi, M. S. Jairajpuri, M. W. Allen, M. W. Khan, N. T. Powell, P. A. A. Loof, R. H. Mulvey, R. M. Sayre, R. S. Pitcher, S. A. Sher, S. I. Husain, S. N. Das, W. R. Nickle, and several others are pioneer nematologists of the world who not only credited for excellent research but have also contributed valuable books. Some of the important books on Nematology which may be available in the market are enlisted in Table 1.

Another major discovery in nematology made at the end of the 19th century was recognition of interaction between nematodes and other pathogens or pests and their management. G. F. Atkinson in 1892 observed that phytonematodes can synergise other pathogens. He demonstrated that the presence of root-knot nematode increased the severity of fusarial wilt on cotton and also broke the wilt resistance reaction of the cultivar (Atkinson, 1892). In 1926, W. M. Carne observed that due to the vector role of *Anguina tritici* in carrying *Clavibacter tritici* to shoot meristem, yellow slime disease of wheat develops (Carne, 1926). W. B. Hewitt, J. D. Raski and A. C. Goheen in 1958 discovered the vector role of some dorylaims to transmit certain plant viruses (Hewitt *et al.*, 1958). Blaire (1964) demonstrated that some plant nematodes need vectors for their own transmission, especially those which attack the tree trunk, for instance *Rhadenaphelenchus cocophilus* that causes red-ring of coconut is transmitted by black palm weevil, *Rhynchophorus palmarum*. The nematode was first discovered separately by Cobb and Nowell in 1919. J. R. Christie in 1941 discovered entomopathogenic nematodes (EPN) and categorized them into three groups.

With the discovery of plant parasitic nematodes and their potential to attack crops, measures to control them were also invented simultaneously. Kuhn (1857) used carbon disulphide to control sugar beet cyst nematode. In 1909, K. Marcinowski recommended the use of hot-water treatment to disinfest the planting material. Discovery of soil fumigants, DD mixture (Carter, 1943) and EDB (Christie, 1945) during the second

Table 1 Important books on plant nematology

Year	Author/Editor	Title and Publisher
1933	T. Goodey	*Plant Parasitic Nematodes and the Diseases They Cause* E. P. Dutton and Co. Inc., U.K.
1934	I. N. Filipjev	*Harmful and Useful Nematodes in Rural Economy.* Moscow, Leningrad, Russia.
1941	I. N. Filipjev	*A Manual of Agricultural Helminthology.* E. J. Brill, Leiden, Netherlands.
1951	T. Goodey	*Soil and Fresh Water Nematodes.* Methuen Co. U.K.
1971	B. M. Zuckerman, W. F. Mai and R.A. Rohde	*Plant Parasitic Nematodes* (Vols. I & II). Academic Press, USA.
1982	J. F. Southey	*Plant Nematology.* Her Majesty's Stationery Off. U.K.
1981	M.B. Zukherman and R.A. Rohde	Plant Parasitic Nematodes (Vol. III) Academic Press, U.S.A.
1981	A. R. Maggenti	*General Nematology* Springer-Verlag, NY, Germany.
1986	J. F. Southey	*Laboratory Methods for Work with Plant and Soil Nematodes.* Her Majesty's Stationery Office, U.K.
1986, 2000	M. R. Siddiqi	*Tylenchida - Parasites of Plants and Insects* (I & II editions). CAB International, U.K.
1987	R. H. Brown and B. R. Kerry	*Principles and Practices of Nematode Control in Crops.* Academic Press, U.S.A.
1988	G. O. Poinar Jr. and H. B. Jansson	*Diseases of Nematodes* (Vols. I & II). CRC Press, U.S.A.
1990, 2005	M. Luc, R. A. Sikora and J. Bridge	*Plant Parasitic Nematodes in Subtropical and Tropical Agriculture.* (Vols. I & II) CAB International, U.K.
1991	W. R. Nickle	*Manual of Agricultural Nematology.* Marcel Dekker, U.S.A.
1991	G. R. Stirling	*Biological Control of Plant Parasitic Nematodes.* CAB International, U.K.
1992	M. S. Jairajpuri and W. Ahmad	*Dorylaimida-Free-living, Predaceous and Plant Parasitic Nematodes.* Oxford & IBH, India.
1993	D. J. Hunt	*Aphelenchida, Longidoridae and Trichodoridae.* CAB International, U.K.
1993	M. W. Khan	*Nematode Interactions.* Chapman & Hall, U.K.
1993	K. Evans, D.L. Trudgill and J. M. Webster	*Plant Parasitic Nematodes in Temperate Agriculture.* CAB International, U.K.
2002	R. Gaugler	*Entomopathogenic Nematology,* CAB International U.K.
2006	E. Geraert	*Functional and Detailed Morphology of the Tylenchida (Nematoda).* Brill Academic Publisher, The Netherlands.
2007	R.N. Perry and M. Moens	Plant Nematology, CABI, U.K.

world war marked the begining of a new era in nematode control. These fumigants were relatively less toxic and cheaper, and easy to apply. Their application provided the farmers with an opportunity to compare yield of crops grown with and without nematicide. In 1950, systemic nematicides were introduced in the U.S.A. and Europe.

By the middle of last century, significance of plant nematodes had been realized and enough relevant information had been generated. This necessitated the formation and introduction of scientific forums and journals to exchange observations and ideas. The Society of European Nematologists, now European Society of Nematologists formed in 1955 was the first association that was solely dedicated to nematology. The first scientific journal on nematology, *Nematologia* was also published from Europe. Important nematological societies and journals are enlisted in Table 2.

In India the first report on plant nematodes was made on root-knot nematode on tea roots by C. A. Barker in 1901. Since then, there had been occassional reports on occurrence of nematode diseases, such as ufra of rice (Butler, 1913), root-knot of vegetables and other crops (Ayyar, 1926; 1933) and white-tip of rice (Dastur, 1936). Research on plant parasitic nematodes in a regularized and organized manner started in India in 1953, when Prof. Abrar Mustafa Khan, basically a plant pathologist became interested in plant nematodes at the Aligarh Muslim University. Postgraduate teaching and research on phytonematology started at the Department of Botany and Zoology, AMU in 1957. Contributions by Drs. M. R. Siddiqi, M. S. Jairajpuri, E. Khan, M. W. Khan and other researchers from 1959 onwards greatly strengthened the nematology at AMU in India. At the Indian Agricultural Research Institute, the Division of Nematology was established in 1965. The Nematological Society of India was established in 1969 by its founder President Prof. Abrar M. Khan presiding for two consecutive tenures.

AGRICULTURAL IMPORTANCE OF PLANT NEMATODES

Plant parasitic nematodes are major destructive pests of agriculture. For centuries they have been causing severe damage to crop plants. They attack all kinds of plants and can infect roots, stems, leaves, crowns, inflorescence, flowers and developing seeds. The degree of damage to a particular crop is influenced by the plant species or cultivar, nematode species, level of soil infestation and the prevailing environment around host and nematode. For the infection of potato by *Globodera rostochiensis*, soil moisture and low temperature are essential. Similarly, ear-cockle dis-

Table 2 List of nematological societies and journals

Year	Society/Association	Journal
1955	Society of European Nematologists (now European Society of Nematologists)	
1956	-do-	Nematologica
1961	Society of Nematologists, U.S.A.	
1969	-do-	Journal of Nematology
1969	Nematological Society of India	
1971	-do-	Indian Journal of Nematology
1971	Organization of Tropical American Nematologists, U.S.A.	Nematropica
1970	Commonwealth Agricultural Bureau, U.K. (Nematological Abstracts since 1992)	Heliminthological Abstracts Part: B Plant Nematology
1973	Mediterranean Nematological Society, Italy.	Nematologia Mediterranea
1972	Japanese Nematological Society	Japanese Journal of Nematology
1976	Brazilian Society of Nematologists	Brazilian Journal of Nematology
1978		Review de Nematologia, France (Fundamental of Applied Nematology, now Nematology)
1982	Pakistan Society of Nematologists	
1983	-do-	Pakistan Journal of Nematology
1983	Nematological Society of South Africa	
1990	Afro-Asian Society of Nematologists	Journal of AASN now International Journal of Nematology
1994	Russian Society of Nematologists	Russian Journal of Nematology English, biannual, 1993

ease develops under humid and cool climate when a fine film of water is formed on the plant surface which is necessary for nematode movement.

When susceptible crops are planted in heavily infested fields and environmental conditions are conducive, nematodes cause severe reduction in the plant growth and yield, both quantitatively and qualitatively (Luc et al., 2005). Root-knot of vegetables and pulses (*Meloidogyne* spp.), decline of fruit trees (*Radopholus* spp.), molya of wheat (*Heterodera avenae*), ufra of rice (*Ditylenchus angustus*) and cyst disease of potato (*Golobodera* spp.) are some of the diseases which cause tremendous economic loss to growers (Sasser, 1989). Nematodes also cause serious decline or death of highly prized ornamentals and turf, e.g., *D. dipsaci* to tulip, *A. ritzemabosi* to chrysanthemums, *Meloidogyne* spp. to seasonal ornamentals, *Belonolaimus longicaudatus* and *Hypsoperine graminis* to turf (Singh and Sharma, 1998). Despite tremendous adverse impact on agriculture, nematodes had been neglected for centuries and even today majority of the growers do not consider them major pests. This is probably because of the fact that the damage caused by nematodes is less obvious than that caused by fungi or chewing insects on leaves; consequently it remains unrecognized. From the economic point of view, the hidden damage is more important, because due to absence of visible disorders, control measures are not adopted, consequently nematodes keep causing damage to crops.

Nematodes do not always cause hidden damage. When fields are heavily infested, characteristic symptoms appear on roots or shoots. Symptoms develop more frequently on roots because usually nematodes are root feeders. Specific symptoms are root-lesions, root-rot, root pruning, root-galls, stubby roots and cessation of root growth (Sasser, 1989) (Fig. 2B–D). The roots damaged by nematodes cannot absorb water and nutrients efficiently from soil. Nonspecific or general symptoms of

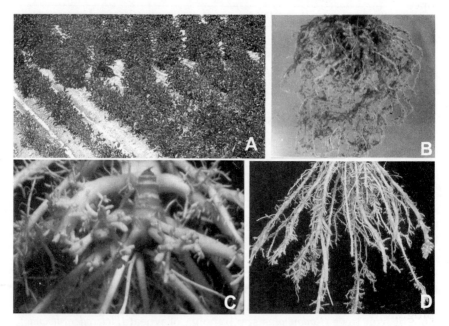

Fig. 2 Symptoms of nematode attack. A- Plants in a field showing stunted growth with chlorotic foliage, B- Root-knot (*Meloidogyne* spp.), C- Cessation of root growth (*Belonolaimus* sp.), D- Stubby root (*Trichodorus* sp.). (Courtesy: Sasser, 1989).

nematode attack are characterized by patches of plants irregularly distributed in a field showing stunted growth, sparce foliage and mild chlorosis on leaves (Fig. 2A). During the periods of accelerated rate of transpiration, plants show incipient wilting despite adequate soil moisture. Hence, the general symptoms resemble nutritional deficiency, the difference being that latter is farely uniform in the field. Further, roots so weakened and damaged by nematodes are easily invaded by many bacteria and fungi, leading to rapid decay of roots. This secondary damage also does not draw immediate attention because foliage of the crop remains healthy and symptoms appear at maturing stage. Another serious impact is that nematode infection in the field renders resistant cultivars susceptible to the pathogen.

Agricultural importance of plant parasitic nematodes comes immediately after insects, fungi, bacteria, and viruses. Nematodes cause about 10-12% yield loss when various crops are considered. The yield losses vary greatly depending on inoculum level and host species. The severe infection may result in as much as 80-90% yield loss in an individual field and sometimes plants fail to give yield of any economic value. Crop loss due to nematodes is greater in developing countries than the developed countries. It is probably due to unplanned agricultural practices and unawareness of the farmers towards nematodes. Despite the implementation of management practices in developed countries, nematodes do cause considerable crop losses. In the U.S.A. alone annual monetary loss due to nematodes is above $6.0 billion (Agrios, 2005). The average yield loss inflicted by nematodes to agricultural crops on worldwide basis are summarized in Table 3.

In India about 10-20% crop losses occur due to nematodes. This includes loss of Rs.70 million to wheat and barley due to molya disease (*Heterodera avenae*). Yield losses to important crops in India are 25-90% (vegetables), 10-20% (legumes), 8-21% (banana), 10-21% (cotton), 7-28% (tobacco), 8-16% (coffee) and 5-19% (citrus).

Monetary annual losses due to nematodes on worldwide basis considering 21 important crop species were estimated over US $80 billion (Agrios, 2005). The real figure when all crops are considered may exceed US $100 billion annually. It has been estimated that around US $125 million (i.e., less than 0.2% of the total loss) are spent on nematology teaching, research and extension. The imbalance between the crop loss and expenditure incurred on nematode management training programmes appears to contribute towards negligence of farmers to nematode infestations, especially in developing countries. To protect the crops from nematodes in order to get higher yields, it is essential to invest more money on nematode management programmes and extension

Table 3 Global loss to agricultural crops caused by nematodes

Crop	Yield Loss (%)	Crop	Yield Loss (%)
Vegetables	5-43	Soybean	11
Pulses	10-20	Pea	12
Cereals	7-22	Maize	10
Fruit Crops	9-28	Wheat	7
Sugar Crops	7-19	Rice	10
Commercial Crops	10-28	Grapes	8-28
Fiber Crops	8-37	Citrus	14
Ornamentals	11-32	Banana	20
Okra	20	Pineapple	16
Tomato	21	Sugar beet	11-19
Eggplant	17	Sugarcane	15
Potato	12	Tea	8
Pepper	12	Coffee	15
Chickpea	6-19	Tobacco	10-28
Mungbean	23	Cotton	10-15

Table 4 Ten most important nematode genera, their susceptible crops and associated yield loss

S. No.	Nematode Genera	Susceptible Crops and Associated Yield Loss
1.	Root-knot nematode, *Meloidogyne* spp.	Vegetables (5-43%), fruits (9-29%), pulses (10-37%), fiber (11-37%), tobacco (10-28%).
2.	Root-lesion nematode, *Pratylenchus* spp.	Tobacco (23-60%), vegetables (8-39%), sugarcane (10-16%), pineapple (9%), coffee (13%).
3.	Cyst nematode, *Heterodera* spp.	Cereals (10-50%), pulses (7-24%), sugar beet (11-19%).
4.	Stem and bulb nematode, *Ditylenchus* spp.	Onion and garlic (16-42%), tulip (5-17%), rice (10-30%).
5.	Potato cyst nematode, *Globodera* spp.	Potato (13-45%).
6.	Citrus nematode, *Tylenchulus* sp.	Citrus (10-20%).
7.	Dagger nematode, *Xiphinema* spp.	Vegetables and citrus (5-17%).
8.	Burrowing nematode, *Radopholus* spp.	Banana (13-60%), black pepper (5-21%), citrus (40-70%).
9.	Spiral nematode, *Helicotylenchus* spp.	Vegetables (8-13%), papaya (4-10%), cotton (7-10%).
10.	Reniform nematode, *Rotylenchulus* sp.	Vegetables (5-26%).

services. Special attention should be given to reduce the population of the nematodes belonging to 10 most economically important genera, and cultivation of their susceptible crops as summarized in Table 4 should be rotated with tolerant/resistant/nonhost crops.

2

Methodology

SAMPLING FOR NEMATODE ASSAY

The major purpose of sampling for nematodes is to estimate the nematode population for research, advisory and predictive programmes. Population estimation is also essential for disease diagnosis and management. The specific objective of sampling should be known before the sampling is done. This is particularly important for the selection of sampling design, type and amount of nematodes required. The decision whether to collect soil or root or both for nematode assay is also important. This actually depends on the nematode, host plant and purpose of sampling. Polyspecific populations that contain both ecto and endo-parasitic nematodes may require both soil and root samples. For instance, to assay ectoparasites, only soil is collected. The specific time of sampling is also important as nematode populations vary greatly with the season and age of the host plant.

Soil Sampling Technique

The technique for soil collection is simple and it does not require any special device. The classical, cylindrical tube type sampler and auger are standard equipments for the sampling. For soil collection, the top soil of 5-6 cm is discarded as it does not contain many nematodes. A good population generally occurs at 5-15 cm depth, so the soil of this zone is collected for nematode assay.

Sampling Pattern

There are different patterns of soil sampling which depend on the purpose and requirement. The relative size of sample area and number of cores collected per area for nematode assay differ greatly with the pur-

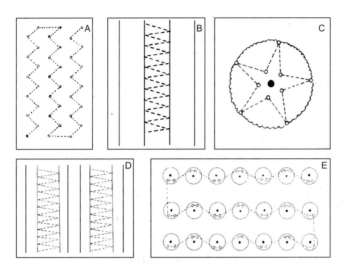

Fig. 3 Various patterns of soil sampling for nematode assay from large plot (A), four row plot (B), single plant plot (C), multirow plot (D) and perennial orchard (E) (Courtesy: Barker, 1985).

pose. For the advisory purpose, a sample area of 1-2 hectare or less is recommended, whereas for regulatory purpose a sample area of 1/3 hectare, and for research purpose, the entire research plot is sampled.

Different patterns are used for collecting soil samples. It is particularly important to obtain reliable data of a field or an area. Nematode population usually occurs in patches, and if such patches are missed, the average population may be too low. In a reverse situation the population will be too high. To minimize such variations and to have a reliable average of nematode population, different patterns of soil collection are used, of course depending on objective, size of sampling area and type of vegetation, which are as follows:

Large plot

In large plots of 1-2 ha, 20-30 cores are made to collect soil samples. The cores can be made in randomized or in stratified random manner (Fig. 3A). In stratified manner, cores are made in lines, while in random, cores can be made anywhere in the plot, as desired by the researcher.

Rowed vegetation

In the field where plants are grown in rows (e.g., tobacco, cotton, etc.), cores are made in two central rows in the stratified manner if it is a four-row plot (Fig. 3B). In multi-rowed plot, cores should be made after two consecutive rows (Fig. 3D).

Perennial orchard

To collect soil samples from feeder root zones of established perennial plots, e.g., a plot having trees of citrus, peach, populus, etc., a pair of cores are made around each tree in an alternate manner (Fig. 3E).

Single plant plot

To collect soil for nematode assay from a single plant plot, a total of ten cores are made in two circles around the plant in a pattern giving an impression of a star (Fig. 3C), which is also known as star sampling.

A sample of soil collected from a specific site or soil of a single core is called a simple sample, whereas a mixture of samples of soil collected from different sites in a field or locality is known as composite sample.

STORAGE OF SAMPLES

Soil Samples

Since plant parasitic nematodes are thin-walled short-lived organisms, improper handling such as exposure to sun, temperature extremes, droughtness or poor storage, result in high mortality rates, leading to erroneous information on the nematode population. After collection, each sample should be sealed in a plastic bag and stored at a place away from direct sun exposure and preferably at 20-25°C.

Plant Material

To isolate endo or semi-endoparasitic nematodes, part of the plant invaded by the nematode is collected, e.g., roots (*Meloidogyne* spp., *Rotylenchulus reniformis*), bulbs, tuber, rhizome (*Ditylenchus* spp., *Radopholus similis*), leaves (*Aphelenchoides* spp.), trunk (*Rhadinaphelenchus cocophilus, Bursaphelenchus xylophilus*), seeds (*Anguina tritici, Aphelenchoides besseyi*), and fruits (*Schistonchus caprifici*). The root samples should be gently rinsed in water before storage.

The plant material containing nematodes should usually be kept cool and moist and examined as soon as possible. Intact plants (free of soil) remain fresh for longer periods during storage. Shoots decompose much faster than roots, hence they should be kept in separate polyethylene bags. The plant material should be moistened before packing in polybags and stored at 10-15°C especially in subtropical and tropical regions.

EXTRACTION OF NEMATODES FROM SOIL

There are various methods of separating nematodes from soil samples. Although vermiform nematodes and cysts can be isolated by a single method, recovery is, however, good if a specific technique is used. Hence, a technique which is appropriate for the type of nematodes to be expected or isolated, is selected.

Extraction of Vermiform Nematodes

Cobb's Decanting and Sieving Method (Modified)

This is a most simple and versatile method which is modified from Cobb's sifting and gravity method. It is useful in extracting all kinds of nematodes for inoculation purpose and routine assay when combined with the Baermann trays or funnels. Soil from the composite or simple sample (500 g or cm^3) is taken in 5-6 liters of water in a bucket. The lumps are thoroughly broken and the suspension is left for 2 minutes to allow settling of crumbs, debris, etc. The suspension is decanted onto a 35 mesh sieve placed over a 350 mesh sieve. In case the soil is rich in humus, silt or clay, a 100 or 200 mesh sieve is placed before the 350 or 400 mesh sieve (Fig. 4). If very fine nematodes, e.g., larvae of root-knot or cyst forming nematodes are to be isolated, a finer sieve of 400 mesh is recommended. If the specimen is dirty, it can be cleaned by pouring water onto a 500 mesh seive. In fact, combination of the seives is selected on the basis of the nematode species or the stage which is to be isolated. Commonly used sieves are 10, 20, 35, 60, 100, 200, 300, 400 and 500 mesh/linear inch. Mesh size or pore aperture with regard to mesh number of sieves is provided in Fig. 4.

Mesh	Mesh size µm
10	2,000
20	840
25	710
35	500
60	250
100	150
200	75
300	53
350	45
400	38
500	26

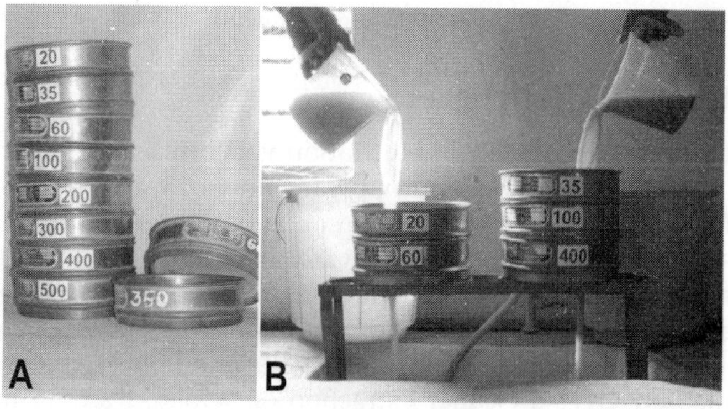

Fig. 4 A- Brass sieves of varying mesh. B- Isolation of nematodes using Cobb's decanting and sieving method.

The catch from the finest seive is poured over a wire gauze sieve or coarse seive (7-8 cm diameter) having two layers of tissue paper at the bottom. The sieve is kept on the modified Baermann funnel holding sufficient amount of water to remain in contact with the bottom of the wire gauze (Fig. 5). The sieve is kept in the funnel in such a way that large air bubbles do not form in between the water surface and sieve bottom. The assembly is left overnight, during this period the nematodes due to their wriggling mobility and random movements to find minute pores or gaps pass through the tissue paper and gradually settle down in the lower end of the funnel, while the fine silt particles remain on the tissue paper. The nematode suspension is collected from the beak of the funnel through the stop valve.

Fig. 5 Various versions of Baermann funnel

Elutriation Techniques

Dictionary meaning of elutriation is a process to separate lighter and heavier particles under an upward flow of water or gas. This technique for nematode extraction from soil is based on the principle that a measured flow of water in an upward direction will allow the nematodes to suspend (since they are light weight), and allow heavier soil debris to settle downward. Elutriation techniques use an upcurrent of water to separate nematodes from soil particles and hold them in suspension. This method gives a cleaner extraction of nematodes than that obtained by direct decanting and seiving method. The chief use of an elutriator is to isolate nematodes from a large amount of soil. In addition, nematodes such as *Criconema, Criconemoides* and *Hemicriconemoides,* which are not easily extracted by decanting and sieving method, are readily isolated by an elutriator. Three elutriators being presently used work on a similar principle of elutriation but differ markedly in structure and processing technique.

Oostenbrink's elutriator

Structure: The elutriator consists of two major parts i.e., floatation apparatus and a funnel having a coarse sieve which are made of brass (Fig. 6). The main body of the elutriator is called a floatation apparatus, cone column or can, which is conical in shape, comprising two cones one larger than the other. The larger cone is fitted with a glass tube on its body, bearing markings of three water levels. The lower end of the larger cone terminates into a smaller cone. Slightly above the junction point of the cones, an L-shaped outlet pipe is fixed. The terminal lower end of the

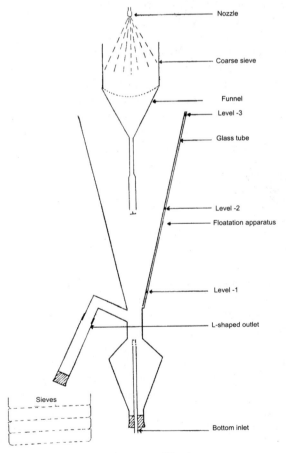

Fig. 6 Line diagram of Oostenbrink's Elutriator

smaller cone is provided with a bottom inlet. The top of the larger cone holds a metallic funnel with a coarse sieve (1 mm). At the top an inlet pipe with nozzle is fitted over the coarse funnel. The whole apparatus is mounted on the wall of a washbasin or sink.

Functioning: The bottom inlet plug is pulled out to remove the remnants of the previous isolation. The plug is then fixed and the floatation body is filled with water upto level-3. The rubber plug of the L-shaped outlet is removed to drain off the water. When the water stops flowing out, the bottom inlet plug is pulled out again for the complete drain off. These steps are precautionary to avoid any contamination. Before loading the sample, the funnel and coarse sieve should also be rinsed with water.

The floatation body is filled to level-1 by a constant flow of 1,000 ml water/minute. The moist soil sample 100-500 g is placed on the coarse sieve and the sample is washed into the can via the funnel by means of the nozzle delivering about 700 ml water/minute. Until two-thirds of the column is filled (upto level-2). Thereafter, water is allowed to flow through the bottom inlet at 600 ml/minute until the water reaches the level-3 of the can. If small and slender nematodes, such as *Tylenchus*, *Paratylenchus*, etc., are to be extracted, the water flow rate is reduced. For larger nematodes, such as *Longidorus*, *Trichodorus* and *Dorylaimus*, a flow rate of 1,500-2,000 ml/minute is maintained. The plug of the L-shaped outlet is pulled out to gently pour the suspension onto sieves of appropriate mesh number, such as 100, 200, 300 and 400. The catch is washed and poured onto a wire gauze sieve having two layers of filter paper placed in a Baermann funnel, and is left overnight.

Seinhorst's elutriator

This is an efficient elutriator and can separate nematodes of different categories according to their size. Isolation of nematodes by this technique is, however, cumbersome and time consuming as 30-40 minute/sample, are needed. Also, as it is made of glass, it is extremely fragile and may prove very costly if it breaks.

Structure: The elutriation system consists of a 2-litre capacity Erlenmeyer flask, three bigger funnels and a container (Fig. 7). The Erlenmeyer flask is provided with a stopper that opens into a funnel (F1) which has a 32 cm long neck (N1) of 3.2 cm diameter. The F1 has an overflow pipe which discharges into a small funnel. The neck of F1 is connected with the second funnel F2 that is smaller than the F1 but possesses a similar shape. Its neck is 24 cm long and 2 cm wide. The F2 is connected with a third funnel (F3) which has a very short neck (N3) but wider than N2. The lower portion of the necks of the three funnels have outlet pipes (O1, O2, O3), which are provided with rubber sleeves and stopcocks into ves-

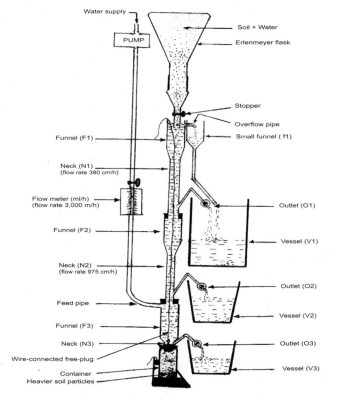

Fig. 7 Line diagram of Seinhorst's Elutriator

sels V1, V2 and V3. The bottom funnel, i.e., F3 is connected to a container. The junction of N3 and the container is provided with a wire-connected free plug, which is used to close the passage between F3 and the container. Internal diameter of the neck is in the order of N3>N1>N2. A feed pipe is connected to the upper end of the funnel F3 to supply water at a uniform and measured flow rate. The whole assembly is made primarily of glass.

Functioning: The Erlenmeyer flask is rinsed and funnels are filled with water. The wire-connected free-plug is removed to allow the settling of remnants of previous sampling. After 5-10 minutes, the passage is closed by pulling the wire, and the water and other remnants from the container are removed and thereafter it is refixed with F3. The outlet valves of O1, O2 and O3 are opened to drain off the water completely from the funnels. The valves are closed and vessels are rinsed.

The soil sample (500g) is stirred in 750 ml of water and is filtered through a coarse sieve. The filtrate is poured into the Erlenmeyer flask and is fixed over the funnel F1. The stopper of the Erlenmeyer flask is removed to regulate the water flow through the feed pipe at a flow rate such that it rises in N1 @ 380 cm/hour and in N2 it should be 975 cm/hour. The required water rise rates are obtained by setting the inflow of water at a particular rate with the help of a flow meter fitted to the feed pipe. By knowing the cross sectional areas of N1 and N2, the water flow rates can be calculated for the required water rise rate according to the following formulae. N1 and N2 are designed with such a diameter that both the required rates of water rise are obtained at a single water flow rate.

Flow rate (ml/hour) = Current speed (cm/hour) × Cross sectional area (cm^2)

Cross sectional area = πr^2

Current speed: N1 = 380 cm/hour, N2 = 975 cm/hour

Diameter: N1 = 3.2 cm, N2 = 2.0 cm

Cross section area of N1(πr^2) = 3.14 × 1.6 × 1.6
= 8.04 cm^2

Cross section area of N2 (πr^2) = 3.14 × 1 × 1
= 3.14 cm^2

Flow rate in N1 (speed × area) = 380 × 8.04
= 3,055 ml/hour

Flow rate in N2 (speed × area) = 975 × 3.14
= 3,061 ml/hour

When the inflow rate is adjusted at 3 L/hour or 50 ml/minute in the flow meter fitted to feed pipe, the counterflow of water keeps the nematodes and small particles into a suspended state (settle very slowly) while heavier soil particles (over 50 μ size) settle relatively rapidly in the container. Small nematodes with a settling rate below 380 cm/hour remain in F1 + N1, whereas in F2 + N2 the nematodes with the settling rate of 380-975 cm/hour are separated from particles of over 100 μ size. The larger nematodes of more than 2 mm length settle in 7-9 minutes in F3 and larger soil particles are collected in the container.

The inflow rate of 3 L/hour is maintained for 20-30 minutes. The overflow of water is collected in the vessel V1 through the overflow pipe and small funnel (f1). In the meantime, the soil particles of over 50 μ size settle from the Erlenmeyer flask into the container. The flow rate is maintained for another 10-15 minutes, thereafter, the clamp of outlet, O1 is removed and the contents are poured into the attached vessel, V1. When N1 is emptied, O2 is opened to collect the suspension in V2.

Finally, the clamp of O3 is removed to collect the contents into V3. After this, the container is emptied.

The contents of vessel V2 are sieved through a set of 100-150 mesh and a set of seven sieves of 150 mesh. If seven sieves are not available, the suspension is re-sieved several times through the 150 mesh sieve. The residue (filtrate) of the sieve is washed into vessel V1. The contents of vessel V1 are poured on a set of seven sieves of 300 mesh and the catch is collected in a small beaker. The contents of vessel V3 are sieved through a set of seven sieves of 60 mesh. The residue is washed several times on a 300 mesh sieve. The nematode suspension (catch) is poured on the wire gauze containing two layers of filter paper in a Baermann funnel and is left over night.

Fluidizing column

In addition to vermiform nematodes, saccate females and cysts of *Heterodera, Globodera* etc., can also be efficiently isolated using a fluidizing column. Much faster flow rates are, however, required for their extraction from soil. Initially a flow rate of 3.5 L/minute is maintained for the first 3 minutes, followed by 7 L/minute for the next 3 minutes. The overflow is caught at 20 and 60 mesh sieves, respectively.

Structure: This is a simple, robust and versatile elutriator. Its structure resembles the measuring cylinder of 1.5-2 litre capacity (Fig. 8). It has an internal diameter of 7.5 cm and column height of 42 cm above the disc of 10 cm, i.e., total height of around 52 cm. It is constructed from a transparent cylinder (perspex) which fits tightly into a short cylindrical disc of 10 cm height. A continuous column of around 50 cm without any disc may also be used. The column is fixed on a bottom plate heavy enough to support the column without any risk of disbalance or tilting. The base of the column contains an inlet pipe at a height of 2-3 cm from the bottom plate. An outlet protrudes from the column at a height of 36 cm, which delivers nematode suspension to an appropriate set of sieves.

Functioning: The column is rinsed with clean water, and is placed in an upright position.

The soil sample (200 g) is mixed in 500 ml of water and is passed through a coarse sieve. The column is filled one third with water, and the soil suspension is poured into it. Water is allowed to inflow through the inlet of the column at a rate of about half of that required to wash over the nematodes with the help of a flow meter for about 3 minutes to mix and fluidize the suspension. Thereafter, the water is run at the full required rate for 3 minutes. The overflow from outlet of the column is collected on appropriate sieves. Inlet flow rates vary with the nematodes which are as follows:

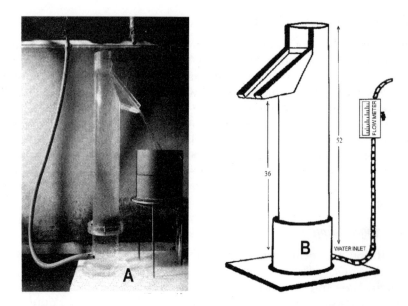

Fig. 8 A- Fluidizing column. B- Line diagram of the column with dimension in cm (Courtesy: Hooper, 1990).

29 ml/minutes: Juveniles of *Heterodera, Meloidogyne, Globodera, Pratylenchus*

291 ml/minutes: *Xiphinema, Longidorus, Trichodorus*

7,000 ml/minutes: Females/cysts of *Heterodera, Globodera*

In practice, about twice these flow rates should be used to ensure good recovery of nematodes. It means that the apparatus should be run for 30 and 300 ml/minute for the first 3 minutes and 60 and 600 ml/minutes for the next 3 minutes for extraction of juveniles of Heteroderidae and Longidoridae or Trichodoridae, respectively.

Extraction of Cysts of *Heterodera* and *Globodera*

Simple Method

Due to the flat size and shape, the cysts float on water. Their isolation from soil is different, infact easier, in comparison to vermiform nematodes. The soil sample should be dried in air before the isolation, so as the cysts could become lighter in weight. A small amount of air dried soil (50 g) is put on a 60 mesh sieve and washed carefully with a slow stream of water. The catch at this sieve is poured into a white bowl with sloping edges. The cysts float along the edge of the bowl and are picked with a fine camel-hair brush.

Fenwick Can Method (Modified)

Structure: This is a simple but specially designed metallic can, made of brass (Fig. 9). The can is 30 cm high, tapering towards the top and with an internal sloping base. Inside the can at the lowest point of the slope, there is a drain hole which is closed with a rubber plug. A sloping collar is moulded just below the rim of the can with a 6 cm upright height of the rim. The collar tapers towards the outlet of the rim (4 cm wide). An inlet of 5 cm from the top is attached to the water supply for filling the can. A large funnel of 20 cm diameter and a stem of 20.5 cm length are supported above it. A coarse sieve (1 mm aperture) is fitted in the funnel.

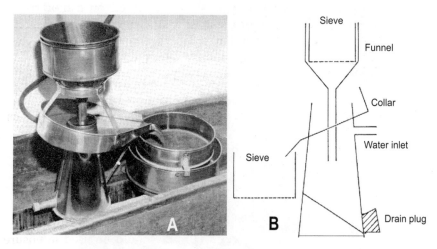

Fig. 9 A- Fenwick Can (Courtesy: Shepherd, 1975). B- Line diagram of the Can

Functioning: Before isolation, the can should be cleaned with water. It is then filled with water and 100 ml air-dried soil is placed in the top sieve of 1 mm pore size. The sample is washed into the apparatus via the funnel. The coarse material is retained on the top. Heavy soil particles like sand sink to the bottom of the apparatus. The floating cysts are carried off over the overflow collar. The cysts, root debris and other particles are collected on a 60 mesh sieve. The catch of this sieve is poured into a white bowl with sloping edge. The catch can also be poured on a white filter paper and the cysts are collected by using a fine camel-hair brush under stereomicroscope.

Semi-Automatic Elutriator

This extraction technique is a combination of Oostenbrink's elutriator fitted with a sample splitter and sieve shaker. It may be used in combination with Baermann trays or funnels. The major advantage of this elutriator is that migratory endoparasites, eggs of *Meloidogyne*, *Heterodera*, *Globodera*, etc., can also be extracted from roots along with soil samples.

Structure: Three to four Oostenbrink elutriators are mounted on a steel tray **(Fig. 10)**. Equal number of sample splitters are fitted in a position such that they receive the soil suspension from the elutriator. The splitter, infact, has two funnels. In the lower funnel there are various outlets which discharge suspension to fine sieves mounted on a motorized shaker. On the upper funnel of the splitter a smaller sieve of desired mesh is kept to separate root parts, larger nematodes (*Longidorus*, *Xiphinema*, etc.) on 35 mesh sieve, and females and cysts on 60 and 400 mesh sieves for smaller nematodes.

Fig. 10 Semi-automatic elutriator (Courtesy: Barker, 1985).

Functioning: Desired flow rate of water is regulated to the elutriator, then 500 g soil is added to it. The elutriator is run for 3-4 minutes, thereafter root parts (along water) are taken to a 35 mesh sieve over the sample splitter. A sieve (10 or 20 mesh) can be placed over 35 mesh sieve to isolate larger nematodes, such as *Longidorus* and *Xiphinema*, which can be collected on 35 mesh sieve. Soil suspension from the sample splitter falls on the 400 mesh sieve mounted on the motorized shaker. A coarser sieve (100, 200 or 300) can supplement the 400 mesh sieve taking into consideration the nematode species to be collected. The catch is rinsed and transferred to a wire guaze with tissue paper in a Baermann funnel.

Roots collected on a 10, 20 or 35 mesh sieve are processed for extraction of nematode larvae or eggs of *Meloidogyne* using an appropriate technique described earlier. Females or cysts of *Heterodera* and *Globodera* can be collected on a 60 mesh sieve under a 10 or 20 mesh sieve placed on the sample splitter.

EXTRACTION OF NEMATODES FROM PLANT TISSUE

Semi-endo and endoparasitic nematodes are extracted from plant tissues. On the basis of nematode/stage and amount of material available, an extraction method is selected.

Direct Examination of Plant Material

If a small amount of material is available, it can be examined directly under a stereomicroscope using transmitted and/or incident light for nematodes assay. The gently washed tissue is taken in petri dishes (without cover) and is teased with a slout needle under the microscope, as a result nematodes are released and collected. The material can be re-examined after a few hours, often deeply embedded nematodes tend to migrate from the damaged tissue in 2-3 hours.

To isolate egg masses and females, the washed roots are placed under the stereomicroscope with incident light. Egg masses (*Meloidogyne* spp.) and females (*Heterodera* and *Globodera*) are picked up with forceps. To collect females of *Meloidogyne* spp., the root tissue beneath the egg mass is gently torn to excise out the creamy or transparent looking shiny females.

Baermann Funnel Technique

Plant material containing nematodes is chopped and placed in muslin or on a wire gauze with tissue paper. The muslin sieve is immersed in water in a Baermann funnel (Fig. 5). A few drops of 0.15% methyl -p-hydroxybenzoate solution is added to water to inhibit microbial activity. The funnel is incubated for 12 hours at room temperature. During incubation the larvae migrate from the tissue, pass through the muslin or tissue paper and aggregate in the beak of the funnel, from where they are collected. This technique is effective in recovering nematodes from seeds of onions, clover (*Ditylenchus* spp.), rice (*Aphelenchoides besseyi*), wheat (*Anguina tritici*), etc. It is advisable to soak seeds in water for a couple of hours before their placement in Baermann funnel. Recovery of *Aphelenchoides* and *Ditylenchus* species from shoot tissue of chrysanthemum and strawberry would be better if 0.15% H_2O_2 water is used in the funnel, which improves oxygenation.

Maceration-Filtration Technique

This technique is more efficient for extraction of nematodes from the plant material and quicker to perform. Roots (5 g) are chopped into 1 cm long pieces and are macerated with 100 ml water in an electric blender for 10-15 seconds. The suspension is poured on a wire gauze sieve having a layer of tissue paper. The sieve is placed in a dish of water for 24-48 hours. During this period the larvae migrate down through the paper and are collected from the dish. Addition of 10 ml of 30% H_2O_2/L water may improve nematode recovery. Baermann funnel can also be used in place of the dish. This technique is used to extract *Radopholus similis*, *Pratylenchus penetrans* and other migratory endoparasitic nematodes from underground parts.

Mistifier Extraction Technique

Nematodes extracted by this technique are often more active than those isolated from other methods because of good oxygenation and washing-off of toxic decomposition products (Fig. 11). A fine mist of water is applied continuously for 1-2 minutes in a 10-20 minute cycle for 4-5 days on the plant material which facilitates migration of nematode larvae from internal tissue to the surface. The plant material is chopped into 3-4 mm pieces and is placed on a wire gauze (with tissue paper) or a 80-100 mesh sieve (without paper). The sieve is kept inside a funnel which is placed on a 250 ml beaker or a large boiling tube. The stem of the funnel should be near the bottom of beaker or tube. Large mesh-wire netting stretching on a tray or a multifunnel holder may support several funnels in the tray. One or two nozzles passing around 4.5 L water/hour at 2.8 kg/cm^2 pressure may provide mist for all funnels. An oil-burner nozzle or gas-jet can also be used. Due to misting, nematode larvae migrate to the surface and then gradually reach the beaker and settle at the bottom. A slow rate of

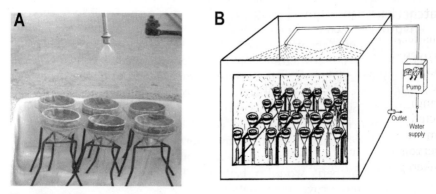

Fig. 11 Mistifier technique. A- Lab arranged model. B- Standard model.

misting is regulated so that nematode larvae do not flow out along the overflowing water from the beaker. The funnel-beaker sets can be fixed in a closed chamber or arranged in a washbasin (Fig. 11). Nematode suspension from beakers can be collected after every 12-24 hours. Volume of the suspension can be reduced by sieving through a 500 mesh sieve. The technique is useful for extracting nematodes from the tissue of roots, stems, leaves, bulbs, seeds, litter, mushroom compost, etc., but is not suitable for the recovery of *Rhadinaphelenchus* and some *Aphelenchoides* spp., as these nematodes swim and are lost in the overflowing water from the beaker or tube.

Root Incubation Technique

Root pieces 5-10 cm long are cut, large diameter roots are first split longitudinally to help nematode emergence. The root pieces are put into screw-capped jars, airtight polyethylene bags or closed petri dishes. The pieces are either wetted well or immersed in shallow water (inside a container and are incubated at 20-25°C) for a few days to weeks. Water from the container is poured off daily and the roots are rinsed, fresh water is added, and containers are recapped. The collected water and rinset are poured on 35 and 500 mesh sieves. The catch from the finer sieve is examined for nematodes. This process is continued for some days. Most of the nematodes are recovered within 4-7 days. *Radopholus* may be recovered in 12 days, *Pratylenchus* may take 5 days to 4 weeks for emergence.

Shoot Incubation Technique

This technique is not very effective for the extraction of nematode larvae from leaves. Wetted chrysanthemum leaves are incubated in polyethylene bags for 3 days. The emerging larvae of *A. ritzemabosi* are collected in condensation droplets on the inner surface of the bag.

Matchstick Extraction Technique

This technique is as efficient as the Baermann funnel or sieving. The recovered specimens are also clearer. It is used to isolate *Aphelenchoides* and *Ditylenchus* species from the fungus-agar culture. Wet matchsticks are erected in a chopped fungus-agar culture medium. A wick (piece of string), from a dish having water and placed at a higher level, is connected to the top of the stick to supply water. As a result the sticks remain wet and receive sufficient water to attract nematode larvae. At intervals of 24 hours the sticks are immersed in water in a beaker and shaken gently to release (or detach) the larvae in water.

Extraction of Cysts and Egg Masses

Cysts, egg masses and/or eggs of sedentary endo and semi-endo parasitic nematodes (*Heterodera, Globodera, Meloidogyne, Rotylenchulus,* and *Tylenchulus*) can be recovered by washing the roots over a strong jet of water. The roots are first gently washed to make them free from soil. Thereafter, the roots are washed vigorously with a strong jet of water over a set of 2 or 3 sieves (80 mesh). To isolate eggs a finer sieve (500 mesh) is used. Several washings by water may make the specimen cleaner.

Extraction of Eggs of Root-Knot Nematodes

Gently washed roots are immersed in 1.0% NaOCl or commercial bleach for 2-3 minutes and are shaken vigorously. The solution is suddenly (within 5 minutes) passed through a 200 mesh sieve and the roots are also rinsed on the sieve several times to detach the partially attached egg masses on the roots. The total time of NaOCl treatment and washing should not exceed 5 minutes, otherwise toxicity may occur to eggs. The collected egg masses are further rinsed and blended with 1% NaOCl for 1 minute to liberate the eggs. The egg suspension is poured on a set of 100 and 400 or 500 mesh sieves. The catch of the finer sieve is washed with distilled water several times to remove any residue of NaOCl. The egg suspension in water is then transferred to a 50 ml centrifuge tube and centrifuged at 1,000 g for 5 minutes. The supernatant is discarded and the pellet and eggs are resuspended in sucrose solution (454 g/L) and centrifuged at 1,000 g for 40 seconds. The supernatant is quickly decanted on a 500 mesh sieve and washed repeatedly with distilled water.

EXTRACTION OF ENTOMOPATHOGENIC NEMATODES FROM SOIL

Different methods such as floatation, Baermann funnel, mist extraction technique, naturally occurring infected insect host (Beaver *et al.*, 1983) and insect trap method can be used to isolate entomopathogenic nematodes (EPN) from soil. The insect trap method is, however, an efficient, standard, handy and widely used technique to isolate EPN from soil. The extraction involves collection of soil, trapping of nematodes and their isolation from the host insect.

Collection of Soil Samples for EPN

To extract (EPN), soil is collected from the site having continously good moisture contents. Soil sample (500 g) is collected from a depth of 2-4 cm after removal of 1 cm top layer or skin of soil. The soil sample can be stored in a polybag under shade at room temperature.

Isolation of EPN

Insect Trap Method

In this technique a susceptible insect host is allowed to be infected by EPN. Thereafter, the insect larva is incubated to facilitate nematode reproduction. A layer of soil sample is put in plastic jars of 500 ml capacity and onto this layer 5-6 *Galleria mellonella* larvae are placed as bait for EPN. The jar is filled to half its capacity with the soil and closed with a lid (Fig. 12A). Perforations are made in the lid for exchange of air, and the jar is incubated at room temperature for 2 weeks. The soil sample is examined after every 24 hours to check mortality of the insects. During the entire process adequate moisture is maintained of adding distilled water. The dead larvae are collected for further processing for the isolation of EPN from the cadaver, using the following technique.

White trap technique

The EPNs from the dead insect larvae are isolated using the White trap technique (White, 1927). In this technique the dead larvae are placed on an inverted watch glass lined with Whatman filter paper placed in a petri dish having sterile distilled water in the surrounding of watch glass (Fig. 12B). The dish is incubated at 22.5°C. The filter paper is kept wet by adding 5-10 drops of sterilized distilled water around the dead larvae daily.

The insect cadavers are observed after every 24 hours to check the nematode activity and emergence from the insect. When nematodes are detected emerging from the cadaver, they are harvested by adding 1-2 ml distilled water on the whatman paper with out disturbing the cadaver. The nematode suspension is collected from the petri dish and transferred to a culture flask for storage upto 3 months at 22.5°C. Harvesting of nematodes is continued till nematodes keep emerging from the insect cadaver. It is strongly recommended to verify Koch's postulates to ensure entomopathogenic status of the isolated nematodes.

Rearing of EPN

The EPNs are generally reared on live insect host. Numerous insects have been found suitable for mass production of EPNs. The final instars of *Galleria mellonella* and *Helicoverpa armigera* are efficient hosts to mass culture the EPNs (Woodring and Kaya, 1988; Kaya and Stock, 1997; Ali *et al.*, 2005). Infective juveniles (100-200) of the nematode are applied to a water soaked filter paper in the petri dish on which 5-10 *Galleria* larvae are transferred. The dish with covered lid is incubated at 22.5°C. Once

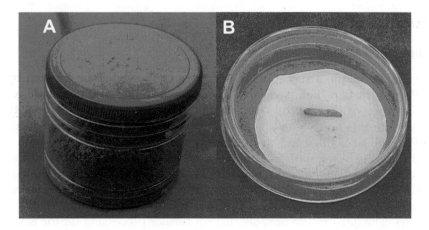

Fig. 12 A plastic jar filled with the soil sample along with *Galleria mellonela* larvae to isolate enotomopathogenic nematodes (A); and a dead larva of *G. mellonella* placed on the White trap in a petri dish (B).

mortality is achieved in the insects, the cadavers are transferred to the White trap and incubated at 22.5°C. After 10-14 days the infective juveniles migrate out from the cadaver in the surrounding water from where they are collected and stored in the nematode culture flask at 22.5°C.

Bioassay of EPN

The infective juvenile suspension obtained from an insect cadaver is cleared and cleaned by washing several times in water on a 500 mesh sieve, through 100 mesh sieve. The cleaned suspension of the nematodes is used in the bioefficacy test.

A single final instar larva of the test insect is kept on the soil layer in 100 ml capacity earthen pots of 8×8 cm dimension (Fig. 13A-C) and the nematode suspension containing 25, 50, 75, 100, 125, 150, 175 and 200 infective juveniles/ml is applied directly to the body surface of the larvae. For each treatment 10-15 replicates should be maintained. Thereafter, the pots are filled almost completely with the moistened and sterilized soil. The pots are covered with muslin to avoid the exit of the test larvae (Fig. 13D). The whole process is carried out in the lab at room temperature. After 24 hours of inoculation the insect larva in each pot is monitored for mortality every 12 hours. The dead larva is transferred to the White trap to further study nematode emergence from the cadaver.

Fig. 13 Bioefficacy test of EPN. A larva being placed in the half filled earthen pot of (8 x 8 cm size) (A); the larvae being inoculated with the EPN suspension (B); a pot filled with the soil after the EPN inoculation on the insect larva (C); and the pot covered with muslin to avoid exit of the test larva (D).

PROCESSING OF NEMATODES

For microscopic study, nematodes extracted from soil or plant material are picked, killed, fixed and mounted. The method is, however, selected on the basis of purpose and type of nematode species to be handled, and involves various steps.

Picking of Vermiform Nematodes

The collected nematode suspension is poured into a watch glass and the desired nematode species is picked with the help of a fine pointed needle. The fine steel needle, bamboo splinter, eyebrow, eyelash hair or horse tail hair are very effective tools for picking nematodes from water suspension. Nematode suspension in a cavity block is placed under a stereomicroscope. The picking device (needle) is held in the right hand, whereas the left hand is placed on the focussing knob. The needle is brought under the nematode larvae perpendicular to its body, then an upward push is given to bring the larvae to the surface. With the left hand the focussing knob is rotated to keep an eye on the larvae. When the nematode has come on the surface, the needle is again brought under perpendicular to the nematode body and with a jerk it is lifted. The nematode is held under surface tension of water on the needle. The picked larva is transferred to a drop of water on a glass slide and is further processed.

Anaesthetizing and Killing of Nematodes

Nematodes are to be anaesthetized or killed to make them immobilized for microscopic examination. Nematodes are damaged if transferred alive to a fixative. Some structures are more easily seen and clearly visible in a temporary mount of an anaesthetized specimen than in dead, fixed or processed nematodes. Sodium azide (0.01 M) is very effective for anaesthetizing nematodes within a few minutes; 0.5-1% solution of propylene phenoxetol in water can also be used to immobilize nematodes. Some of the important methods are as follows:

Direct flame method: The best method to kill nematodes is gentle heating on a flame. A glass slide having a specimen in a drop of water is exposed to a spirit lamp flame for a few seconds. The temperature is assessed by touching the slide with the skin of hand (opposite to palm). It is heated intermittently to the extent that it becomes unbearable to touch. During the heating care must be taken not to let the water drop evaporate. This method is very handy when a few specimens are to be killed.

Hot water method: A safer method of killing a larger sample is the hot water treatment. The volume of suspension is decreased by centrifugation at 1,500 rpm. The concentrated specimen in a test tube is shaken and the tube is plunged in a hot water bath at 65°C for 2-5 minutes or until the nematodes are killed. Thereafter, an equal volume of double strength fixative is added.

Hot fixative method: One of the best methods to kill and fix nematodes by one single process is the use of a hot fixative. Formal-acetic fixative (4:1) is heated at 100°C in a test-tube or dropper in boiling water. The hot fixative is quickly added to nematode suspension in double volume. This method fixes glands and gonads well, and nuclei are clearly visible; use of propionic acid in place of acetic acid followed by 4% formaldehyde may yield better killing and fixing.

Chemical method: Addition of an equal volume of potassium-iodine solution (0.1g I_2 + 0.2 g KI + 100 ml water) to live nematode suspension can provide effective killing. This method is useful when activity of nematodes makes counting difficult. This technique is not recommended for specimens which are to be mounted permanently or examined critically.

Fixing of Nematodes

This is a process which prevents distortion or decay of the nematode body and preserves the organs/structures intact and in the original form. Nematodes can be stored in a fixative indefinitely. Vials containing nematodes should be labelled with identity, if known, locality, fixative used and date of fixation. Fixatives containing alcohol should be avoided as they cause some shrinkage to the body. Well-fixed specimens have a smooth outline. Various fixatives are used to fix nematodes, some of the commonly used fixatives are as follows:

Formalin-glycerol: Solution of 5-10% formalin plus 2% glycerol (glycerin) is widely used. If a small amount of $CaCO_3$ is added to stock solution of formalin, it will prevent darkening and granulation of tissue.

Formalin (40% formaldehyde)	8 ml
Glycerol (glycerin)	2 ml
Distilled water upto	100 ml

Formal-acetic (FA) or formal-propionic (FP) acid: This is another widely used fixative for nematodes. Addition of 2% glycerol helps in preventing drying of specimens if the fixative evaporates slowly, hence the nematodes stored in vials will eventually end up in glycerol which is the best known mounting material for nematodes.

Formalin (40% formaldehyde)	10 ml
Glacial acetic acid or propionic acid	1 ml
Glycerol (glycerin)	2 ml
Distilled water upto	100 ml

Formal aceto-alcohol (FAA): This fixative contains 20% alcohol, as a result a certain amount of shrinkage of the specimen always occurs. This quality is sometimes desired to have the nematode structures plain such as lateral lines and annulations.

Ethanol 95%	20 ml
Formalin (40% formaldehyde)	6 ml
Glacial acetic acid	1 ml
Distilled water upto	100 ml

Triethanol amine formalin (TAF): The TAF has an advantage over FA or FP. With this fixative nematodes retain the life-like appearance for several hours. However, during longer storage in TAF some degeneration in the cuticle may occur. To prevent further degeneration, the specimens should be transferred to FA or FP plus 2% glycerol. Addition of these fixatives to TAF with specimens in excess also works satisfactorily.

Formalin (40% formaldehyde)	7 ml
Triethanol amine	2 ml
Distilled water upto	100 ml

Fixatives for Specific Purpose

Cuticular structures, stylet tip, knobs, etc.: To study cuticular structures, stylet, etc. specimens are first fixed in 4% formaldehyde. A drop of aqueous $KMnO_4$ is added to 4 ml of nematode suspension in formalin and left for 12 hours. The specimens are mounted temporarily in formalin plus $KMnO_4$ for microscopic examination.

Stylet and spicule protrusion: Nematode stylet or spicule is more easily examined and measured if it is protruded. Protrusion can be induced in live specimens by adding aqueous ammonia (density 0.88 g/ml) in nematode suspension (in water) in the ratio 1:50. To a few nematodes it can be done by adding two or three drops of diluted ammonia solution to a drop of water containing nematodes on a glass slide. The spicule and/or stylet will protrude within 1-2 minutes, the effect can be increased by adding more ammonia. Once the desired protrusion has been achieved the specimen should be immediately killed and processed further.

Fixation of Soil Sample before Extraction

During a prolonged storage, the nematode population may increase (egg laying and hatching) or decrease (death and decay). To obtain the actual population occurring at the time of collection, the soil can be fixed by transferring it to formalin-glycerol fixative (100 ml of 40% formaldehyde + 10 ml glycerol + 890 ml distilled water). The fixed soil can be extracted using centrifugal floatation method. This method is also helpful in avoiding the quarantine restrictions.

Processing of Nematodes for Mounting

Internal body details, especially gonad structure of nematodes in a fixative, may gradually diminish or become obscure due to granular appearance of the intestine. The specimen can be cleared by processing with a mountant such as lactophenol or glycerol. It is quicker to mount in lactophenol, and the specimen remains clear for several years but it may deteriorate eventually, especially when kept in light. However, some nematologists have claimed to have the specimen clearer upto 16 or even 30 years. A well-prepared specimen in glycerol remains intact and clear without diminishing any structure infinitely. Some of the common mounting methods are as follows:

Rapid lactophenol method: Lactophenol mountant is prepared from the following ingredients. A cavity slide is filled with the mountant and heated at 65-70°C.

Phenol (liquid)	500 ml
Lactic acid	500 ml
Glycerol	1,000 ml
Methyl blue (cotton blue)	A pinch
Distilled water	500 ml

The heating should be done under an exhaust-hood as the fumes containing phenol are highly toxic. Already fixed nematodes are transferred to the hot lactophenol. The slide is kept hot for 2-3 minutes and is checked under a stereomicroscope for clearing the specimens. The heating can be prolonged to get the desired clearance. Heating should, however, be avoided until the mountant thickens as this may cause distortion or collapse of the specimen. If cotton blue stain is used, the mountant should be heated carefully until the desired depth of stain is achieved. Hot lactophenol should always be used as the cold one may cause irreversible distortion to the nematode body. Some nematodes, such as the members of Hoplolaimidae and Criconematidae, get cleared but do not stain blue. Instead of cotton blue, other stains, e.g., acid fuchsin or picrofuchsin can also be used. Live nematodes can also be processed by lactophenol. Hot lactophenol of double strength is added to a nematode suspension in the cavity slide (half filled) or a test tube. The suspension is heated at 65-70°C for 3-5 minutes on a hot plate or in hot water.

Glycerol methods: Processing of nematodes by glycerol method is time taking but gives excellent specimens which retain originality indefinitely. There are a few methods which take a fairly short time (24-36 hours) but the fact is that the slower the process the better the specimen. Some of the common glycerol methods are as follows:

Glycerol-ethanol method: Nematode specimens from a fixative are transferred into a cavity block containing 0.5 ml glycerol-ethanol mountant (20 ml 96% ethanol + 1 ml glycerol + 79 ml distilled water). The cavity block is placed in a desiccator containing 96% ethanol (one-tenth of the desiccator volume) and is left for 12-16 hours at 35-40°C. This removes almost all the water and leaves the nematodes in a larger volume of mixture of glycerol and ethanol. Thereafter, another glycerol-ethanol mixture (5:95) is added in the cavity block which is placed in a partially covered petri dish to allow a slow evaporation of ethanol which is achieved in 3-4 hours at 40°C, with the result the nematodes are left in pure glycerine. The obtained specimens can be mounted in pure and dehydrated glycerol.

Slow glycerol method: After fixation, nematodes are transferred to diluted glycerol (1.5% in distilled water or in 75% ethanol) in a cavity block. A trace of copper sulphate or a little thymol is added to prevent microbial growth. The cavity block is kept in partially closed petri dishes inside a desiccator. The diluted glycerol (1.5%) evaporates very slowly (3-5 weeks). After desiccation, the specimens remain in pure glycerol. A rapid evaporation may lead to collapse of the nematodes. An oven at 30-60°C can substitute the desiccator.

Staining for Mortality Determination in Nematodes

Nematode eggs or larvae are soaked in 0.05% aqueous solution of blue R dye for 1-24 hours, and cyst nematode eggs upto 7 days. Dead specimens become darkly stained, whereas live ones are unstained. Treatment with 50 ppm chrysoidin results in yellow staining to dead, and orange to live nematodes. Phloxin B (5%) can also differentiate dead (rapidly staining blue) and live (slowly staining blue) nematodes; results are, however, not uniform. Dead nematodes take an amber to dark brown stain with 0.5% $KMnO_4$ (1-2 hours exposure), whereas live ones do not accept the colour.

Mounting of Nematodes

Temporary Mounts

Various external and internal structures are more clearly visible in live, anaesthetized or even a freshly fixed nematode than in an older specimen. In addition, temporary mounting takes very little time and is particularly effective for teaching, diagnostic and predictive purposes. A few ml of nematode suspension is taken in a cavity block or watch glass. The desired number of nematodes are picked and transferred to a large drop of water in the centre of a clean glass slide. Generally water is used to mount unprocessed specimens, otherwise lactophenol or glycerol is used as a mountant. Around the drop, 3-4 glass fibres or small pieces of cover-slip are placed. After ensuring, under a stereomicroscope, that the specimens are near the centre of the drop and are not floating, but are resting on the glass surface, the cover-slip is applied gently. Excess of water, if any, is removed by soaking with a blotter and the cover-slip is sealed firmly from the edge using molten wax or nail polish in such a manner that a wax-ring (circular cover-slip) or wax-square (squarish cover-slip) is formed.

Permanent Mounting

Nematodes can be permanently mounted in lactophenol or dehydrated glycerol, later is more frequently used. Glycerol can be easily dehydrated by keeping it inside a desiccator for 1-2 weeks. Glycerol should not be shaken or stirred vigorously as this may cause formation of enormous small air bubbles. A drop of glycerol is kept on a clean glass slide. The drop should not be spread, instead it should be kept as convex as possible to avoid air bubbles. The drop should be of such a size that it just spreads to the edge of cover-slip without any excess. This, however, depends on the size of cover slip and nematode thickness that comes with practice. Nematodes are picked and mounted in glycerol in the same manner as described in temporary mounting. It is recommended to gently warm the cover-slip over a flame before applying to the drop, because a warmed glass settles more quickly. The cover-slip can be sealed by glyceel, DPX mountant or Canada balsam. The sealing should be done 2-3 times at intervals of 6-12 hours.

Wax-ring Method of Sealing a Mount

In this method a wax-ring is formed on the glass slide with the purpose of surrounding and confining a drop of mountant and holding the cover-slip. To make a wax-ring, a 1.5 cm diameter metallic cork-borer is warmed on a flame and dipped in molten paraffin wax or wax mixture (8 parts wax plus 3 parts petroleum jelly) and applied to the centre of the slide, as a result a hollow circle of wax with raised circumference is formed. For square cover-slips, molten wax can be applied to the slide by a brush to match with the cover-slip. A drop of mountant, e.g., glycerol, is put in the centre of the wax circle or square to which nematode specimens are transferred and 3-4 supports (glass fibre or cover-slip pieces) are placed. A cover-slip is applied on the drop and the slide is placed on a hot-plate at 65°C for a few seconds or in an oven at 70-80°C for a few minutes. In place of a wax-ring, 3 lumps of wax slightly bigger than the mountant drop can also be used. The cover-slip is placed on the lumps and the slide is heated. The wax melts allowing the cover-slip to settle down and confines the glycerol to the centre of the mount. As soon as the wax melts, the cover-slip should be pressed gently with a needle to ensure that it has settled sufficiently. A thick mountant will prevent the use of oil immersion objective. A secondary seal with glyceel, DPX mountant or Canada balsam can be done to ensure proper sealing.

Remounting of Old, Broken or Dry Slides

Partially damaged or dried specimens can be remounted. The sealing ring of the slide is removed and excess mounting fluid is added along the edges of the cover-slip. The cover-slip is carefully lifted by a fine needle. The cover-slip should not be moved laterally as this may cause rolling or skidding of the specimen, and eventually result in extensive damage. The cover-slip is finally lifted and placed on a clean slide in an upside down position. Excess of mounting material is added on the cover-slip as well as on the old slide (from where the cover-slip has been removed) and both are examined under a stereomicroscope. The specimen is picked with an eye-brow or horse-tail hair and heated with excess mounting fluid at 50°C for several hours. Deteriorated specimens often improve considerably if put into hot cotton blue or lactophenol and then remounted in glycerol.

PREPARATION OF ENFACE VIEW, PERINEAL PATTERN AND VULVAL CONE

Enface View

Hard glycerol jelly is prepared by soaking 9 g gelatin in 40 ml distilled water for 2 hours. Thereafter, 50 ml glycerol and 1 ml phenol is added and the mixture is stirred in hot water bath at 50°C till it is homogenized. The mixture is allowed to cool and solidify. A bit of the hardened jelly (sufficient for formation of a drop while melted) is taken on a clean slide and is melted over a flame to form a drop. The drop is spread on one side by a needle. A nematode specimen (processed with glycerol) is placed on the drop (molten jelly) with its head pointing towards the spread out part of the jelly drop, and the jelly is allowed to harden. The nematode head end is cut off by a sharp scalpel, Burradaile needle or a blade at about one head width from lips under a stereomicroscope. The section is transferred to a drop of molten glycerol on another slide. The section is carefully oriented in an upright position. Three glass fibres are arranged radially around the drop and a cover-slip is applied. The cover-slip is adjusted properly on the hardened jelly by gently pressing by a heated needle tip and is fixed by small drops of the ringing material first at three points. After ensuring a proper upright position of the section, the cover-slip is properly sealed with an appropriate material.

Perineal Pattern

Cuticular markings surrounding the vulva and anus are called as posterior cuticular or perineal pattern which is used to identify a mature female of *Melodoigyne* spp. Females in fresh or preserved roots are stained in cotton blue, lactophenol or lactoglycerol, and excised. Use of fresh root material should be preferred as the body content of females are easily removed and the cuticular structures are clearer. The head of the female is removed with a sharp scalpel or a similar tool and the body contents are pressed out through the opening. The posterior part, one-fourth of the body, is cut and transferred to 45% lactic acid for 30 minutes to several hours. The lactic acid facilitates removal of body tissues that adhere to the cuticle. The section is removed from lactic acid and cleared and trimmed to a square containing perineal pattern. The trimmed section is transferred to another drop of 45% lactic acid and the remaining tissues are removed. After proper clearing, the patterns are transferred to a drop of glycerol on a clean glass slide and arranged in rows orienting anus (ventral side) downward and outer cuticle facing the objective of the microscope. A cover-slip is applied gently and sealed by a ringing material. A well-processed pattern in dehydrated glycerol can also be mounted in glyceel, DPX mountant or Canada balsam.

Vulval Cone or Cone Top

The posterior end of cysts have vulva, fenestra and other associated structures which help in the identification of cysts of heteroderids. Dry cysts should be soaked in water for 24 hours. A moist cyst is placed on a glass slide and its posterior end (one-fifth of the body) is cut by a sharp scalpel or blade under a stereomicroscope. The posterior end can be differentiated based on the fact that it is opposite to the neck and may be shorter, wider and straighter. The section is bleached by transferring to 45% H_2O_2 solution for 5-10 minutes in a watch glass or a large drop on a glass slide. The section in H_2O_2 should be examined under a microscope to terminate or prolong the treatment to achieve proper bleaching or to avoid over bleaching. The section can also be cleared by a fine brush and washed in distilled water. The clear section is dehydrated in ethanol series (30, 70, 95 and 100%, 3 minutes for each) and transferred to clove-oil for 30 minutes to get a clearer section. The section can also be stored in clove oil. The section is trimmed to make uniform edges so that it can stand in an upright position. Initially a pinch of Canada balsam is applied in the centre of a glass slide and at three places around which the cone and glass fibres or cover-slip pieces are placed, respectively. A drop of Canada balsam is added to the cone and the cover-slip is applied. Orientation of the cone top must remain in an upright position which can be adjusted by a slight movement of cover-slip with the help of fingers under the microscope. The slide is dried in an oven at 40-50°C. Proper positioning of the cone should be ensured atleast two times after every 6-8 hours. The slide may dry in 3-4 days. Vulval cones may also be mounted in Euparal after passing through 70% ethanol and isobutanol, or directly in glycerol jelly and sealed with an appropriate material.

PRESERVATION AND STAINING OF NEMATODES IN PLANT MATERIAL

Fixation and Preservation of Plant Material

Roots, stems, leaves, flowers, etc., containing nematodes are fixed and stored for subsequent examination and/or staining. The method of fixing and preserving the plant material, however, depends on the ultimate treatment of the tissues. For subsequent staining in lactophenol it is recommended to use lactophenol in the first instance because the tissues get soft and nematodes (especially *Meloidogyne* spp.) are more easily excised. The fixing/preserving in hot (70-80°C) formalin (5-10% formaldehyde), formal-acetic acid (FA) or triethanol amine formalin (TAF) also gives better results. The whole plant or its parts can also be stored in these fixatives; the colour fades at first but returns later. Slow bubbling of SO_2 in a container may clear the cloudiness in the fixative.

Staining of Nematodes in Plant Tissue

To distinguish nematodes from plant cells, the infested tissue is processed with certain stains. Some of the commonly used methods are as follows:

Cotton blue or acid fuchsin in lactophenol method

The plant tissue is rinsed and made free of soil particles and other debris. Small pieces, 3-5 cm long are cut and plunged into boiling cotton blue (0.1% cotton blue) or acid fuchsin stain for 3 minutes in a deep beaker. Multiple samples can be treated in one operation by wrapping them separately in muslin. This also helps to immerse the leaves which tend to float. After 3 minutes the tissue is removed, washed in running water and placed in petri dishes with plain lactophenol. Liquid phenol is used to decolourize leaves and stem. Care should be taken to avoid fumes from phenol. The treated tissue is left for several hours to 2-3 days to allow the tissue to differentiate. Nematodes stain blue or red, whereas plant tissues, except meristematic tissue, remain largely stainless. Differentiation of root tissues can be accelerated by autoclaving the tissue with clear lactophenol in an autoclave at 15 lbs for 10 minutes. The small parts of plant tissue may be pressed gently between two glass slides to make the surface plain and is examined under a microscope.

Lactoglycerol method

To avoid use of toxic phenol, after staining with cotton blue or acid fuchsin, the plant tissue is boiled for 3 minutes in a mixture of glycerol lactic acid and distilled water, and transferred to 50% glycerol in distilled water, acidified with a few drops of lactic acid.

Sodium hypochloride-acid fuchsin method

In this method the plant tissue is first bleached and then stained. Washed roots are chopped into 1-2 cm pieces and are immersed in 1.5% NaOCl solution (chlorine bleach) in a beaker for 4 minutes with intermittent shaking. The material is taken out and rinsed with water for 15-20 minutes to remove residues of the bleach. The material is then transferred to a beaker containing 30 parts water and 1 part stain (3.5 g acid fuchsin, 250 ml acetic acid and 750 ml distilled water) and heated to boil for 30 seconds, thereafter allowed to cool at room temperature. The material is washed with water to remove the excess of stain and transferred to acidified glycerol (a few drops of 5N HCl), heated to boil and cooled. Root pieces are pressed gently between two glass slides and examined microscopically.

Acetic acid-acid fuchsin method

This method is particularly useful for staining nematodes in young root tissues. The roots are fixed in a mixture of acetic acid- acid fuchsin for 4-24 hours. The roots are then transferred to a saturated solution of chloral hydrate for 2-12 hours and examined in lactophenol or lactoglycerol.

Picric acid and aniline blue method

This method can stain the nematodes (blue coloured) in hard and woody roots. Washed fresh roots are cut into small pieces/sections upto 1 cm thickness and fixed in FAA for 48 hours. The roots are taken out, washed with water, cut into thin strips or slices and immersed in water for 2-3 hours. The roots are transferred to 1% safranin solution for 1-2 minutes. Excess of stain is washed in water and the material is immersed in a mixture of aniline blue and picric acid (100 ml saturated picric acid plus 25 ml saturated aqueous solution of aniline blue) in a beaker and heated near boiling point. The pieces sections are taken out and washed in water.

Flemming's solution for staining shoots and roots

A mixture of 1% chromic acid, 2% osmic acid and glacial acetic acid in the ratio of 15:4:1 was prepared by C.C. Flemming. The mixture stains nematode (black) in shoot or root tissue. Direct treatment of green shoot material with Flemming's solution sometimes does not give satisfactory staining as the tissue becomes darker. It is recommended to treat the leaves first with boiling 80% acetone in a hot water bath (acetone boils at 63°C) and let them boil for a few minutes, then leave for 3-4 hours until the green colour disappears. Acetone is drained off and replaced with water 2-3 times. The shoot is transferred to hot water (70-80°C) for 2-3 minutes. The roots are directly treated with the hot water for 3-4 minutes. The shoots or roots are then treated with Flemming's solution for 4-5 minutes. The degree of depth of staining can be controlled by shortening or prolonging the duration of treatment. The treatment must be carried out in an airtight container as the osmic acid vapours are highly toxic to eyes. The stained shoots/roots are washed in running water for 4-12 hours.

ELECTRON MICROSCOPY OF NEMATODES

The electron microscope is a type of microscope that uses electrons to create an image of the target. It has much higher magnification or resolving power than a normal light microscope. A high resolution electron microscope may magnify an image upto million times, thus enabling to see the minutest objects and details. The first electron microscope prototype was built in 1933 by the German engineers Ernst Ruska and Max Knoll. It was based on the ideas and discoveries of French physicist Louis de Broglie. Although it was a primitive model and not fit for practical use, the instrument was still capable of magnifying objects by 400 times. Reinhold Rudenberg, the research director of Siemens, had patented the electron microscope in 1931; the company, however began developing the electron microscope in 1937. The first practical electron microscope was built at the University of Toronto in 1938 by Burton and his students (Burton *et al.*, 1939). Siemens produced the first commercial EM in 1939.

Modern electron microscopes are still based on the Ruska and Knoll prototype. The electron microscope is an integral part of many laboratories and is used to examine biological materials (microorganisms and cells), a variety of large molecules, medical biopsy samples, metals and crystalline structures, and the characteristics of various surfaces. There are basically four types of electron microscopes, among them transmission electron microscope and scanning electron microscope have wide use and application in biological research including nematology.

Transmission electron microscope (TEM): This machine is the original form of electron microscope that involves a high voltage electron beam emitted by a cathode formed by magnetic lenses. The electron beam that has been partially transmitted through the very thin specimen (hence semitransparent for electrons) carries information about the inner structure of the specimen. The spatial variation in the image is then magnified by a series of magnetic lenses until it is recorded by hitting a fluorescent screen, photographic plate, or light sensitive sensor such as a CCD camera. The image detected by the CCD may be displayed in real time on a monitor or computer (Fig. 14).

Resolution of the TEM is limited by spherical aberration and chromatic aberration. A new generation of aberration correctors has been able to overcome spherical aberration. Software correction of spherical aberration has allowed the production of images with very high resolution to show even carbon atoms in diamond separated by only 0.89 angstrom and atoms in silicon separated by 0.78 angstrom at a magnification of 50 million times.

Scanning electron microscope (SEM): Unlike the TEM, where electrons are detected by beam transmission, SEM produces images by detecting secondary electrons which are emitted from the surface due to excitation by the primary electron beam. In the SEM, the electron beam is rastered across the sample, with detectors building up an image by mapping the detected signals with beam position. Generally, the TEM resolution is about an order of magnitude better than the SEM resolution, however, because the SEM image relies on surface processes rather than transmission, it is able to image bulk samples and has a much greater depth of view, and so can produce images that are a good representation of the three-dimensional structure of the sample (Fig. 15).

Reflection electron microscope (REM): Like TEM, this microscope involves electron beams incident on a surface, but instead of using the transmission as in TEM or secondary electrons as in SEM, the reflected beam is detected. This technique is typically coupled with reflection high energy electron diffraction (RHEED) and reflection high energy loss spectrum (RHELS).

Scanning transmission electron microscope (STEM): This is a type of transmission electron microscope. With it, the electrons pass through the specimen, but, as in SEM, the electron optics focus the beam into a narrow spot which is scanned over the sample in a raster. By using a STEM and a high-angle detector, it is possible to form atomic resolution images where the contrast is directly related to the atomic number. This is in contrast to the conventional high resolution electron microscopy technique, which uses phase-contrast, and, therefore, produces results which need interpretation by simulation.

Transmission Electron Microscopy of Nematodes

The transmission electron microscopy (TEM) provides detailed anatomical information, hence needs a fine section of the nematode body or its part to be viewed (Fig. 14). TEM has been found highly helpful in studying the details of minutest structures including sensory organs, ganglion, etc. The basic steps of processing nematodes for TEM are fixation of the specimen, dehydration, infiltration, embedding, section cutting and staining which are as follows.

Isolation of Nematodes for TEM

An appropriate method of isolation of nematodes from soil or plant material should be used which must not be harsh in order to obtain nematode specimens free of aberrations or artifacts. Use of bleach, sucrose, flocculating agents, etc., should be avoided. Nematode specimens should be individually picked by a fine metallic wire, horse tail hair, bamboo splinter or micropipette.

Fixation and Relaxation for TEM

The nematodes are fixed in a cold glutaraldehyde buffer followed by osmium tetraoxide; however, before the latter treatment the nematode

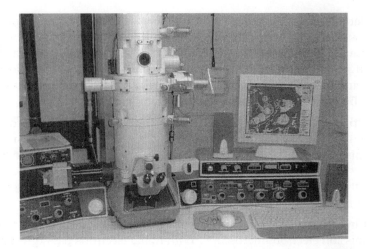

Fig. 14 A complete assembly of transmission electron microscope (TEM)

specimens should be rinsed with cold water (4°C) several times. Researchers have also achieved satisfactory fixing of the specimen with osmium tetraoxide applied prior to glutaraldehyde. It has also been found that addition of formaldehyde to glutaraldehyde followed by osmium tetraoxide gives superior results over the other methods. Fixation of heterodorids has been best achieved with the addition of hydrogen peroxide in glutaraldehyde. Dimethyl sulphoxide (DMSO) can also increase the rate of penetration of fixative. The duration of fixation may vary with the fixative and nematode. The killing of nematodes can be very fast and can be achieved in 5-6 minutes at 4°C or 2 seconds at 70°C with 3% formaldehyde, but it may take one to several hours for complete fixation by the aldehyde.

Nematodes have different body postures like coiled, twisted, semispherical, etc. For proper section cutting it is necessary that the larvae should be relaxed and straightened during the fixation. Application of 1% propylene phenoxetol for 1 minute with acrolein and glutaraldehyde is effective to straighten nematodes at the begining of fixation. Iodine (0.001 N) at room temperature before osmium tetraoxide treatment can also relax nematode larvae. Sodium azide treatment is also effective for relaxation of nematode larvae.

Puncture and Penetration of Fixatives

After proper fixation the nematode specimens are required to be punctured for speedy penetration of the fixative into various organs. Care should be taken during puncturing to avoid pulling of tissue near the puncture. Eye knives may give proper puncturing of cuticle. Similarly, a three sided swing needle point or electrochemically sharpened tungsten needle can help to puncture the cuticle during glutaraldehyde fixation. Heating to aid penetration of chemical fixatives should not be done. The time of puncturing is important and it may vary with the nematode stage. The puncturing in heteroderid adults should be done atleast 2 hours after fixation, whereas the puncture time for their second stage larvae is 30 minutes. Puncturing of eggs is difficult to achieve with a needle or knife. The broken pieces of cover-slips or cellulose acetate sheeting have been found effective in puncturing nematode eggs. Likewise, application of lasers or chitinase enzymes may also cause puncture in egg shells. Thereafter, the specimens are rinsed several times with an appropriate buffer. Cacodylate buffer appears to be an ideal solution to rinse the unbound fixative from nematode body/organs. Its application also reduces osmotic shrinkage in nematode body that may result due to glutaraldehyde fixative. Phosphate buffer has been widely used but sometimes its higher concentration (above 0.1 M) may cause artifactual deposits.

Transfer of Specimens to the Blocks

After fixation, the nematodes are transferred for dehydration and other processes. This can be done by tucking the individual specimens into a shallow agar groove followed by overlaying with a drop of warm agar (45°C). The agar should be made in salt solution of osmolarity similar to the fixative. Sometimes, nematodes after the killing are placed in the agar followed by fixation before or after osmium treatment. The specimens can also be transferred in standard centrifuge tubes, small polypropylene microcentrifuge tubes, small sieves, glass tubes or teflon rings.

Orientation of the Specimens

The position or orientation of nematodes on agar sheets or blocks is important. A safer and precise method is placing nematode specimens in a thin layer of epoxy on a slide treated with a releasing agent or inside a gelatin capsule. Proper orientation of the nematode can also be achieved by placing the specimen between two epoxy sheets or under a small epoxy drop on a prepolymerized plate. Eggs or larvae can be oriented into a distal, compact pellet at the end of a small rod. This can be done by microcentrifuging in a microcapillary containing a tiny polyethylene tube with nematode specimens in unpolymerized epoxy.

Embedding of the Specimens

The embedding involves infiltration of epoxy resin which is a good infiltration agent. The 50% epoxy acetone is added dropwise for 90 minutes.

A puncture in the soft and delicate nematode specimen may improve the infiltration; in a tougher specimen puncturing is of no use. Infiltration may also be achieved with unaccelerated Epon CY212 resin for 3 days at room temperature and in accelerated resin for 12 hours at 30°C. These embedding materials are sensitive to oxygen, hence a gelatin capsule must be used in place of a flat embedding mold.

Sectioning of the Specimens for TEM

For proper section cutting, the embedded specimen should be of desired hardness. This can be achieved with an appropriate adjustment in the epoxy components, proper orientation of the specimen near the block surface and extra polymerization time. If the specimen is not hard enough, it may tear up or fall out of the embedded block during section cutting. Excessive electrostatic current generated during the section cutting may make the serial ribbons stick one above the other. The charge can be neutralized by antistatic guns, ionizing polonium bars or a copper earthing connection with the knife loader. The block is first trimmed for ease in the sectioning. The block grids are coated with parlodion for the serial sectioning. This coating prevents falling out or tearing of larger nematodes, such as saccate females.

Staining of the Sections for TEM

Staining of nematode specimens can be done during fixation or after section cutting. Treatment of serial sections with uranyl acetate and lead citrate give adequate stain to the nematode tissue. However, contrast may be achieved if the blocks before sectioning are stained with aqueous or alcoholic uranyl acetate followed by the staining after the sectioning. Pre-treatment of uranyl acetate may also be given immediately after fixation; while this is done, the specimens should be washed with water several times to rinse cacodylate or phosphate buffers to avoid precipitation due to uranyl acetate treatment. After staining the sections are loaded in the TEM and are viewed as per instructions of the model.

Scanning Electron Microscopy of Nematodes

The scanning electron microscope (SEM) provides high resolution of structures present on the surface of the nematode body. The resolution in comparison to light microscopes is thousand times greater in SEM (Fig. 15). In recent times the SEM has become an indispensable tool in morphological and taxonomical studies of nematodes. The SEM has been found immensely useful to study the minute details of cuticular striations, sensory structures, lip region, posterior end, spicules, sperm, etc., and the resulting informations have helped to overcome several confusions in the taxonomic positions of nematode taxa. This tool is, however, expensive (US$ 0.1-0.15 million) and also involves a lengthy procedure of preparing nematode samples for SEM study which is described in the following paragraphs:

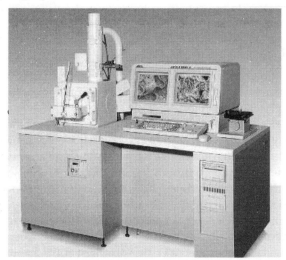

Fig. 15 A complete assembly of scanning electron microscope (SEM)

Isolation of Nematodes for SEM

An appropriate method is used to isolate nematodes both from soil as well as plant tissue. Proper precautions, as indicated for TEM, should be taken to avoid any aberration or injury to nematode specimens.

Killing and Fixing of Nematodes for SEM

There are several methods of killing and fixing of nematodes but an appropriate method is that which performs the two processes in one action and maintains natural integrity and quality of all structures. Glutaraldehyde buffer is one of the best fixatives and yields good results while used at 4% solution (2 ml of 70% glutaraldehyde + 33 ml sodium phosphate buffer). Sequential fixation in the cold is superior to rapid fixation at room temperature. To achieve this, nematode specimens are placed in a known volume of cold tap water (4°C). This is a lethal low temperature and causes death to the nematode in 5-6 minutes. The 4% glutaraldehyde buffer is added drop by drop to the nematode suspension every 15 minutes until the concentration of the suspension is increased to 2%. The suspension is incubated for 1-2 days at 4°C to achieve complete fixation.

Thereafter, the specimens are washed with cold water several times to rinse unbound glutaraldehyde and are treated with 2% solution of osmium tetraoxide for 4-12 hours at 4°C. After the treatment, the specimens are rinsed several times with cold water (4°C) and are allowed to gradually attain room temperature. Another good fixative to prepare nematodes for SEM study is sodium cacodylate buffer. Sodium cacodylate buffer (0.1 M) is prepared by mixing 21.4 g of sodium cacodylate with distilled water to make a final volume of 500 ml and a neutral pH is adjusted with concentrated hydroch/loric acid. However, sodium cacodylate is toxic and may be hazardous to the user.

Dehydration

The step after fixation is dehydration of nematode specimens, which is achieved with a gradual exchange of 5, 10, 25, 50, 75, 90, 95 and 100% ethanol for 15-20 minutes duration in each concentration, whereas 100% ethanol is exchanged three times. For dehydration, the nematodes may be placed in a nucleopore filter envelope (non-fibrous porous paper), and the envelope may be transferred directly from one concentration to another. It is recommended to use the ethanol for 100% treatment from a sealed bottle. The specimens may be sonicated for 1-5 minutes in 30% ethanol to detach the adhered particles of soil, organic matter, fixative, etc., from the nematode body. A longer duration or harsh sonication may damage the specimen.

Drying of Nematodes

After dehydration, the nematode specimens are dried to remove the inter and intra-cellular fluid. The drying can be achieved by freeze-drying or critical point drying. Both the methods are equally effective and can be used depending on the availability and personal choice.

Freeze-drying

The freeze-drying equipment is handy and less expensive. The technique involves rapid freezing of fluids in the nematode specimen followed by sublimation of ice under a high vacuum. The specimens kept in a nucleopore envelope are rapidly immersed in liquified propane cooled by liquid nitrogen for 2 hours. The dried nematode specimens are either immediately mounted on SEM stubs or may be stored in an apendrof tube or nucleopore filter envelope kept inside a desiccator for several weeks to months.

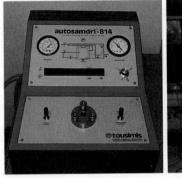

Fig. 16 Critical point dryer (left) and sputtering machine (right)

Critical Point Drying

Critical point drying is an expensive technique but it has a slight edge over the freeze-drying. It is based on the principle that at a certain temperature and pressure a solution reaches a critical point where liquid and gaseous phases are indistinguishable. For this purpose liquid CO_2 is used in the critical point dryer (Fig. 16).

Mounting on SEM Stubs

Before the mounting of nematodes on SEM stubs, electrostatic charge on the nematode specimens is first neutralized with the help of an antistatic gun. An adhesive, silver paint, glue or double sided adhesive cellotape is applied to the SEM stub (Fig. 17). The nematode specimen is very carefully picked from the nucleopore envelope and transferred to the SEM stub with the help of a fine needle having a pinch of adhesive. The nematodes are mounted on the stub in varied positions to view the different structures. A piece of aluminium foil may also be erected on the stub, with the help of some adhesive, and nematode specimens, are mounted on the upper edge of the foil.

Sputtering

Sputtering is the final step of preparation which involves coating of the specimen with 300 Å gold for about 3 minutes. The coating is achieved by using a sputtering machine (Fig. 16). A fine coating of approximately 20 nm of gold is adequate. A thick coat may diminish the surface details. After sputtering, the stub is taken out and mounted inside the vacuum chamber of the SEM, as per instructions of the machine, to view the specimen. The stubs may also be stored in the storage box for viewing later (Fig. 17).

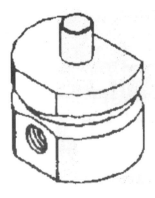 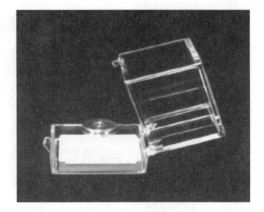

Fig. 17   The SEM stub (left) and SEM mount storage box (right)

BIOCHEMICAL AND MOLECULAR TECHNIQUES

Methodological developments have provided an opportunity to aid the conventional morphological methods to precisely identify nematodes at generic and specific levels and also to characterize races or pathotypes. Certain biochemical and molecular characters, especially proteins and deoxyribo nucleic acid (DNA), are stable characters that do not change under varied environmental conditions, hence can be used in nematode identification. DNA is the genetic code of an organism, whereas proteins are products of the genetic code. A protein analysis reflects amino acid composition, which is determined by the sequence of DNA. Although numerous techniques are used to study proteins and DNA, this chapter, however, describes the techniques which have potential application in nematode identification. Proteins are mainly assayed by electrophoresis whereas DNA by PCR tests followed by gel electrophoresis.

Biochemical Techniques

Protein Electrophoresis

Electrophoresis is a technique that separates molecules in a matrix such as starch, agar, acrylamide or cellulose acetate. Total proteins can be analyzed by staining the matrix containing the electrophoresed protein extract with a protein specific stain. The electrophoresic pattern of total proteins are complex due to the presence of thousands of different proteins. However, specific proteins, such as enzymes or isozymes, may give simple or less complex patterns which vary greatly interspecifically. Isozymes are enzymes that catalyze the reactions, which differ considerably in electrophoresic mobility and are more sensitive to stains than total proteins. This relative sensitivity allows identification of individual nematode in a mixed population. Isozyme analysis has been extensively used to identify root-knot nematode species because of the larger size of a female that yields enough protein to undergo electrophoresis. However, isozyme electrophoresis has not been effective to characterize races of root-knot nematodes (Williamson et al., 1997). Species of *Meloidogyne* from numerous populations have been identified using esterase isozyme patterns (Vovlas et al., 2005). Other isozymes, such as malate dehydrogenase or glucose phosphate isomerase, have also been found effective in identifying *Meloidogyne* spp. Isozyme patterns have also been used to identify *Heterodera glycines, Radopholus similis, Anguina tritici, A. agrostis* and other nematodes. The following procedure is used to study isozyme patterns:

Sample Preparation for Isozyme Study

Nematodes are isolated from soil or excised from roots, and are processed immediately or can be stored in 0.9% saline solution at room temperature for upto 3 weeks. An individual nematode, e.g., female of root-knot nematode is crushed in a small volume of extraction solution (20% sucrose, 2% Triton X-100). The crushed nematode is immediately processed or can be stored at -15°C, as esterase activity remains stable for several months.

Casting of Gels

The gels may be prepared following a simple method. A thin (less than 0.7 mm) acrylamide or agarose gel is used in the electrophoresis system. Finer gels give better resolution matrix for separation of isozymes. The starch or cellulose acetate paper, which is less expensive, can also be used as an alternative to gels. A detailed description on gel electrophoresis is given in DNA assay.

Running of Electrophoresis

The cassette along with the gel is placed in the electrophoresis apparatus, and the upper and lower tanks are filled with the electrophoresis buffer. A pipette is used to remove any air trapped in the gel wells so that the wells become filled with the buffer. The sample tubes containing nematode extracts are removed from the freezer and are placed into a centrifuge at 13,000 g for 15 minutes at 5°C. The centrifugation results in three fractions, the insoluble pellet at the bottom, clear aqueous layer in the middle and opaque layer at the top in the sample tube. The middle layer is transferred carefully into a sample well in the stacking gel. This sample contains sucrose or glycerol, and is more viscous and dense than the electrophoresis buffer, hence remains in the well without mixing with the buffer. Thereafter, a drop of bromophenol blue dye solution is added to the bridge buffer in the upper tank and the electric current at 80 volts is allowed to run for 30 minutes followed by 200 or 300 volts for another 15 minutes. By this time a definite band of bromophenol dye should be visible in the gel. When this band has migrated approximately 100 mm, the power supply is turned off and the cassette is removed from the electrophoresis apparatus.

Staining of Gels

A mixture of buffer substrates, chelators and coenzymes is poured in the cassette. The dye and other photosensitive chemicals are added to the mixture just prior to the staining of the gels. The gel showing bands is cut from the unused/excess portion and is washed under a slow stream of water in a glass dish. The cassette is held sloping towards the dish to allow the gel to slide into the dish without tearing. The water from the dish is poured out and the staining solution is filled in the dish to cover the gel thoroughly. The dish is covered with a polysheet and placed inside an incubator on a gentle shaker (1 horizontal shaking/10 minutes) under darkness at 37°C for 30-60 minutes, depending on proper staining of enzyme bands. Thereafter, the gel is washed in a slow stream of water to remove the staining solution and is stored in 7% acetic solution in an airtight container for 1-2 hours. The enzyme pattern can now be studied and photographed. Isozyme stain for esterase is 25 ml 0.1% sodium phosphate buffer, pH 7.4 + 7.5 mg EDTA + 15 mg fast blue R R salt + 10 mg á napthyl acetate in 0.5 ml acetone, and the stain for malate dehydrogenase is 38 ml water + 5.0 ml solution A (10.6 g sodium bicarbonate + 1.3 g L - malic acid + water to 100 ml final volume) + 7.5 ml Tris HCl (pH 7.1) + 25 mg NAD + 15 mg tetrazolium + 1.0 mg phenazine methosulphate.

Molecular Techniques (DNA Study)

The molecular characterization of nematodes involves isolation of DNA from a nematode followed by its amplification in a polymerase chain reaction(PCR) machine or Thermocycler, and the gel electrophoresis. These tests initially require very small amounts of DNA, that can be sufficiently obtained from a single nematode larva, as small as *Pratylenchus* spp. or a second stage juvenile of *Meloidogyne* spp. The DNA information has enormous potential to provide precise identification of nematodes and to test their phylogenetic relationships. DNA is a bipolymer of complementary strands of nucleotides. There are 4 nucleotides in DNA and their coding in a genome determines identity of an individual, e.g., genome of *M. incognita* comprises about 51 million DNA sequences.

DNA is mainly present in the chromosome (called nuclear, genomic or chromosomal DNA, ch DNA) and also inside the mitochondria in the cytoplasm (mitochondrial DNA, mtDNA). The mtDNA in nematodes have relatively shorter DNA sequence (14,000-36,000 nucleotides) compared to chromosomal or nuclear DNA. The mtDNA is present in hundreds of copies per cell, whereas nuclear/genomic DNA is without a duplication (single copy). The sequence of the mtDNA is evolved more rapidly that makes it useful for observing differences between closely related organisms. Recent biotechnological developments have enabled isolation and amplification of genomic DNA as well as mtDNA and to sequence them. The sequence variability provides huge information that can be used to diagnose nematode genera, species and even their races. In addition to sequencing, use of DNA primers and markers greatly help in characterizing nematode species and races (Williamson et al., 1997). The technique involves mainly three steps, isolation of DNA from a single nematode individual followed by its amplication in a PCR machine, and gel electrophoresis of the amplified DNA. These steps are detailed below:

Isolation of DNA from Nematodes

Isolation of genomic DNA (chDNA)

The genomic DNA can be isolated from a single juvenile or an adult. A nematode larva is crushed by a minute pin in 2.5 µl PCR buffer (10 mM Tris HCl, pH 8.3; 50 mM KCl, 15 mM $MgCl_2$, 0.1% Triton X-100, 0.01% gelatin) and 90 µg/ml proteinase K in a 0.5 ml microfuge tube. The sample is frozen for 10 minutes at -70°C and immediately stored at -70°C or used immediately. For amplification, the sample is overlaid with mineral oil and incubated for 1 hour at 60°C and 15 minutes at 95°C. Amplification is carried out in 25 µl PCR buffer with 0.1 mM dNTP, 0.1 µM primer, and 1.25 units *taq* polymerase under the following programme: 2 minutes at 94°C, followed by 35 cycles of 30 seconds at 94°C, 1 minute at 58°C, and 1 minute at 72°C, followed by 5 minutes at 72°C (Williamson et al., 1997).

Isolation of mitochondrial DNA (mtDNA)

The mtDNA is isolated from the eggs of nematodes. Eggs of a nematode species are isolated from soil (if pure cultured in sterilized soil), plant tissue, egg mass or cyst. The egg suspension is passed through a series of 200-mesh and 500-mesh sieves. Eggs are further separated from the plant and soil debris by centrifugation in a 20 and 40% (w/v) stepwise sucrose gradient at 1,250 g for 15 minutes. The eggs are thoroughly washed on a 500-mesh sieve to remove the sucrose, concentrated by gravity sedimentation at 4°C, and stored in 1 ml portions in labelled vials at –20°C. The frozen eggs are ground to a paste using a 15 ml mortar-pestle. 1-2 ml of extraction buffer (0.1 M Tris HCl, 0.1 M EDTA, 0.2 M NaCl; pH 8) is added to the paste and the suspension is refrozen in the mortar. The frozen material is scraped gently with a rapid motion using a metallic spatula, reground, mixed with 1-2 ml extraction buffer and refrozen. This procedure is repeated three times to ensure sufficient disruption of the eggshells and the cuticles of juveniles in embryonated eggs. Sodium dodecyl sulphate is added to a concentration of 4% (w/v), and the suspension is incubated at 37°C for 3 hours with occasional agitation. For further lysis of the cell and tissue, the homogenate is incubated with 60 mg elastase, 0.5 ml cesium trifluoro acetate (134 g CsTFA in 100 ml stock solution) per ml eggs, and sufficient sarkosyl is added to obtain a final concentration of 8% (w/v) at 50°C for 20 minutes. The CsTFA is added 10 minutes after the start of incubation. Ethidium bromide (40-50 mg/ml) and 3.5 ml of the preparation are added onto 1.5 ml of cushion of CsTFA in extraction buffer (density of 1.60 g/ml) in a polyallomer centrifuge tube, and ultra centrifuged at 150,000 g in a swinging bucket rotor for 16 hours at 20-25°C. Thereafter, the UV-fluorescent DNA band at the interface between the cushion and the suspension are removed with a 16-gauge syringe needle. The covalently-closed circular mtDNA will be precipitated with two volumes of 95% ethanol, washed with 70% ethanol, resuspended in 0.12 M sodium phosphate containing 0.001 M EDTA, pH 6.8, and is stored at –20°C. The DNA concentration can be measured using a fluorometer. DNA amplification will be done using a random decamer primer or with mtDNA primers in a PCR machine.

Amplification of DNA Isolated from Nematodes

Polymerase Chain Reaction Machine or Thermocycler

The thermocycler is a machine that amplifies a derived DNA sequence, irrespective of its origin, through polymerase chain reaction and is commonly known as PCR or PCR machine (Fig. 18). The DNA can be amplified by millions of times within a few hours. PCR is especially valuable because the polymerase chain reaction is highly specific, easily automated and capable of amplifying minutest amount of the sample. Other methods of DNA amplification, such as recombinant technology, are time consuming. The PCR also has a great potential for use in the diagnosis of pathogens including nematodes. A PCR machine, infact, is a thermocycler that adjusts the temperature as per requirement of the process. A PCR machine can work in the range of 4–99°C temperature. Generally, the thermocycler of a PCR machine can increase and decrease the temperature by 3-4°C/second. The change in temperature helps to

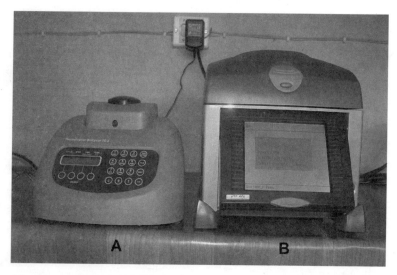

Fig. 18 Thermocyclers or PCR machines: Nongradial (A), and Gradial (B).

perform the three most important steps of PCR (denaturation, annealing and extension) (Fig. 19) in the same vial and no shift of material is required from one machine to another machine. PCR machines are of two types. In a Gradial PCR different thermocycles can be stored which run one after the other, whereas in a Nongradial PCR one cycle is stored and operated (Fig. 18).

Restriction Amplified Polymorphic DNA (RAPD) Analysis

Standard or nematode species-specific primers are used to determine RAPD bands that are eluted from the gel using UI-trafree-MC Filter Unit followed by the cloning using a TA cloning kit. Plasmid DNA from isert-containing clones are extracted using QIA prep-spin Plasmid kit. The DNA sequence of clones is determined by automated sequencer. The DNA sequence is used to design forward and reverse 20-mer primers. The primers do not include the original RAPD primer sequence, and are chosen to have atleast 50% GC content and to lack obvious secondary structure. Multiplex reaction condition are the same as for single primer pairs. To assess the diagnostic utility of nematode specific primers, DNA isolated from a juvenile is subjected to PCR amplification using specific primers. Prior to the assessment, mtDNA isolated from the J_2 is subjected to PCR amplification using mtDNA primers for comparative purpose (Williamson et al., 1997).

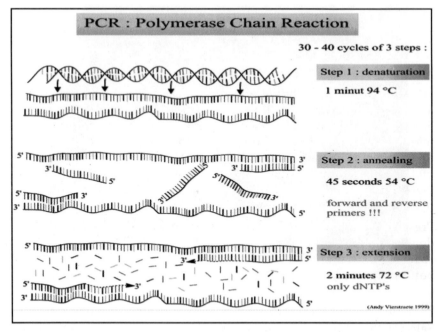

Fig. 19 Schematic presentation of DNA amplification in a PCR machine

Restriction Fragment Length Polymorphism (RFLP) Analysis

Eight restriction endonucleases are used to digest the nematode DNA, which are Msp I, Hpa II, Hha I, Cfo I, Ava I, Bcl I, and Taq I. Enzymatic reactions (15 µl), containing 1.5 µg nematode DNA and 30 units of the enzyme, are carried out in the appropriate enzyme buffer for 8 hours at an optimal temperature. The digestion of nematode DNA under these reaction conditions for 16 hours does not change restriction patterns. Agarose gels are run at 60 V for 105 minutes and stained with ethidium bromide at 0.5 µg/ml. DNA is visualized with a TM-36 transilluminator (302 Km) and photographed on a polaroid type film.

Amplified Fragment Length Polymorphism (AFLP) Analysis

The AFLP procedure for nematodes, according to the original method referred by Zabeau and Vos (1993) and Vos *et al.* (1995), now has some minor modifications, such as choice of the rare cutter enzyme, amount of DNA restricted enzyme and final volume of enzymatic reactions, cycling programmes, and number of amplification cycles in both the preselective and selective PCR reactions. *Hind*III and *Mse*I adapters are used and preselective and selective primers are derived from primers H+0 and M+0, which are complementary to the core of the adapter sequences.

Preselective primers will have one additional A nucleotide at their 3' end (H + A, M + A). Selective primers have three additional nucleotides at their 3' end (H+AAA, M+AAA, M+AAC, M+AAG, M+AAT, M+ACA, M+ACC, M+ACG, M+ACT, M+AGA, M+AGC, M+AFF, M+AGT, M+ATA, M+ATC, M+ATG, M+ATT).

DNA restriction and ligation of adapters: DNA (1 µg) from the sample are double digested with 10 units each of *Hind*III and *Mse*I for 3 hours at 37°C in RL buffer (10 mM Tris HCl, pH 7.5, 10 mM MgAc, 50 mM Kac, 5 mM DTT, 50 Kg/µl BSA) in a final volume of 80 µl. Then, 10 µl of a mixture containing 5 pmoles of *Hind*III adapter, 50 pmoles of *Mse*I adapter, 1 unit of T4 DNA ligase, 1 mM ATP in RL buffer are added to 50 µl of the digestion and incubated for more than 3 hours at 37°C. After ligation, the reaction mixture is diluted 10-folds in $1 \times$ TE buffer and stored at –20°C.

Preselective amplification: 5 µl of primary template resulting from double digestion and adapter ligation is mixed with 75 ng of H + A and 75 ng of M + A primers, 0.2 mM of all four dNTPs, 0.5 units *Taq* in $1 \times Taq$ buffer (appligene) in a final volume of 50 µl. Amplifications are performed in a PCR machine for 28 cycles following the cycle programme of 60 seconds at 94°C, 80 seconds at 60°C, 60 seconds at 70°C.

Selective amplification: The H + AAA primer (5 ng) is labelled using ^{a}P-dATP and T4 polynucleotide kinase for 45 minutes at 37°C and 15 minutes at 70°C. Amplification is performed as described above using 5 g of labelled and 30 ng of unlabelled selective primer (M + three additional nucleotides) and 1 µl of pre-amplified DNA previously five-fold diluted in $1 \times$ TE buffer. The cycle profile is as follows: 60 seconds at 90°C, 80 seconds at 65°C and 60 seconds at 72°C. The annealing temperature is subsequently reduced in each cycle by 0.7°C for the next 12 cycles, and is continued at 56°C for the remaining 24 cycles.

GEL ELECTROPHORESIS

There are two general kinds of gel electrophoresis, i.e., agarose gel electrophoresis and polyacrylamide gel electrophoresis, which are commonly used in DNA and protein analysis. These gels can be poured in a variety of shapes, sizes and porosities that depends primarily on the sizes of fragments being separated.

Agarose Gel Electrophoresis

Agarose gels have a lower resolving power than polyacrylamide gels, but they have a greater range of separation of DNAs from 50 bp to several megabases. Various types, concentrations and configurations of agarose are used that depend on the objective (Table 5). Small DNA fragments

(50-20,000 bp) are best resolved in agarose gels which are run in a horizontal configuration in an electric field of constant strength and direction (Sambrook and Russell, 2001, Brody et al., 2004). Under these conditions the velocity of the DNA fragment decreases as the length of a fragment increases, and is proportional to the electric field strength. Gelatin of agarose results in a three-dimensional mesh of channels with a diameter range of 50 - 200 nm. Agarose is a linear polymer composed of alternating residues of D and L- galactose joined by $\alpha - (1 \rightarrow 3)$ and $\beta - (1 \rightarrow 4)$ glycosidic linkage. The L-galactose residue has an anhydro bridge between the three and six positions. Chains of agarose form helical fibres that aggregate into super coiled structures with a radius of 20-30 nm. Different types of agaroses and buffers are used in the agarose gel electrophoresis.

Table 5 Different types of agaroses

Type of Agarose	Gelling Temp.	Melting Temp.
Standard agaroses Low EEO isolated from *Gelidium* spp.	35-38°C	90-95°C
Standard agaroses Low EEO isolated from *Gracilaria* spp.	40-42°C	85-90°C
High-gel-strength Agaroses	34-43°C	85-95°C
Low melting/gelling temp. (modified) agaroses		
Low melting	25-35°C	63-65°C
Ultra low melting	8-15°C	40-45°C
Low viscocity, low melting/gelling temp. agaroses	25-38°C	70-85°C

Electrophoresis Buffers

Gel loading buffer: A suitable buffer is an important part of gel electrophoresis. The buffer serves three important purposes. It increases density of the sample, ensuring that the DNA sinks evenly into the gel well, adds colour to the sample, thereby simplifying the loading process, and it contains dyes, which in an electric field move towards the anode at predictable rates. Different types of gel loading buffers are used which are enlisted in Table 6.

Electrophoresis buffer: Several different buffers are available for electrophoresis of native double-stranded DNA. These contain *tris*-acetate (TAE), EDTA (pH 8.0, also called E buffer), and *tris*-borate (TBE) or *tris*-phosphate (TPE) at a concentration of ~50 mM (pH 7.5-7.8). Electrophoresis buffers are usually prepared as a concentrated solution

Table 6 Gel Loading Buffers (6x buffer)

Buffer	6x Buffer	Storage Temperature
I	0.25% bromophenol blue 0.25% xylene cyanol FF 40% (w/v) sucrose in H_2O	4°C
II	0.25% bromophenol blue 0.25% xylene cyanol FF 15% Ficoll (Type 400; Pharmacia) in H_2O	Room temp.
III	0.25% bromophenol blue 0.25% xylene cyanol FF 30% glycerol in H_2O	4°C
IV	0.25% bromophenol blue 40% (w/v) sucrose in H_2O	4°C

and stored at room temperature. Important electrophoresis buffers are enlisted in Table 7.

Table 7 Electrophoresis Buffers

Buffer	Working Solution	Stock Solution/Litre
Tris-acetate buffer (TAE)	1×40 mM Tris-acetate 1mM EDTA	50×242 g of Tris base 57.1 ml of glacial acetic acid 100 ml of 0.5 M EDTA (pH 8.0)
Tris-phosphate buffer (TPE)	1×90 mM Tris-phosphate 2 mM EDTA	10×108 g of Tris base 15.5 ml of phosphoric acid (85%, 1.679 g/ml) 40 ml of 0.5 M EDTA (pH 8.0)
Tris-borate buffer (TBE)	0.5×45 mM Tris-borate 1 mM EDTA	5×54 g of Tris base 27.5 g of boric acid 20 ml of 0.5 M EDTA (pH 8.0)

Casting of Gels

Special equipments, viz., clean, dry horizontal electrophoresis apparatus with chamber and comb, or clean dry glass plates with appropriate comb, gel sealing tape, common type of lab tape, microwave oven or boiling water bath, power supply device capable of generating upto 500 V and 200 mA, water bath preset at 55°C, etc., are needed to cast the gel. First of all, the edges of a clean, dry glass plate or the open ends of the plastic tray supplied with the electrophoresis apparatus are sealed with cellotape to form a mould, which is positioned horizontally on a bench. The electrophoresis tanks are filled with electrophoresis buffer (usually 1 × TAE or 0.5 × TBE). The solution is prepared in electrophoresis buffer at a concentration appropriate for separating the particular size fragment expected in the DNA sample(s). An estimated amount of powdered agarose is added to a measured quantity of electrophoresis buffer in an Erlenmeyer flask or a glass bottle and the neck is loosely plugged. The flask is heated in a boiling water bath or microwave oven until the agarose gets dissolved and the suspension becomes clear. The flask is then transferred to a water bath at 55°C; after 10-15 minutes ethidium bromide is added to make its final concentration as 0.5 µg/ml, and is mixed thoroughly by gentle swirling. In the mean time, an appropriate comb is selected to make well-formed sample slots (wells) in the gel. The comb is positioned 0.5-1.0 mm above the plate and the warm agarose solution is poured into the mould. The gel is left for 30-45 minutes at room temperature to allow its complete solidification. Thereafter, a small amount of electrophoresis buffer is poured on top of the gel and the comb is removed carefully. More buffer is added and the tape is also removed carefully.

Some other chemical substances, viz., ethidium bromide, SYBR Gold, acrylamide solution and acrylamide N,N', methylene-bis-acrylamide are used in agarose and polyacrylamide gel electrophoresis. These solutions are prepared according to following methods:

Ethidium bromide solution (10 mg/ml): Ethidium bromide (1 g) is added to 100 ml of distilled water and stirred on a magnetic stirrer for several hours to ensure complete dissolution of the chemical. The solution is either stored in a dark coloured bottle at room temperature or the bottle is wrapped in aluminium foil (States, 2003).

SYBR Gold staining solution: SYBR Gold (molecular probes) is supplied as a stock solution of unknown concentration in dimethylsulfoxide. Agarose gels are stained in a working solution of SYBR Gold, which is a 1:10,000 dilution of SYBR Gold nucleic acid stain in electrophoresis buffer. The working stock solution of SYBR Gold is prepared daily and stored in a dark coloured bottle at regulated room temperature (Tuma *et al.*, 1999).

Acrylamide solution (45% w/v): The solution having 434 g acrylamide (DNA sequencing grade), 16 g N,N'-methylene-bis-acrylamide and 600 ml distilled water is heated at 37°C to dissolve the ingredients and the final volume is made to 1 litre with the water. The solution is filtered through a nitrocellulose filter (0.45 µ pore), and stored in a dark coloured bottle at room temperature.

Acrylamide and N,N'-methylene-bis-acrylamide solution: A stock solution containing 29% (w/v) acrylamide and 1% (w/v) N,N'-methylene-bis- acrylamide is prepared in deionized warm water to assist the dissolution of the bisacrylamide. Acrylamide and bisacrylamide are slowly converted to acrylic acid and bisacrylic acid during storage. This deamination reaction is catalyzed by light and alkali. It should be checked that the deamination pH of the solution is maintained less than 7.0, and the solution is stored in a dark coloured bottle at room temperature.

Ammonium persulfate solution: Ammonium persulfate provides the free radicals that drive polymerization of acrylamide and bisacrylamide. A small amount of a 10% stock solution (w/v) should be prepared in deionized water and stored at 4°C. Ammonium persulfate decomposes slowly, hence fresh solutions should be prepared weekly. The polymerization reaction is driven by free radicals generated by an oxido-reduction reaction in which a diamine (e.g., TEMED) is used as the adjunct catalyst.

Running of Electrophoresis

The casted gel is mounted in the electrophoresis tank and enough electrophoresis buffer is added to cover the gel to a depth of ~1 mm. The DNA sample is mixed with 0.20 volume of the desired 6x gel-loading buffer and the sample mixture is slowly and carefully loaded into the slot (well) of the submerged gel using a disposable micropipette, drawn-out Pasteur pipette or glass capillary tube. The gel tank is covered with the lid and the current is run to allow migration of the DNA towards the positive anode (red lead) (Fig. 20). A voltage of 1-6 V/cm is applied. If the leads have been attached correctly, bubbles should be generated at the anode and cathode (due to electrolysis), and within a few minutes, the bromophenol blue should migrate from the wells into the body of the gel loaded in the cassette.

The current is run until the bromophenol blue and xylene cyanol FF migrate to an appropriate distance through the gel. Thereafter, power is turned off and lid from the gel tank is removed. If ethidium bromide is added to the gel and electrophoresis buffer, the gel is examined under UV light and photographed directly. Otherwise, the gel is stained by immersing in the electrophoresis buffer or water containing ethidium bromide (0.5 µg/ml) for 30-45 minutes at room temperature or by soaking in a 1:10,000-fold dilution of SYBR Gold stock solution in electrophoresis buffer (Suenga and Namura, 2005). Several factors, viz., molecular size of the DNA, concentration of agarose, conformation of the

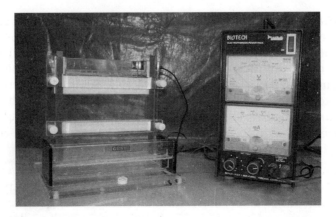

Fig. 20 The gel electrophoresis unit showing gel loaded on the assembly and power system

DNA, presence of ethidium bromide in the gel and electrophoresis buffer, applied voltage, type of agarose, the electrophoresis buffer, etc., may affect the rate of migration of DNA through agarose gels. The DNA separation also depends on the percentage of agarose in the gel (Table 8).

Table 8 Range of separation of DNA fragments through different types of agaroses

	Size Range of DNA Fragments Resolved by Various Types of Agaroses			
Agarose (%)	Standard	High Gel Strength	Low Gelling/ Melting Temp.	Low Gelling/ Melting Temp. Low Viscosity
0.3	-	-	-	-
0.5	700 bp to 25 kb	-	-	-
0.8	500 bp to 15 kb	800 bp to 10 kb	800 bp to 10 kb	-
1.0	250 bp to 12 kb	400 bp to 8 kb	400 bp to 8 kb	-
1.2	150 bp to 6 kb	300 bp to 7 kb	300 bp to 7 kb	-
1.5	80 bp to 4 kb	200 bp to 4 kb	200 bp to 4 kb	-
2.0	-	100 bp to 3 kb	100 bp to 3 kb	-
3.0	-	-	500 bp to 1 kb	500 bp to 1 kb
4.0	-	-	-	100 bp to 500 bp
6.0	-	-	-	10 bp to 100 bp

Polyacrylamide Gel Electrophoresis (PAGE)

Polyacrylamide gels are highly effective for separating small fragments of DNA (5-500 bp). Their resolving power is extremely good, and fragments of DNA that differ in size by as little as 1 bp in length or by as little as 0.1% of their mass, can be separated from one another. They run faster and accommodate much larger quantities of DNA than agarose gels. Upto 10 µg of DNA can be applied to a single slot (1 cm × 1 mm) of a typical polyacrylamide gel without significant loss of resolution. The DNA recovered from polyacrylamide gels is extremely pure and can be used for other demanding purposes. They can be run very rapidly and can accommodate comparatively large quantities of DNA. Polyacrylamide gels, however, have the disadvantage of being more difficult to prepare and handle than agarose gels. Polyacrylamide gels are run in a vertical configuration in a constant electric field. In the presence of free radicals, which are usually generated by reduction of ammonium persulfate by TEMED (N,N,N',N'-tetramethylethylene diamine), vinyl polymerization of acrylamide monomers results in the formation of linear chains of polyacrylamide. When bifunctional cross-linking agents (e.g., N,N' methylene-bis-acrylamide) are included, the copolymerization reaction generates three-dimensional ribbon-like networks of cross-linked polyacrylamide chains with a statistical distribution of pore sizes.

Cross-linked chains of polyacrylamide are used as electrically neutral matrices to separate double-stranded DNA fragments according to size, and single stranded DNAs according to size and conformation. Two types of polyacrylamide gels, viz., denaturating and nondenaturating gels are widely used. The denaturating gels are used for the separation and purification of single stranded fragments of DNA, whereas the nondenaturating polyacrylamide gels are used for the separation and purification of fragments of double-stranded DNA.

Casting of the Gels

Various materials, viz., buffer solutions, acrylamide:bisacrylamide (29:1 w/v), ammonium persulfate (10% w/v), ethanol, 6x gel loading buffers, KOH/methanol, siliconizing fluid (e.g., sigmacote or acrylease), TEMED, electrophoresis apparatus, glass plates, combs, spacers, gel sealing tape (time tape or VWR tape), gel temperature monitoring strips, micropipette with drawn out plastic tip or Hamilton syringe, petroleum jelly and syringe (50 cc) etc., are needed to cast the gel.

The glass plates and spacers are cleaned with KOH/methanol followed by their rinsing in warm detergent solution, first in tap water and then in deionized water. The plates are held by the edges or gloves, to avoid deposition of oils from the hands on the working surfaces of the plates. The plates are rinsed with ethanol and set aside to dry. It is optional but better to treat one of the two plates with siliconizing fluid. The plate is placed on a pad of paper in a chemical fume hood and a small quantity of siliconizing fluid is poured on to the surface. The fluid is wiped over the surface of the plate with a pad of Kimwipes, rinsed in deionized water and then dried with paper towels.

Assembling of glass plates with spacers is done. The larger (unnotched) plates are layed flat on the bench, and the spacers arranged on each side parallel to the two edges. A minute dab of petroleum jelly is applied to keep the spacer bars in position. The inner (notched) plate is positioned, resting on the spacer bars, the plates are clamped together with binder or "bulldog" paper clips, and the entire length of the two sides and the bottom of the plates are bound with gel-sealing tape to make a watertight seal. The acrylamide gel solution of a desired concentration is prepared (Table 9).

Table 9 Volume of reagents used to cast polyacrylamide gels

Reagents to Cast Polyacrylamide Gels of Indicated Concentrations in 1 x TBE [a]

Polyacrylamide Gel (%)	29% Acrylamide +1% N,N'-Methlylene (ml)	Water (ml)	5 x TBE (ml)	10% Ammonium Persulfate (ml)
3.5	11.6	67.7	20.0	0.7
5.0	16.6	62.7	20.0	0.7
8.0	26.6	52.7	20.0	0.7
12.0	40.0	39.3	20.0	0.7
20.0	66.6	12.7	20.0	0.7

[a]*Some investigators prefer to run acrylamide gels in 0.5 x TBE. In this case, the volumes of 5 x TBE and H_2O are adjusted accordingly.*

Since polyacrylamide is poisonous, gloves should be used and casting should be done quickly before the gel starts polymerizing. TEMED is added @ 35 µl/100 ml of acrylamide:bis solution, and is mixed by gentle swirling. The gel solution is taken into a 50 cc syringe. The syringe is inverted and any air that entered the barrel (syringe) is expelled. The nozzle of the syringe is introduced into the space between the two glass plates, and the acrylamide gel solution from the syringe is expelled to fill the space almost to the top. The glass plate is then placed against a test tube rack at an angle of ~10° to the bench top, and an appropriate comb is inserted immediately into the gel, being careful not to allow air bubbles to become trapped under the teeth. The top of the teeth should be slightly higher than the top of the glass clamp. The comb is clamped with bulldog paper clips. The gel mould should be completely filled with the remaining acrylamide gel solution in the syringe, but acrylamide solution must not leak from the gel mould. The gel is left for 30-60 minutes at room temperature for polymerization, and molten acrylamide:bis gel solution may be added if the gel retracts significantly. After complete polymerization, the comb and the top of the gel are surrounded with paper towels soaked in 1x TBE. The entire gel is then sealed in Saran Wrap and stored at 4°C until needed. When ready to proceed with electrophoresis, squirt 1x TBE buffer around and on top of the comb. The comb is then carefully pulled from the polymerized gel. A syringe is used to rinse out the wells with 1x TBE. The gel sealing tape is removed from the bottom of the gel with a razor blade or scalpel.

Running of Electrophoresis

The gel is attached to the electrophoresis tank, using large bulldog clips, and the notched plate should face inward towards the buffer reservoir. The electrophoresis tank is filled with electrophoresis buffer prepared from the same batch of 5x TBE used to cast the gel. Any air bubbles trapped beneath the bottom of the gel are removed with the help of a bent Pasteur pipette or syringe needle. The wells are flushed out with 1x TBE again by using a Pasteur pipette or a syringe. The DNA sample is mixed with an appropriate amount of 6x gel loading buffer (Table 6), and the mixture is loaded into the wells using a Hemilton syringe or a micropipette equipped with a drawn-out plastic tip. The electrodes are connected to the power pack, and current is allowed to flow until the marker dyes have migrated the desired distance as described in agarose gel. Separation of DNA on the gels depends on the composition of the ingredients (Table 10). Thereafter, power is turned off, and the glass plates are detached and the gel sealing tape is removed by a scalpel or razor blade. The glass plates are laid on the bench (siliconized plate uppermost), and spacer or plastic wedge is used to fill a corner of the upper glass plate. Care should be taken to keep the gel attached to the lower

Table 10 Effective range of separation of DNAs in acrylamide gels

Concentration Acrylamide Monomer (%)[a]	Effective Range of Separation (bp)	Xylene Cyanol FF[b]	Bromophenol Blue[b]
3.5	1,000-2,000	460	100
5.0	80-500	260	65
8.0	60-400	160	45
12.0	40-200	70	20
15.0	25-150	60	15
20.0	6-100	45	12

[a] N,N'-methylene-bis-acrylamide is included at 1/30th the concentration of acrylamide.
[b] The numbers given are the approximate sizes (in nucleotide pairs) of fragments of double stranded DNA with which the dye comigrates.

plate. The upper plate, is pulled smoothly and spacers are removed. The gel, together with its supporting glass plate, is wrapped with Saran Wrap, and any air bubbles are smoothed out. To align the gel and the film, adhesive dot labels marked with radioactive ink or with chemiluminescent markers are attached to the surface of the Saran Wrap. The radioactive ink labels are covered with cellophane tape to prevent contamination of the film holder or intensifying screen. The gel is inverted

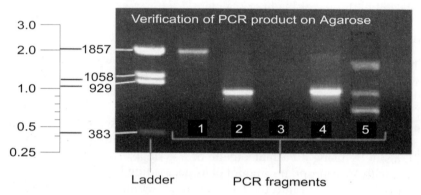

Fig. 21 The ladder is a mixture of fragments with known size to compare with PCR fragments. Lane1: PCR fragment is approximately 1,850 base long. Lanes 2 & 4: Fragments are approximately 800 base long. Lane 3: No product is formed, so PCR failed. Lane 5: Multiple bands are formed because one of the primers fits on differrent places.

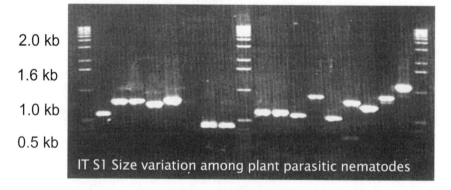

Fig. 22 ITS1 size variation among Secernentean and Adenophorean plant parasites. Here amplicon is varying from 0.5 to 1.2 kb in size. The size variation can be used to identify plant parasitic nematodes, for species identification RFLPs are used.

and exposed to X-ray film (Wilmes and Bond, 2004). A nematode species is characterized on the basis of DNA bands that appear on the gel (Figs. 21-23). Different PCR-based tests, such as AFLP of isolates of *M. hapla* (Liu and Williamson, 2006); *Steinernema feltiae* (Campos-Herrera *et al.*, 2006), ITS 1 of *B. longicaudatus* (Han *et al.*, 2006), and mDNA-16S rRNA of five *Meloidogyne* species (Jayaprakash *et al.*, 2006) can aid in precise characterization of nematodes. Takeuchi *et al.* (2005) and Takeuchi and Futai (2007) have used nested PCR method to identify *Bursaphelenchus xylophilus* in pinewood.

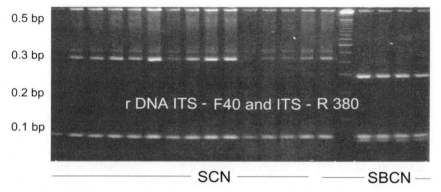

Fig. 23 Gel showing RFLP variation between Soybean Cyst Nematode (SCN) and Sugar Beet Cyst Nematode (SBCN)

3

Morphology of Nematodes

GENERAL STRUCTURE OF NEMATODES

Nematodes are elongated, thread-like (derived from the Greek words, *nema* i.e., thread, and *oides* i.e., form or like), unsegmented (without any segment on the body wall, although in some species intense cuticular modification may confuse), bilaterally symmetrical (the two longitudinal halves are identical, although the anterior region is radially symmetrical), triploblastic (three germ layers, i.e., ectoderm, mesoderm and endoderm) and pseudocoelomic animals (body cavity, the coelom is not true as it is devoid of mesodermal lining). The size of nematodes vary considerably, plant and soil nematodes may be 0.2-11 mm long and 20-100 micron wide (Fig. 24). The most common length of plant parasitic nematodes is 0.5-1.5 mm with the exception of female *Paratylenchus* 0.2-0.3 mm long

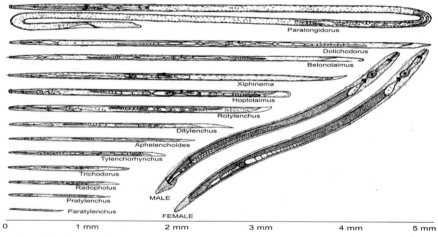

Fig. 24 Schematic presentation of plant nematodes of different size

and *Paralongidorus* 11 mm long. Animal nematodes are much bigger, *Dracunculus medinensis* may attain a metre length, whereas the female giant larvae, *Placentonema gigantissimum* from whales may reach a 8.5 m length and 0.5 m width.

Nematodes have four sides, i.e., dorsal, ventral and two lateral sides. Anteriorly, the body is terminated by an oral opening surrounded by lips, and posteriorly by anus or cloaca and tail (Fig. 25). On the body surface transverse and longitudinal striations or markings are present. The striations are prominently visible, except at anterior and posterior ends. The longitudinal section of the entire length shows two continuous tubes, outer tube — the body cavity, and inner tube — the digestive tract. The digestive system is well-developed consisting of stoma, oesophagus, intestine and rectum. The excretory system is primitive type. The nervous system consists of circum-oesophageal or circum-intestinal commissure (nerve ring) and longitudinal nerves. Circulatory and respiratory systems are absent. The sexes are generally separate, and reproductive organs are tubular and independent. Females have a separate genital pore. Males have a well-developed copulatory organ (spicules) located in the cloaca.

BODY SHAPE

Nematodes are generally eel-like, worm-like or thread-like, and the body tapers at both ends. All nematodes are curved atleast slightly on the ventral side, which is characterized by the presence of genital and anal openings. During movement, it is very difficult to distinguish the body shape; however, upon killing by a gentle heating, nematodes are relaxed giving specific postures. A careful observation represents considerable variation in body shape of nematodes. They may be extremely slender and appear as a glass fibre (*Ecphyadophora, Ecphyadophoroides*), straight (*Pratylenchus*) or arcuate ventrally (*Hoplolaimus*), curved to resemble the "C" alphabet (*Tylenchorhynchus, Belonolaimus*), circular (*Rotylenchus*) or spiral (*Helicotylenchus*) (Fig. 26).

Some nematodes especially females become sedentary after the root penetration. These nematodes incite nurse cells and feed upon them during the immobile phase of life (Fig. 26). At maturity, they assume obesity and become thick (*Cacopaurus pestis*), kidney shaped (*Rotylenchulus reniformis*), pear shaped (*Meloidogyne*), lemon shaped (*Heterodera*), and globular (*Globodera, Nacobbus*). Male individuals of these nematodes, however, remain or become vermiform. This character of variation in body shape and size among males and females is known as sexual dimorphism.

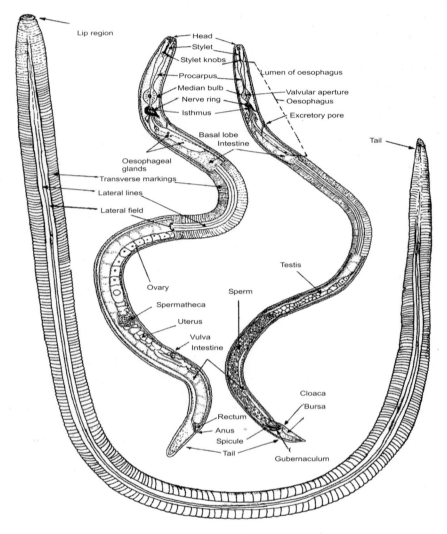

Fig. 25 Schematic diagram of male and female nematodes, showing general structures

BODY WALL

The longitudinal section of a nematode reveals two continuous tubes, one inside the other. The body wall forms the outer tube, and encloses the inner tube (digestive tract) and also reproductive organs. The body wall consists of three layers: outermost is cuticle, hypodermis in the middle, and somatic muscle layer or musculature is the innermost layer.

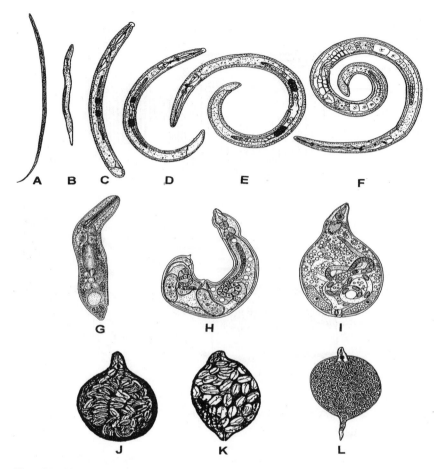

Fig. 26 Variation in the body shape of vermiform (A-F) and saccate (G-L) nematodes. A- *Ecphyadophoroides*, B- *Pratylenchus*, C- *Hoplolaimus*, D- *Tylenchorhynchus*, E- *Rotylenchus*, F- *Helicotylenchus*, G- *Cacopaurus*, H- *Rotylenchulus*, I- *Meloidogyne*, J- *Globodera*, K- *Heterodera*, and L- *Nacobbus*.

The transverse section of the nematode, somewhere in the region from the oesophagus to rectum, reveals four hypodermal chords, i.e., dorsal, ventral and two lateral chords (Fig. 27). The lateral chords are prominent and wider.

CUTICLE

The cuticle is the outermost covering of the body. It is a tough, noncellular and nonliving layer and is secreted by ectodermal cells (hypodermis). It protects the body from unfavourable environmental conditions, acts as

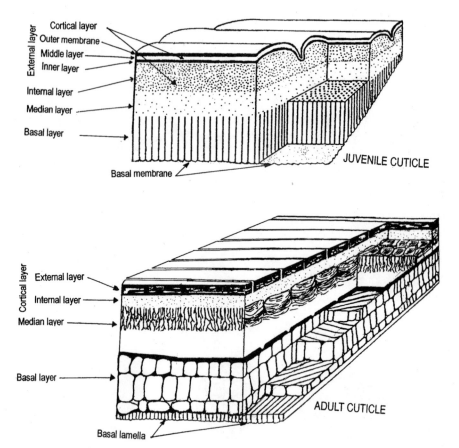

Fig. 27 Schematic three-dimensional diagram of juvenile and adult cuticle of nematode

a barrier to regulate permeability, maintains turgor pressure, and serves as an exoskeleton to maintain body shape and size. The cuticle that covers the outer surface of the entire body is called external cuticle. The cuticle also invaginates into the body through natural openings (oral opening, excretory pore, anus and vulva) and is called internal cuticle. The internal cuticle lines the stoma, oesophagus, excretory duct, vagina, rectum, cloaca, etc. During moulting, the external cuticle and lining of various organs (internal cuticle) shed and are replaced by a new cuticle. The basal part of stylet (feeding apparatus) of tylenchids gets dissolved during moulting.

Structurally, both external and internal cuticles are identical and, indeed, complex showing tremendous variation. The cuticle is multilayered, the number and pattern of the layers, however, may vary with the nematode genera or species. Generally, the cuticle consists of three basic layers (Fig. 27). The outermost is cortical layer, the middle one is median layer or matrix and the innermost is basal layer or fibrillar layer. Bird (1980) recognized four layers, the outermost epicuticle followed by cortical zone, median zone and basal zone. In some nematodes there may be five more layers in the cuticle. They may however, be considered as sublayers of the basic three layers. The sublayers are easily recognizable in juvenile stages. The cuticle may also consist of a delicate thermolabile membrane possibly made of wax or sterol (lipid), which disappears when the nematode is heated above 65°C. The thermolabile layer makes the cuticle quite impermeable. The external cuticle of *Hemicycliophora*, *Hemicriconemoides*, etc., consists of an outer loose fitting sheath and the true external cuticle lies underneath. The number of layers in both cuticles vary. The cuticle of female *Hemicycliophora arenaria* contains seven layers which are in the sheath and five layers in true cuticle, whereas *Ditylenchus dipsaci* has six and five layers, respectively. The male of *Meloidogyne arenaria* has four and six layers in the sheath and body cuticle.

Cortical layer: It consists of two layers viz., external layer and internal layer (Fig. 27). The external layer has three sublayers: (1) outer membrane, (2) middle layer, and (3) inner layer. These layers can be recognized in the cuticle of a juvenile, whereas at maturity they are diminished. The outer membrane is three layered and resembles the plasma membrane. The middle layer is non-osmophilic, whereas the inner layer is osmophilic. Chemically, the entire external layer is rich in keratin, quinone, polyphenols, etc. The cuticle of *Heterodera* cyst has a very high accumulation of tannin. The internal layer has a fibrous structure. Its thickening varies, it is thinner in the preparasitic or juvenile stage than adult. This layer contains RNA and enzymes.

Cortex or median layer: This layer is structurally simple. In the preparasitic forms, it appears electron transparent and contains globular electron dense bodies. In adults, it traverses by thin columns of material which connect this layer to the basal layer. It contains higher concentration of collagen like proteins, lipids, mucopolysaccharides, etc.

Basal layer: It consists of regularly arranged vertical rods or striations. The depth varies with species and stage, e.g., 125 µm thick in *Meloidogyne javanica* larvae, 250 µm in *Globodera rostochiensis* larvae, and 500 µm in *M. javanica* adult male. In adults it becomes multilayered. This layer is composed of a protein with very high linkage between molecules, which is responsible for resistance and protection to nematodes from environmental fluctuations. The basal layer is followed by a fine basal membrane in larvae and basal lamella in adults.

Hypodermis

The layer immediately underneath the cuticle is hypodermis or epidermis. Since this layer is not the outermost, it is not appropriate to refer to it as epidermis. Hypodermis is metabolically one of the most active regions of the nematode body and it secretes as well as maintains the cuticle. Hypodermis is either syncytial (ample cytoplasm with many nuclei) or made up of a single layer of normal cells. The thickening varies, being thicker at chords, especially at lateral chords. The chords provide four interchordal areas on which muscle cells are arranged (Fig. 28A, B). Nuclei of hypodermal cells and nerve chords are located in the chords. In some nematode groups, lateral excretory canal is located in either lateral chords. In marine nematodes (Mermithidae), small additional four chords may be present, hence giving rise to eight interchordal zones

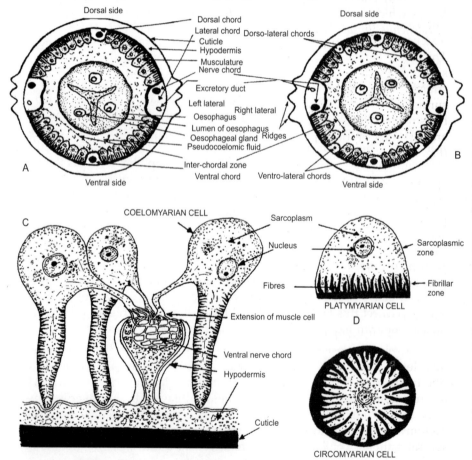

Fig. 28 A, Transverse section of nematode body from oesophageal region to show general structures. B. 8 chords in Mermithidae. C-D, various types of muscle cells and their attachment to nerve chord.

(Fig 28B). Some secretory glands, viz., hypodermal glands, caudal glands, etc., are also present in the hypodermis. The hypodermal glands help to regulate osmotic or ionic concentration of pseudocoelomic fluid. Caudal glands produce adhesive secretions which serve to anchor nematode during mating.

Somatic Musculature

Muscle cells are arranged longitudinally (vertically) beneath the hypodermis in the interchordal zone (Fig. 28). The nematode muscles are unique, i.e., unlike other animals, the muscles send innervations (branches) to nerves rather than having nerves extended from the central nervous system to muscle cells. The shape of muscle cells may vary from rhomboidal (opposite sides equal) to spindle. A muscle cell has two zones, sarcoplasmic zone and fibrillar zone (Fig. 28C-D). The sarcoplasmic zone contains nucleus and sends out protoplasmic protuberances (extensions) to longitudinal nerves at dorsal or ventral chords (Fig. 28C). This zone usually remains towards pseudocoelom. The fibrillar zone is situated next to the hypodermis and consists of numerous oblique (not straight) ribbon shaped contractile fibres lying perpendicular to the cell surface. The fibres are separated from each other by sarcoplasm. On the basis of cell shape and arrangement of contractile fibres, the muscle cells are of three types: platymyarian, coelomyarian and circomyarian (Fig. 28C-D).

Platymyarian type of muscle cell has a wide flat base with contractile elements limited only to the base lying close to the hypodermis (Fig. 28D). The fibres do not extend to the side of the cell that lies towards the pseudocoelom. These cells are considered basic (primitive) and are found in nonparasitic nematodes. Coelomyarian cell has a narrow base. Contractile elements extend to the sides of the cell towards pseudocoelom and fibres partially enclose the sarcoplasm (Fig 28C). Infact, coelomyarian cells have strong neuromuscular relationship. Circomyarian cell is rounded and has contractile fibres all along its circumference (Fig. 28D). This type is rare and is found among specialized muscles of vulva, spicule, etc.

The platymyarian type is considered primitive or basic type of muscle cell which has modified into coelomyarian type by narrowing (the base) and upward elongation of fibrillar zone. Coelomyarian type has further modified to circomyarian type. It is not rare to find muscle cells into the intermediate stage, i.e., platymyarian-coelomyarian, coelomyarian-cricomyarian type. In plant parasitic nematodes, few to several muscle cells may be present in the interchordal zone. Based on the number, there may be different arrangements of muscle cells, viz, holomyarian type (1-2 platymyarian cells), meromyarian type (2-5 platymyarian cells), and polymyarian type (more than 5 coelomyarian cells). The nematodes of primitive origin (nonparasitic, weak parasites) have holo or meromyarian type of arrangement with platymyarian cells, whereas evolved nematodes (*Meloidogyne, Heterodera*) have polymyarian arrangement of coelomyarian cells.

Body Cavity/Pseudocoelom

Body cavity, the coelom is not true as it is devoid of mesodermal lining, hence called pseudocoelom. The body cavity is well-developed and is extended full length from head to tail. The cavity is filled with pseudocoelomic fluid which is rich in protein and other dissolved substances. In addition, pseudocoelomocytes, pseudocoelomic membranes and mesentries are also present in the body cavity fluid. The high osmotic value of pseudocoelomic fluid is responsible for the turgescent state of the nematode body.

CUTICULAR MARKINGS

The external cuticle of nematodes undergoes certain modifications, as a result markings appear on the surface which are also called cuticular striations, annulations or lines. The markings are of three types, i.e., transverse striations, longitudinal striations (Fig. 29), and perineal or cyst wall patterns (Fig. 31).

Transverse Striations and Annulations

The outer cuticle of tylenchid nematodes is marked with transverse grooves which are known as striae (singular stria), and the space between two striae is called interstrial zone. It is generally epicuticle and exocuticle which are involved in the formation of striae.

The stria is continuous (except broken at the lateral field) forming a ring around the nematode body. Striations are of different types depending on the prominence and shape of striae.

Fine striae: Members of a number of families, Tylenchidae, Psilenchidae, Anguinidae, etc., have shallow and narrow striae with a small interstrial zone. Such striae are called fine striae, e.g., *Tylenchus*. They are almost absent in *Polenchus* and *Allotylenchus* (Fig. 29).

Coarse striae: In Hoplolaimidae, e.g., *Hoplolaimus*, the striae are fairly deep, prominent and easily resolvable with comparatively longer interstrial zone (Fig. 29). In Tylenchorhynchinae, e.g., *Tylenchorhynchus*, it is difficult to differentiate between fine and coarse striae.

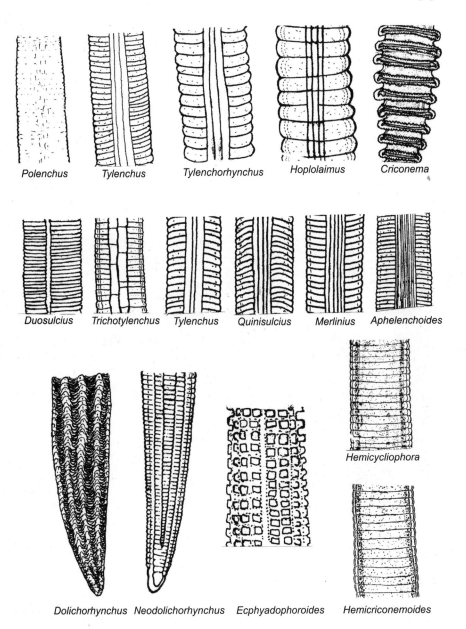

Fig. 29 Schematic representation of variation in transverse lines, lateral lines, longitudinal lines, (*Dolichorhynchus, Neodolichorhynchus*) and body sheath (*Hemicycliophora, Hemicriconemoides*).

Deep striae or annulations: In the members of Criconematoidae, striae are very deep with a wide interstrial zone, dividing the body into superficial segments. These striations are called annulations. The space between two markings is called annule. In *Criconema* and *Ogma*, annules are very prominent (Fig. 30). Annules are provided with cuticular outgrowth which may be spine or scale-like, and are formed due to posterior elongations of annules. The shape and size of annules vary greatly and are of immense taxonomic value. The annulations are present only in females. They lack lateral lines, and the individuals move in a unique fashion, the telescopic movement unlike other nematodes. During motion, annules provide strong body grip with the surface and prevent backward slip.

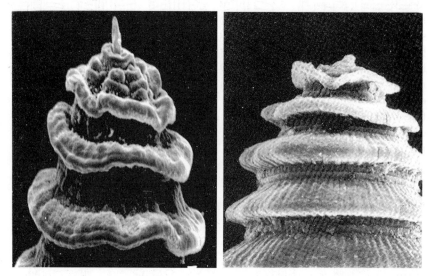

Fig. 30 Scanning electron micrograph of *Criconema* and *Ogma**, showing annules (*Courtesy: M.R. Siddiqi)

Longitudinal Striations

Lateral lines

The area on the cuticle overlaping lateral epidermal (hypodermal) chords is referred to as lateral fields, which are clearly visible under a microscope with some exceptions. The area is usually provided with longitudinal cuticular ridges, running nearly the entire length of the body (Figs. 25, 29). The ridges are most conspicuous at mid-body and tend to diminish posteriorly. Ridges play an important role in nematode movement by preventing bending laterally. In the front-view, ridges do not appear raised structures but look as lines and are called lateral lines. The total number of lateral lines is always one more than the number of ridges in the lateral field. Lateral lines are considered an important diagnostic character and their number varies 2-6. The most common number of lateral lines is 4, followed by 6, 3, 5 and 2, with the exception of 8-16 (Aphelenchida), 2 in the members of Duosulciinae; and 3 in *Trichotylenchus* (Fig. 29). Sometimes lateral ridges are widely separated, hence the number of lines may not be clear. To overcome such doubt, it is advisable to cut a section from mid-body of the nematode and count the ridges. In *Hoplolaimus,* the ridges are low, almost inconspicuous, but lines are present. *Hemicycliophora* and *Belonolaimus* have a simple grooved lateral field. In heterodorids (*Heterodera, Meloidogyne,* etc.), the adult female loses ridges because of obesity of the body, while larval stages do possess ridges and are helpful in identifying a species. Lateral lines are usually straight, but often crenate (notches or like round teeth) corresponding to the transverse striae of the body. Usually only the outer lines are crenate, but sometimes the inner ones may also be crenate. Transverse striae normally do not intrude into the lateral fields. However, sometimes they enter into the field, and it is called areolation. The areolation may be confined to a particular region, e.g., oesophageal region in *Quinisulcius capitatus,* and phasmidial region in *Areolaimus* spp. Areolation may be regular or irregular.

Longitudinal Lines

Ridges, Striations, Lamellae and Grooves

These are present outside the lateral field in some nematode groups (some members of Tylenchorynchinae, Hemicycliophorinae, etc.). In addition to transverse striations, longitudinal lines (grooves) run along the entire length of the nematode body, e.g., *Scutylenchus*. The number of these lines along the body circumference varies with species. In some genera of Tylenchorhynchinae (*Dolichorhynchus, Neodolichorhynchus, Prodolichorhynchus,* etc.) longitudinal lines are few but are raised above the body surface (contour) and are called ridges or lamellae (Fig. 29). If both transverse and longitudinal lines present together, they cross each other and form a series of tiny squarish or rectangular blocks as in *Ecphyadophoroides* (Fig. 29).

Perineal and Cyst Wall Pattern

The external cuticle of adult females of *Meloidogyne* spp. becomes smooth due to the obesity at maturity. As a result, transverse striations and lateral lines disappear except at the posterior end around the anus and vulva, where they, along with phasmids, form a fingerprint like pattern

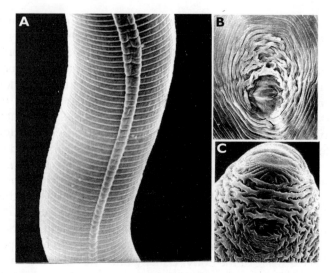

Fig. 31 SEM showing transverse and lateral lines (A); perineal pattern of *Meloidogyne* (B); and cyst wall pattern of *Heterodera* species (C); (Courtesy: Eisenback, 1985)

called perineal pattern (Fig. 31). The pattern consists of transverse striations (dorsal arch and ventral arch), lateral lines, anus, vulva, punctation bodies, etc. Perineal pattern is extremely important to identify a species at adult female stage.

The cuticle of cysts of *Heterodera, Globodera* etc., shows a zig-zag pattern on the surface and punctate pattern immediately on the subsurface (Fig. 31). These patterns may help in the identification of species. In addition, the fenestrae, vulval pattern, under bridge, bullae, etc., are present at the posterior end, and are collectively called as cone top, especially in *Heterodera* spp. (Fig. 31). The cone top provides a reliable diagnostic character at species level. Cysts of some *Heterodera* spp. undergo deposition of chalky looking waxy material which form the subcrystalline layer. The material is produced by a yeast-like symbiotic fungus.

HEAD

It is also called as lip region or cephalic region. The lip region can be best seen at an en-face view of the anterior end of the nematode. This region is hexaradiate with the mouth in the centre surrounded by six lips or labia, two lateral and four submedian in position (two subdorsal and two subventral) (Fig. 32). The lips with regard to shape and size show enormous variation, which form an important diagnostic character at genus or species level.

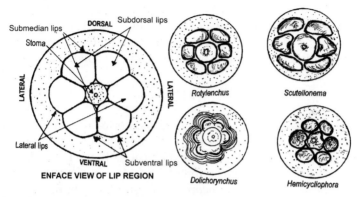

Fig. 32 Schematic diagram of *en-face* view of nematodes and variation in lips

There may be four major types of lip arrangements, i.e., circular, squarish, radiated and raised (Fig. 32). In *Hemicycliophora* all the six lips are somewhat circular in shape. Two lateral lips of *Rotylenchus* are squarish in shape, whereas the submedian lips are flat. The six lips of *Dolichorhynchus* are irregularly lined by transverse striae showing radiated structure with six arms. The lateral lips of *Scutellonema* are large and raised, whereas submedian lips are narrow and flat. In addition, other types of variations may frequently be seen in the head region of nematodes.

Lateral view of head region shows continuity or demarcation (set-off) of lip region from the rest of the body (Fig. 33). The continuous head may be merged (*Paraphelenchus, Polenchus*) or truncate (as cut-off at the

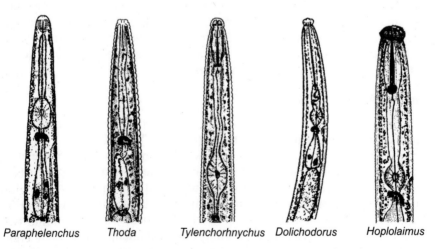

Fig. 33 Lateral view of nematodes showing variation in head region

tip, *Thoda*). The set-off lips may be mildly demarcated (*Histotylenchus*) or the demarcation is moderate (*Tylenchorhynchus*). The lip region in *Dolichodorus, Belonolaimus,* etc., show prominent demarcation, whereas in *Hoplolaimus* and *Rotylenchus* it is annulated and lens like (Fig. 33).

TAIL

The tail is a post-anal elongation or the posterior region of the body beyond anus or cloaca. It is present in all stages of nematodes, juveniles or adults, being smaller in the former. Occasionally the tail is drastically reduced (*Heterodera, Nacobbus*) or absent (*Meloidogyne, Globodera*). In some nematodes (*Belonolaimus, Dolichodorus*), the intestine may be extended in the tail region forming a post-anal sac or blind sac. The tail is essential for locomotion, swimming and copulation. Various structures, such as phasmids, papillae, pores, bursa, caudalids, mucros, etc., are present in the tail region. There are innumerable types of tails which vary in size and shape (Fig. 34). The tail may be filiform, elongated, cylindroid or short, with further variation within each type. Filiform tail, the very long and thread or filament-like, may be acute, sharp and pointed (*Polenchus, Cephalenchus, Tylenchus*) or the tip may become clavate (club-shaped or swollen) as in *Psilenchus* (Fig. 34). Elongated tails may be conoid, hooked or mucronate. In *Ditylenchus,* the tail is like a long cone (conoid), whereas in *Halenchus,* the terminal part of the tail bends ventrally or laterally to become hooked. A mucro or spine (mucronate) may be present on the tail tip of *Hirschmanniella.* Cylindroid tail is shorter but longer about two times the anal body width, e.g., Tylenchorhynchinae (*Tylencho-rhynchus, Dolichorhynchus*). When the length of a short tail is equal or shorter than the anal body width, the tip may be bluntly conoid, (*Rotylenchus, Helicotylenchus*), hemispheroid (*Hoplolaimus, Scutellonema*), or clavate (*Trophurus, Macrotrophurus*). The clavate (club) shape develops due to cuticular thickening at the terminal region of the tail (Fig. 34).

Caudal Alae or Bursa

It is a wing-like cuticular expansion arising from both lateral sides around the cloaca of male nematodes (Fig. 35). The caudal alae is made up of cortical and median layers of cuticle. The main function of the alae or bursa is to provide some protection to the spicule and to help the male nematode to grasp a female during copulation. The shape and size of the bursa vary considerably and form an important taxonomic character. The length of the bursa usually decreases with the increase in tail length and vice versa (Fig. 35). Hence, depending on the tail length, the bursa may be short, subterminal, terminal and notched. In filiform tail the bursa is

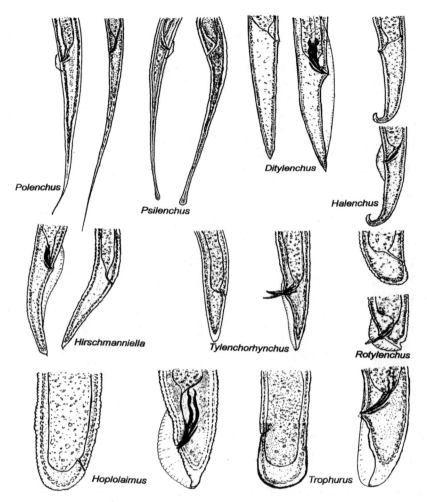

Fig. 34 Schematic diagram of different types of nematode tails, filiform acute (*Polenchus*), clavate (*Psilenchus*), long conoid (*Ditylenchus*), hooked (*Halenchus*), mucronate (*Hirschmanniella*), cylindroid (*Tylenchorhynchus*), short conoid (*Rotylenchus*), short hemispheroid (*Hoplolaimus*) and short clavate (*Trophurus*).

short (*Tylenchus, Psilenchus, Cephalenchus*), whereas in an elongate conoid tail, the bursa is relatively longer and subterminal (*Ditylenchus, Hirschmanniella, Radopholus*). The bursa is large and terminal covering the entire length of the conoid tails as in the family, Tylenchorhynchinae (*Tylenchorhynchus*) and Rotylenchinae (*Rotylenchus*), and in the family Hoplolaiminae it extends a little beyond the tail (*Hoplolamus*). The

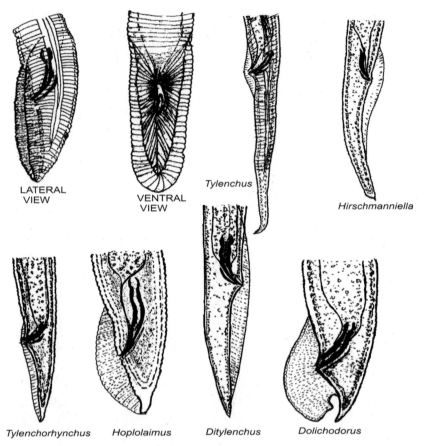

Fig. 35 Lateral and ventral view of bursa or caudal alae and its variations

notched or lobed caudal alae can be seen in *Dolichorhynchus* and *Prodolichorhynchus*. In *Dolichodorus,* the notch is large and deep forming a trilobed alae (Fig. 35).

4

Systematics and Classification of Nematodes

Systematics is a method of classification in a hierarchical system through defining groups and determining their ranking. Taxonomy is a fundamental science which deals with the recognition of taxa. The subject matter covered by both terms overlaps so greatly that they are usually considered as synonyms. More elaboratively, taxonomy is the science of classification including its bases, principles, procedures and rules. It includes the identification of specimens, publication of comparative and descriptive data and the study of diversity and relationships among organisms. The term alpha, beta and gamma taxonomy are frequently used to indicate descriptive, classificatory and evolutionary phases of systematics.

Systematics is a broad scientific study of the principles of classification of organisms, using all variable biological data. It deals with types, diversities, relationships and evolution of organisms and their arrangement into groups and systems. Systematics play an important role in both theoretical and applied biology. It is the only science that deals with the development of a complete understanding of nature and its diversity. Classification, which is the ordering or arrangement of organisms into groups and systems, is based on their relationships that may be phyletic (evolutionary), phenetic (numerical) or both. There are two types of classifications: (1) artificial, and (2) natural; the nature of the classification, however, depends on its purpose. For example, classification of nematodes entirely on the basis of body shape or mode of parasitism is artificial or special classification. The classification based on numerous characters, such as evolutionary, morphological, ecological etc., is a natural classification.

Another term widely used in taxonomy is Nomenclature, which is, infact, naming of a taxa. Naming of nematodes from subspecies to superfamily is governed by the International Code of Zoological Nomenclature (ICZN), which is a collection of rules and recommendations previously adopted by the International Congress of Zoology. The name of a species is binomial, consisting of two words (generic and specific) and the subspecies name is trinomial. The name of a subgenus can be cited in parenthesis between the generic and specific names but is not counted as a part of binomial or trinomial names. The generic and species names are italicized, underlined or in a type face different from that of the text. The generic and specific names begin with upper and lower case alphabets, respectively. All the names of supra generic categories are uninomial, they begin with an upper case letter and are not italicized. Species and suprageneric names are always written in full. Taxonomic categories of nematode nomenclature are enlisted in Table 11.

Table 11 Suffices of International Code of Zoological Nomenclature (ICZN) to assign systemic position to nematodes

Taxonomic Category	Suffix	Name	Name
Kingdom	Variable	Animalia	Animalia
Phylum	-a	Nemata	Nemata
Class	-ea	Secernentea	Secernentea
Subclass	-ia	Diplogasteria	Diplogasteria
Order	-ida	Tylenchida	Aphelenchida
Suborder	-ina	Tylenchina	Aphelenchina
Superfamily	-oidea	Tylenchoidea	Aphelenchoidea
Family	-idae	Tylenchidae	Aphelenchidae
Subfamily	-inae	Tylenchinae	Aphelenchinae
Tribe	-ini	Tylenchini	–
Genus	Variable	*Tylenchus*	*Aphelenchus*
Species	Variable*	*filiformis*	*avenae*
Subspecies	Variable*	*parvus*	–

* If a Latin or Latinized adjectival or participial name, it must agree in gender with the generic name (ICZN Article 34 (b)).

The history of taxonomy of animals or plants is lost in antiquity. Linnaeus (1758, 1767) adopted the most suitable criteria of classification from his predecessors and incorporated them into a workable system of *Systema Naturae*. His binomial system of nomenclature of all kinds of organisms (except viruses, etc.) laid down the foundation of taxonomy. Two genera and three species of nematodes were included in the *Systema*

Naturae. Another early contribution dealing with the taxonomy of nematode was made by J.S. Huxley in 1840 as *The New Systematics*. History of higher classification of nematodes began when L. Orley in 1880 proposed grouping of similar nematode genera into families. However, the first major attempt to classify marine, freshwater, terrestrial and parasitic nematodes above the family rank was made by I. N. Filipjev in 1934; many of his orders stand valid even today. Filipjev in 1934 stated that if classes could be distinguished, he would accept phylum status, nemata proposed by N. A. Cobb in 1919. Chitwood (1937) expanded the classification of Filipjev recognizing the phylum Nematoda and proposed two classes, Phasmidia (Secernentea) and Aphasmidia (Adenophorea). Chitwood (1950) improved his classification of 1937. Hyman (1951) rejected Chitwood's 1950 classification and reassigned nematodes to the phylum Aschelminthes, Grobben, 1909. Neither Chitwood (1950) nor Hyman (1951) agreed with Filipjev's proposal that the most ancestral nematodes were to be found among the marine enoplids. Maggenti (1961) proposed a classification with the basic framework of Chitwood's 1950 proposal, but with certain philosophical differences. The evolution of Secernentea from Araeolaimida was emphasized. De Coninck *et al.*, (1965) proposed two important changes in the hierarchy of Adenophorea, i.e., this class be subdivided into two infraclasses: Enoplia and Chlromadoria. Gadea (1973) proposed that Adenophorea and Secernentea be abandoned and replaced by Enoplimorpha and Chromadorimorpha. Andrassy (1974, 1976) reinstated the basic format that had been emerging prior to Gadea (1973); however, he discarded bifurcation of Adenophorea. Maggenti (1982) in an exhaustive and critical review of phylum-Nemata proposed within Secernentea three subclasses, viz., Rhabditia, Spiruria and Diplogasteria. Within Diplogasteria, Tylenchina was reduced to seven families from more than twenty families. Later, Maggenti in 1983 proposed the subdivision of Enoplia (Adenophorea) into two suborders, Marenoplica and Terrenoplica. The orders, Trichosyringida, Trichocephalida and Mermithida were combined in the single order, Stichosomida. In 1983 an international team comprising A. R. Maggenti, M. Luc, D. J. Raski, R. Fortuner and E. Geraert took up the task of revising the suborder Tylenchina. Absence of Dr. M. R. Siddiqi in the team had been surprising as Dr. Siddiqi was an undisputed authority on the order Tylenchida. The team proposed the classification in 1987 (Luc *et al.*, 1987) and 1988 (Maggenti *et al.*, 1988) that adopted the phylum designation Nemata which was originally proposed by Cobb. A major controversy arose in the order Tylenchida. Maggenti *et al.* (1988) proposed within Tylenchida three suborders: Tylenchina, Aphelenchina and Sphaerulariina. Due to major phyletic and phenetic differences in Tylenchida and Aphelenchida (listed in Table 12), Siddiqi (1980, 1986, 2000) considered them separate orders. Siddiqi agrees that convergent evolutionary similarities exist among Tylenchida and Aphelenchida due to development of a protrusible stylet and its associated muscles and resulting modifications (Siddiqi, 2000). These similarities are far to an extent to merge Tylenchida and Aphenlenchida under a common order. Hunt (1993) have also recognized Siddiqi's order - Aphelenchida in the classification which he proposed for Aphelenchida. Some controversy also exists for the classification of the order Dorylaimida. Jairajpuri and Ahmad (1992) proposed three suborders: Dorylaimina (5 superfamilies, Dorylaimoidea, Actinolaimoidea, Longidoridea, Belondiroidea and Tylencholaimoidea), Nygolaimina (1 superfamily, Nygolaimoidea) and Campydorina (1 superfamily, Campydoroidea), whereas Maggenti (1991) within the order Dorylaimida proposed 3 suborders: Dorylaimina (3 superfamilies, Dorylaimoidea, Actinolaimoidea and Belondiroidea), Diptheraphorinae and Nygolaimina.

Table 12 Comparison of the orders Tylenchida and Aphelenchida (Siddiqi, 2000)

Tylenchida	Aphelenchida
1. Amphids mostly lateral in position.	1. Amphids latero-subdorsal in position.
2. Stylet shaft mostly formed by metarhabdions, lacking innervation; basal knobs formed by telorhabdions, often well developed and marked off.	2. Stylet shaft mostly formed by telorhabdions, anteriorly innervated; basal knobs not a separate entity but represented by thickenings, not marked off from each other at basal region of the shaft.
3. Orifice of dorsal oesophageal gland in procorpus at the base of the stylet or a short distance behind it.	3. Orifice of dorsal oesophageal gland in muscular postcorpus anterior to the central valve-like cuticular thickening.
4. Median oesophageal gland, if present without a muscular valve anterior to the central valve-like cuticular thickening.	4. Median oesophageal bulb always present, with a muscular valve anterior to the central valve-like cuticular thickening.
5. Anus inconspicuous, minute, pore-like, directed outward.	5. Anus conspicuous (except when in a degenerate state), large, crescentic, backwardly directed slit.
6. Sperm usually small-sized, nucleus not showing discrete chromosomes.	6. Sperm large-sized, nucleus showing discrete chromosomes.
7. Male caudal papillae absent; bursa lacking papillary ribs or rays, never present only at the tail tip.	7. Male caudal papillae present; bursa with papillary ribs or rays when it is large and envelopes the entire tail; a short bursa only at the tail tip in several genera.
8. Spicules not thorn-shaped.	8. Spicules mostly thorn-shaped.

To avoid controversies, confusions and dissimilarities among the systems of nematode classification and to provide a universal system of classification acceptable to all those concerned with nematode taxonomy, an attempt in this direction can be initiated by picking and clubbing those systems (at order level) which have received greater recognition and acceptance. In the present chapter a system of nematode classification is proposed, which is based on Maggenti (1991), Jairajpuri and Ahmad (1992), Hunt (1993) and Siddiqi (2000). A valid and healthy criticism is encouraged so as a single system of nematode classification may be developed.

PROPOSED CLASSIFICATION OF NEMATODES

Phylum: Nemata
Class 1: Adenophorea
Subclass : Enoplia
Chromadoria
Subclass A: Enoplia
Superorder: Marenoplica
Terrenoplica
Superorder 1: Marenoplica
Order 1: Enoplida
Superfamily 1: Oxystominoidea
Families: Paraoxystominidae
Oxystominidae
Alaimidae
Superfamily 2: Enoploidea
Families: Enoplidae
Lauratonematidae
Leptosomatidae
Phanodermatidae
Thoracostomopsidae
Order 2: Oncholaimida
Families: Oncholaimidae
Eurystominidae
Symplocostomatidae
Order 3: Tripylida
Suborder 1: Tripylina

Families: Tripylidae
Prismatolaimidae
Suborder 2: Ironina *inquirenda*
Family: Ironidae
Superorder 2: Terrenoplica
Order 1: Isolaimida *inquirenda*
Family: Isolaiimidae
Order 2: Mononchida
Superfamily 1: Mononchoidea
Families: Mononchidae
Mylonchulidae
Cobbonchidae
Anatonchidae
Iotonchulidae
Superfamily 2: Bathyodontoidea
Families: Bathyodontidae
Mononchulidae
Order 3: Dorylaimida
Suborder 1: Dorylaimina
Superfamily 1: Dorylaimoidea
Families: Dorylaimidae
Encholaimidae
Tylencholaimidae
Tylencholaimellidae
Leptonchidae
Belonenchidae
Longidoridae
Superfamily 2: Actinolaimoidea
Families: Actinolaimidae
Brittonematidae
Carcharolaimidae
Trachypleurosidae
Superfamily 3: Belondiroidea
Families: Belondiridae
Roqueidae
Dorylaimellidae
Oxydiridae
Mydonomidae

Suborder 2: Diptherophorina
 Families: Diphtherophoridae
    Trichodoridae
Suborder 3: Campydorina
 Superfamily: Campydoroidea
 Family: Campydoridae
Suborder 4: Nygolaimina
 Families: Nygolaimidae
    Nygolaimellidae
    Aetholaimidae

Order 4: Stichosomida
Superfamily 1: Trichocephaloidea
 Families: Trichuridae
    Trichinellidae
    Trichosyringidae
Superfamily 2: Mermithoidea
 Families: Mermithidae
    Tetradonematidae
Superfamily 3: Enchinomermelloidea
    incertae sedis
 Families: *incertae sedis*
    Enchinomermellidae
    Marimermithidae
    Benthimermithidae

Subclass B: Chromadoria
Order 1: Chromadorida
Superfamily 1: Chromadoroidea
 Families: Chromadoridae
    Hypodontolaimidae
    Microlaimidae
    Spirinidae
    Cyatholaimidae
    Comesomatidae
Superfamily 2: Choanolaimoidea
 Families: Choanolaimidae
    Selachinematidae
    Ethmolaimidae

Order 2: Desmoscolecida
  Families: Desmoscolecidae
  Greeffiellidae
Order 3: Desmodorida
  Superfamily 1: Desmodoroidea
  Families: Desmodoridae
  Ceramonematidae
  Monoposthiidae
  Superfamily 2: Draconematoidea
  Families: Draconematidae
  Epsilonematidae
  Prochaetosomatidae
Order 4: Monhysterida
  Families: Linhomoeidae
  Siphonolaimidae
  Monhysteridae
  Scaptrellidae
  Sphaerolaimidae
  Xyalinidae
  Meyliidae
Order 5: Araeolaimida
  Suborder 1: Araeolaimina
  Families (unassigned): Axonolaimidae
  Camacolaimidae
  Tripyloididae
  Superfamily 1: Araeolaimoidea
  Families: Araeolaimidae
  Cylindrolaimidae
  Diplopeltidae
  Rhabdolaimidae
  Superfamily 2: Plectoidea
  Families: Plectidae
  Leptolaimidae
  Haliplectidae
  Bastianidae

Class 2: Secernentea
    Subclass: Rhabditia
        Spiruria
        Diplogasteria
    Subclass A: Rhabditia
        Myenchidae: *Familium incertae sedis; dubium*
        Order 1: Rhabditida
            Suborder 1: Rhabditina
                Family: Alloionematidae
                    incertae sedis
                Superfamily 1: Rhabditoidea
                Families: Rhabditidae
                      Rhabditonematidae
                      Odontorhabditidae
                      Steinernematidae
                      Rhabdiasidae
                      Angiostomatidae
                      Agfidae
                      Strongyloididae
                      Syrphonematidae
                      Heterorhabditidae
                      Carabonematidae
                      Pseudodiplogasteroididae
                Superfamily 2: Bunonematoidea
                Families: Bunonematidae
                      Pterygorhabditidae
                Superfamily 3: Cosmocercoidea
                Families: Cosmocercidae
                      Atractidae
                Superfamily 4: Oxyuroidea
                Families: Oxyuridae
                      Thelastomatidae
                      Rhigonematidae
                Superfamily 5: Heterakoidea
                Families: Heterakidae
                      Ascaridiidae

Suborder 2: Cephalobina
    Families: Cephalobidae
          Robertiidae
          Chambersiellidae
          Elaphonematidae
    Superfamily 1: Panagrolaimoidea
    Families: Panagrolaimidae
          Alirhabditidae
          Brevibuccidae

Order 2: Strongylida
    Families (unassigned):
          Diaphanocephalidae
          Metastrongylidae
          Maupasinidae
    Superfamily 1: Strongyloidea
    Families: Strongylidae
          Cloacinidae
          Syngamidae
    Superfamily 2: Ancylostomatoidea
    Families: Ancylostomatidae
          Uncinariidae
          Globocephalidae
    Superfamily 3: Trichostrongyloidea
    Families: Trichostrongylidae
          Amidostomatidae
          Strongylacanthidae
          Heligmosomatidae
          Ollulanidae
          Dictyocaulidae

Subclass B: Spiruria
    Order 1: Ascaridida
        Superfamily 1: Ascaridoidea
        Families: Ascarididae
              Toxocaridae
              Anisakidae
              Acanthocheilidae
              Goeziidae

>   Crossophoridae
>   Heterocheilidae
> **Superfamily 2: Seuratoidea**
>   Families: Seuratidae
>   Schneidernematidae
>   Quimperiidae
>   Subuluridae
>   Cucullanidae
> **Superfamily 3: Camallanoidea**
>   Families: Camallanidae
>   Anguillicolidae
> **Superfamily 4: Dioctophymatoidea**
>   Families: Dioctophymatidae
>   Soboliphymidae
> **Superfamily 5: Muspiceoidea** *incertae sedis*
>   Families: Muspiceidae
>   Robertdollfusidae
>   Phlyctainophoridae

Order 2: Spirurida
> **Superfamily 1: Spiruroidea**
>   Families: Spiruridae
>   Thelaziidae
>   Acuariidae
>   Hedruridae
>   Tetrameridae
> **Superfamily 2: Drilonematoidea**
>   Families: Drilonematidae
>   Ungellidae
>   Scolecophilidae
>   Creagrocericidae
>   Mesidionematidae
>   Homungellidae
> **Superfamily 3: Physalopteroidea**
>   Families: Physalopteridae
>   Megalobatrachone-matidae
>   Gnathostomatidae

Superfamily 4: Dracunculoidea
Families: Dracunculidae
Philometridae
Micropleuridae
Superfamily 5 : Diplotriaenoidea
Families: Diplotriaenidae
Oswaldofilariidae
Superfamily 6: Filarioidea
Families: Filariidae
Aproctidae
Setariidae
Desmidocercidae
Onchocercidae

Subclass C: Diplogasteria
  Order 1: Diplogasterida
Families: Diplogasteridae
Odontopharyngidae
Diplogasteroididae
Cylindrocorporidae *incertae sedis*
  Order 2: Tylenchida
Suborder 1: Tylenchina
Infraorder 1: Tylenchata
Superfamily 1: Tylenchoidea
Families: Tylenchidae
Ecphyadophoridae
Atylenchidae
Tylodoridae
Infraorder 2: Anguinata
Superfamily 1: Anguinoidea
Families: Anguinidae
Sychnotylenchidae
Suborder 2: Hoplolaimina
Superfamily 1: Hoplolaimoidea
Families: Hoplolaimidae
Rotylenchulidae
Pratylenchidae

          Meloidogynidae
          Heteroderidae
    Superfamily 2: Dolichodoroidea
    Families: Dolichodoridae
          Belonolaimidae
          Telotylenchidae
          Psilenchidae
Suborder 3: Criconematina
    Superfamily 1: Criconematoidea
    Family: Criconematidae
    Superfamily 2: Hemicycliophoroidea
    Families: Hemicycliophoridae
          Caloosiidae
    Superfamily 3: Tylenchuloidea
    Families: Tylenchulidae
          Sphaeronematidae
          Paratylenchidae
Suborder 4: Hexatylina
    Superfamily 1: Sphaerularioidea
    Families: Neotylenchidae
          Sphaerulariidae
          Paurodontidae
          Allantonematidae
    Superfamily 2: Iotonchioidea
    Families: Iotonchiidae
          Parasitylenchidae
Order 3: Aphelenchida
    Suborder : Aphelenchina
        Superfamily 1: Aphelenchoidea
        Family: Aphelenchidae
            Subfamily: Aphelenchinae
        Family: Paraphelenchidae
            Subfamily: Paraphelenchinae
        Superfamily 2: Aphelenchoidoidea
        Family: Aphelenchoididae
            Subfamily: Aphelenchoidinae
            Subfamily: Anomyctinae

Family: Seinuridae
　Subfamily: Seinurinae
Family: Ektaphelenchidae
　Subfamily: Ektaphelenchinae
Family: Acugutturidae
　Subfamily: Acugutturinae
　Subfamily: Noctuidonematinae n. subfam.
Family: Parasitaphelenchidae
　Subfamily: Parasitaphelenchinae
　Subfamily: Bursaphelenchinae
Family: Entaphelenchidae
　Subfamily: Entaphelenchinae

CLASS I: ADENOPHOREA

Six lip-like structures are present around the oral opening. The buccal cavity or stoma may be equipped with jaws or teeth or a hollow stylet. The amphids are postlabial. Sixteen cephalic sensory structures are present. Somatic setae, hypodermal glands, and somatic papillae are generally present. Hypodermal cells are uninucleate. External cuticular striae are not clearly visible. The excretory system is single celled and usually lacks a duct. Generally, there are six or more coelomocytes in the pseudocoelom. Three caudal glands are present in selected groups. Males generally have two testes and a paired spicule and caudal alae is rarely present. Upon hatching of an egg, first-stage juvenile emerges and first moulting occurs outside the egg.

The class Adenophorea is divided into two subclasses: Enoplia and Chromadoria.

Subclass A: Enoplia

The Enoplia subclass possesses a diversity of characters. In pouchlike amphids, the aperture is transverse or cyathiform. The aperture may be porelike, ellipsoid, or greatly elongated if the internal structure is tubiform. The cephalic sensilla are papilliform, setiform, or a mixture of both. Caudal glands are generally present in marine nematodes. The oesophageal glands (usually 5 in number) open either into the anterior oesophagus, through stomatal teeth, or posterior to the nerve ring. The shape of the oesophagus varies from cylindrical to bottle-shaped. Cuticular striae are usually absent on the external cuticle.

The subclass is divided into two superorders: Marenoplica and Terrenoplica.

Superorder 1: Marenoplica

Marine nematodes and also some soil and fresh water forms are grouped into this superorder. Out of 16 cephalic sensory structures, 6 are labial and 10 are postlabial. The latter are generally setiform. Five oesophageal glands are generally present which open anteriorly into the stoma or into the anterior portion of the oesophagus. A single excretory cell and three caudal glands are present. Prominent preanal supplements in a single ventromedial row are present in males.

There are three orders in the superorder: Enoplida, Oncholaimida and Tripylida.

Order 1: Enoplida

Members of Enoplida have a very thick cuticle (double) of the cephalic region forming a cap or helmet. The stoma may or may not possess movable jaws or teeth. The oesophagus is cylindrical and the cardia is prominent. Oesophageal glands (5 nos.) are located in the posterior region of the oesophagus; however, the orifices for these glands are located anterior to the nerve ring. The amphid apertures may be transverse slits or oval elongate. The cephalic sensilla are most often in two whorls. A single excretory cell is usually present which opens at the level of or anterior to the nerve ring. Males may or may not have supplementary organs and the spicule is paired. Caudal glands and a cuticular spinneret are found in males and females.

There are two superfamilies: Oxystominoidea and Enoploidea.

Superfamily 1: Oxystominoidea

The labial region is not setoff by a distinct groove. The cephalic region is narrowed, and the stoma is weakly developed and surrounded by oesophageal tissue. The cephalic sensilla are arranged into three distinct whorls. The amphidial pouch is generally elongate and the aperture may be transverse slit to a small oval, rarely longitudinally elongate. The oesophagus is elongated and gradually broadens posteriorly. Male supplementary organs are generally absent and spicules are paired.

There are three families: Paraoxystominidae, Oxystominidae and Alaimidae.

Superfamily 2: Enoploidea

The pouchlike amphids have transverse slits as aperture. The cephalic sensilla are in the typical two-whorl pattern. The anterior helmet is punctate, smooth, or penetrated by numerous fenestrae. The stoma is completely surrounded by oesophageal tissue and may be equipped with jaws or teeth (usually three, rarely one). The oesophagus is cylindrical and may have posterior crenations. The oesophageal glands empty through the teeth or into the stoma. Males have one or two tubiform supplementary organs. Spicules are paired.

There are five families in the superfamily: Enoplidae, Lauratonematidae, Leptosomatidae, Phanodermatidae and Thoracostomopsidae.

Order 2: Oncholaimida

Pocketlike amphids have oval or elliptical aperture. The stoma is generally vaselike with heavily cuticularized walls and is equipped with three teeth (one dorsal and two subventral), the stoma walls may additionally have rows of small denticles. The stoma is divisible into a cheilostome and an oesophagostome. The male stoma in some taxa is indistinct or collapsed. There are cephalic sensilla in the typical two-whorl pattern. In some taxa all sensilla are papilliform. The oesophagus is usually conoid to cylindrical; in some taxa the posterior portion is composed of a series of bulbs.

The order contains three families: Oncholaimidae, Eurystominidae and Symplocostomatidae.

Order 3: Tripylida

The pouchlike amphids have apertures that are inconspicuous or transversally oval. The cephalic cuticle is simple and not doubled; there is no helmet. The body cuticle is smooth or sometimes superficially annulated. Cephalic sensilla are in the typical two-whorl pattern. The stoma may be simple, collapsed, funnel-shaped or cylindrical, and may or may not be armed. The stoma is usually oesophagostome, but when it is expanded both the cheilostome and oesophagostome are evident. The oesophagus is cylindrical to conoid. Oesophageal glands open anterior to the nerve ring. Males generally have 3 or more supplementary organs and caudal glands are generally present.

The order has two suborders: Tripylina and Ironina.

Suborder 1: Tripylina

Cephalic sensilla are in typical two-whorl pattern. Amphidial aperture is porelike or oval. This is the only group in which well-developed striae are present on the external cuticle. Stoma may be collapsed or globular and composed of two distinct parts. Oesophageal glands open anterior to the nerve ring; when a dorsal tooth is present the gland opens through it. The oesophagus is nearly cylindrical and cardia is well-developed. Males have papilloid or vesiculate preanal supplements and caudal glands in the posterior region.

There are two families: Tripylidae and Prismatolaimidae.

Suborder 2: Ironina *inquirenda*

The cephalic sensilla are in the typical two whorl pattern. The stoma is cylindrical and elongate with heavily cuticularized wall and is equipped with three teeth. All oesophageal glands open anteriorly. The cardia is small. Males have papillose or setose preanal supplementary organs and paired spicules accompanied by a weakly developed gubernaculum.

Currently only one family is recognized: Ironidae.

Superorder 2: Terrenoplica

This superorder represents the terrestrial Enoplia and is distinguished by the cephalization of the cephalic sensilla. The sensilla are labial, and are porelike and conelike. Oesophageal glands are 3-5 in number which open posterior to the nerve ring. A discrete excretory cell is seldom seen. Numerous preanal supplements are present in males but they are generally weak and tubiform.

There are four orders in the superorder: Isolaimida, Mononchida, Dorylaimida and Stichosomida.

Order 1: Isolaimida *inquirenda*

This is an order of uncertain position because its morphology is poorly understood; it is represented by the single genus *Isolaimium*. Members of this genus are elongate (3-6 mm) and cylindrical with transverse and longitudinal striations anteriorly and posteriorly being most obvious on the tail. The stoma is surrounded by six prominent cuticularized tubes and two whorls of six cephalic sensilla. The amphids are presumed to be the dorsolateral papillae of the second whorl. The stoma is elongate and has thickened walls anteriorly. The oesophagus is clavate. The females are amphidelphic, sometimes with flexures in the ovaries. The spicule is well-developed, but gubernaculum is weak and has two dorsal apophyses. Males have preanal supplements (papilloid). There are paired caudal papillae on the tail of both sexes.

There is one monotypic family: Isolaimiidae.

Order 2: Mononchida

The amphids are small, cuplike and located just posterior to the lateral lips; the amphidial aperture is slitlike or ellipsoidal. Cephalic sensilla (16 nos) are in a typical two-whorl pattern. The stoma is globular, heavily cuticularized and derived primarily from the cheilostome. The stoma bears one or more massive teeth that may be opposed by denticles in either transverse or longitudinal rows. The oesophagus is cylindrical-conoid, and lining of the lumen is heavily cuticularized. The excretory system is atrophied. Males have ventromedial supplements and paired spicules with or without gubernaculum. Females are mono or didelphic. Caudal glands may be degenerated or absent.

There are two superfamilies: Mononchoidea and Bathyodontoidea.

Superfamily 1: Mononchoidea

Taxa of this group have expanded labial region in which labia are generally angular and distinct. The cephalic papillae are in two-whorls; papillae of the outer circle are terminal on each labium. The stoma is barrel-shaped and equipped with a large immovable tooth on the dorsal wall. Subventral teeth may be present as denticles or as teeth as massive as the dorsal tooth. The oesophagus is nearly cylindrical with a heavily cuticularized lumen. Cardia is well-developed. Females are mono or didelphic (reflexed ovaries). In males, preanal supplements are papilloid, spicules are slender with well-developed gubernaculum. Caudal glands and a spinneret are present in both sexes.

The superfamily-Mononchoidea contains five families: Mononchidae, Mylonchulidae, Cobbonchidae, Anatonchidae and Iotonchulidae.

Superfamily 2: Bathyodontoidea

Taxa of this group have rounded or slightly expanded lip region. The cephalic sensilla are conelike. In stoma, the cheilostome is shortened, and hexaradiate and stomatal cavity is tubular posteriorly. The oesophagostome, which forms a larger part of the stoma equipped posteriorly with a ventrosublateral tooth that may be accompanied with a small denticle. The oesophagus is nearly cylindrical with five oesophageal glands. The cardia is well-developed. Females are didelphic, amphidelphic with reflexed ovaries; rarely monodelphic. Males are rare; preanal supplements are papilloid and spicules are paired, without a gubernaculum. In both sexes caudal glands and a spinneret are present.

The superfamily contains two families: Bathyodontidae and Mononchulidae.

Order 3: Dorylaimida

Cuticle is smooth, finely or coarsely striated. The labial region is often set-off from the body contour by a constriction. When there is no constriction, the labial region is defined as the region anterior to the amphids. The labia are generally well-developed and distinct; however, many taxa exhibit a smoothly rounded anterior region. The cephalic sensilla are all located on the labial region. The amphidial pouch is shaped like an inverted stirrup and the aperture is ellipsoidal or a transverse slit. The stoma is equipped with a movable mural tooth or a hollow axial stylet. The oesophagus is divided into an anterior region (slender and muscular) and a posterior region (elongated or pyriform glandular/muscular) with five oesophageal glands, rarely three or seven, having orifices posterior to the nerve ring. Cardia is well-developed. The mesenteron is often clearly divided into an anterior intestine and a prerectum. Females are monodelphic (pro or opisthodelphic) or didelphic with reflexed ovaries. Males have a pair of opposed testes, paired spicules, lateral guiding pieces, and sometimes a gubernaculum; preanal supplements few to numerous, rarely absent.

The order has four suborders: Dorylaimina, Diptherophorina, Campydorina and Nygolaimina.

Suborder 1: Dorylaimina

The characteristic feature of taxa of this suborder is its hollow axial stylet. Cephalic papilla are in a typical two-whorl pattern. Amphids are postlabial with slitlike or ellipsoidal apertures. The spear is bipartite, i.e., the odontostyle (tip), and the odontophore (posterior shaft). The oral opening may be equipped with denticles or other cuticular accessories. The anterior oesophagus is narrow, and the posterior muscular/glandular portion is expanded which exhibits 3-5 oesophageal glands, rarely 7. The excretory cell and pore are absent. A prerectum is usually present. Males have paired spicules with lateral accessory pieces and gubernaculum rarely present.

The suborder contains three superfamilies: Dorylaimoidea, Actinolaimoidea and Belondiroidea.

Superfamily 1: Dorylaimoidea

External cuticle is smooth or transversely striated, rarely with longitudinal ridges or lamella. Cheilostome is tubular and odontostyle is hollow with an oblique dorsal opening. The posterior portion of the oesophagus is one-third the length of the oesophagus. Cardia is well-developed. Females are generally didelphic but some taxa are mono-opisthodelphic. Males have a pair of adanal supplements preceded by a ventromedial series. Spicules are paired.

The superfamily contains seven families: Dorylaimidae, Encholaimidae, Tylencholaimidae, Tylencholaimellidae, Leptonchidae, Belonenchidae and Longidoridae. Since the book emphasized plant crop nematodes, the family Longidoridae is described here because it contains important plant parasitic nematodes. These nematodes are pathogenic as well as vectors of certain plant viruses.

Family: Longidoridae

Taxa includes long and slender nematodes, 1.5 to 12 mm in length. Cuticle smooth. Cephalic region rounded, continous with body contour or offset. Lips are amalgamated with full compliment of cephalic sensilla, which are arranged in two whorls of 6 and 10 papillae. Amphidial apertures may be small pores to broad transverse slits. Amphids large, pouchlike or stirrup-shaped. Dorsal and ventral series of body pores usually present. Odontostyle greatly elongate (50-220 µm) and attenuate. Odontophore elongate, sometimes with three strong, basal flanges. Junction of cheilostome and stomodeum is marked by a strongly sclerotized guide ring varying in position from near the lip region to near the odontostyle base. Oesophagus bipartite (anterior region narrow and tubular, and posterior, short bulb). The bulb contains longitudinal valve plates. Three oesophageal glands — one dorsal and two ventrosublateral. Females are amphididelphic with reflexed ovaries, sometimes pseudomono-delphic or mono-opisthodelphic. Males are diorchic, opposed, and spicules are large, dorylaimoid-type with lateral accessory guiding pieces. Copulatory supplements in the form of an adanal pair, and then a ventromedian series of 1-20 papillae are present. Tail shape variable, but generally similar in each sex.

The family contains three subfamilies: Xiphinematinae, Xiphidorinae and Longidorinae.

Key to Subfamilies of Longidoridae

1. Dorsal gland nucleus elongate, smaller than those of the ventrosublateral glands and located at some distance posterior to its orifice. 2
   - Dorsal gland nucleus round, larger than those of the ventrosublateral glands and located adjacent to its orifice. .... Xiphinematinae
2. Amphidial apertures a minute slit (porelike), odontostyle with furcate base, guide ring located near the odontostyle/odontophore junction, male copulatory supplements few in number (less

than eight) and with an hiatus between the adanal pair and the ventromedian series. .... Xiphdoriinae
– Not with the above combination of characters. ....
.... Longidorinae

Superfamily 2: Actinolaimoidea

This superfamily is characterized by the heavily cuticularized stomatal walls or the cuticular basketlike vestibule, these may also be modified into large teeth or denticles. The axial spear is always present. The labial region is generally rounded, may be distinct because of constriction. The oesophagus is bipartite with two equal parts, i.e., anterior portion and the expanded posterior region. Females have didelphic ovaries.

The superfamily contains four families: Actinolaimidae, Brittonematidae, Carcharolaimidae and Trachypleurosidae.

Superfamily 3: Belondiroidea

Diagnostic characteristic of this superfamily is the thick sheath of spiral muscles surrounding the basal enlarged portion of the oesophagus. The labial region is narrow and may be rounded or angular. Amphidial apertures are very large, often as wide as the head width. The odontostyle is normally shorter than the width of the labial region; the odontophore is rarely flanged. Females are mono-opisthodelphic or didelphic.

There are five families in this superfamily: Belondiridae, Roqueidae, Dorylaimellidae, Oxydiridae and Mydonomidae.

Suborder 2: Diphtherophorina

The labial region is slightly expanded and bears the cephalic sensilla. The amphids have an inverted stirrup shape and the aperture is ellipsoidal. A fully developed stylet is comprised of both cheilorhabdions and oesophorhabdions, most strongly developed on the dorsal side. Dorsal wall is thickened and functions as a movable tooth. The elongate narrow oesophagus expands posteriorly into a short pyriform bulb. An excretory pore and cell are present ventromedially. Females are didelphic. Males have a single testis, a pair of spicules, sometimes accompanied by a gubernaculum and papilloid preanal supplements. A weakly developed caudal alae may be present on the tail.

There are two families in this suborder: Diphtherophoridae and Trichodoridae. The latter contains important plant parasites that vectors plant viruses.

Family: Trichodoridae

Females are cigar-shaped, males straight or J-shaped on heat relaxation. Cuticle thick, smooth; may swell abnormally upon death. Amphidial apertures wide, gaping ellipses; sensilla sac separated from the fovea only by a constriction. Onchiostylet distally solid, dorsally convex, attached to the dorsal wall of the pharynx. Oesophagus consisting of a narrow anterior portion expanding into a posterior bulboid section of spathulate or pyriform shape containing five glands — one dorsal, two anterior ventrosublateral and two posterior ventrosublateral. Distinct excretory pore present, located either within the oesophageal region or slightly posterior. Prerectum is absent. Females are amphididelphic or monoprodelphic with reflexed ovaries. Uterus a simple tube; oviduct consisting of two cells. Spermatheca present or absent. Female anus almost terminal. Males are monorchic and outstretched. Copulatory supplements present in a precloacal, ventromedian position. Spicules straight or curved with the spicule protractor muscles forming a capsule around the proximal half of the retracted spicules. Bursa present or absent. Tail very short and rounded.

Suborder 3: Campydorina

The important distinguishing features of the suborder are the presence of acute, hollow and sub dorsally located mural tooth, muscular oesophagous forming triquetrous chamber in the basal region. Females: amphidelphic with reflexed ovaries extending beyond the vulva, prerectum absent. Males not found.

The suborder comprises of a superfamily: Campydoroidea and a family: Campydoridae.

Suborder 4: Nygolaimina

The principle diagnostic characteristic is the reversible stoma equipped with a protrusible subventral mural tooth. The oesophagus is bipartite and the posterior part has a distinct bulb that appears to be valved. Three cardiac glands are characteristically present at the junction of the oesophagus and intestine. Prerectum is present or absent.

The suborder contains four families: Nygolaimidae, Nygolaimellidae, and Aetholaimidae.

Order 4: Stichosomida

The most distinguishing characteristic is the modification of the posterior oesophagus into a stichosome (a series of glands exterior to the oesophagus proper). The stichosome may be in one or two rows. The early larval stages possess a protrusible stylet that is present in adults. Amphids are postlabial. Taxa of this order are parasites of vertebrates and invertebrates.

There are three superfamilies relegated to this order, one of which requires confirmation: Trichocephaloidea, Mermithoidea and Echinomermelloidea *incertae sedis*.

Superfamily 1: Trichocephaloidea

A unique feature of the superfamily is that in female gonads, the germinal zone extends along the length of the gonad but no rachis is formed. The stichosome is generally in a two rows of stichocytes; however, in the family Trichinellidae forms a single short row. The orifices are posterior to the nerve ring. Males and females possess a single gonad. Spicules are single or paired. The eggs are distinguished by being operculate.

This superfamily contains three families: Trichuridae, Trichinellidae and Trichosyringidae.

Superfamily 2: Mermithoidea

The stichosome in this family is always formed of two rows of stichocytes with orifices posterior to the nerve ring. The female gonads are not modified and the germinal zone is distal. Both sexes generally have two gonads but a single gonad does occur in some taxa. Males may have a single or a pair of spicules. The intestine in adults forms a trophosome. The eggs are not operculate.

There are two families: Mermithidae and Tetradonematidae.

Superfamily 3: Echinomermelloidea

Superfamily *incertae sedis*

The superfamily comprises three families, all *incertae sedis*: Echinomermellidae, Marimermithidae and Benthimermithidae.

Subclass B: Chromadoria

This subclass represents the transitional taxa between the marine and terrestrial nematodes. They are in all environments but are frequently encountered in the freshwaters throughout the world.The subclass contains some of the most elaborately ornamented forms.

The cephalic sensilla are generally in three whorls; the first circlet is always papilliform, whereas the outer circlets may, in varying combinations, be setiform, elongate coniform or papilliform. The subclass exhibits greater variation in amphids, they may be spiral, circular, vesicular or forms derivable therefrom. Fully developed stoma may contain a large dorsal tooth, three jaws or six inwardly acting teeth. There are three uninucleate oesophageal glands. The dorsal gland orifice is anterior to the nerve ring while the subventral glands empty into the posterior corpus. Shape of the oesophagus varies from near-cylindrical to tripartite: corpus, isthmus and posterior bulb oftenly with a valve. The cardia is usually well-developed. In both sexes there may be one or two gonads. Males in the subclass are distinguished by having moderately developed muscles of the *ductus ejaculatorius* only. Caudal glands are nearly ubiquitous.

The subclass contains five orders: Chromadorida, Desmoscolecida, Desmodorida, Monhysterida and Araeolaimida.

Order 1: Chromadorida

The amphid manifestation is relatively less variable. They may be reniform, transverse elongate loops, simple spirals or multiple spirals not seen in other orders or subclasses. The cephalic sensilla are in one or two whorls at the extreme anterior. The cuticle is smooth or annulated and usually exhibits punctuations. Fully developed stoma is oesophagostome and is usually equipped with a dorsal tooth, jaws or protrusible rugae. The corpus of the oesophagus is cylindrical, the isthmus is not seen and the postcorpus is distinctly expanded with heavily cuticularized lumen, which forms the crescentic valve. Cardia is triradiate or flattened. The females are usually didelphic with reflexed ovaries.

The order contains two superfamilies: Chromadoroidea and Choanolaimoidea.

Superfamily 1: Chromadoroidea

External cuticle is usually ornamented with transverse rows of punctuations or other intracuticular designs. Cephalic papillae in three whorls, the second and third whorls may be combined. The amphids are normally located just posterior to the third whorl of cephalic sensilla, may sometimes be located among the four sensilla of the third whorl. The amphids are transversely elongate ovals to loop-shaped, circularly spiral or tightly coiled multispirals. The vestibule leading to the stoma proper (oesophagostome) generally possesses weakly to well-developed rugae. The stoma proper is cylindrical, funnel-shaped or bowl-shaped and may be equipped with a dorsal tooth or the main chamber may have three well-developed teeth that are protrusible; in some taxa there are small denticles at the anterior of the oesophagostome. The postcorpus bulb may be slightly expanded or well-developed. Males may have knoblike preanal supplements; the spicules may be elongate and accompanied by a gubernaculum with or without a caudal apophysis.

The superfamily contains six families: Chromadoridae, Hypodontolaimidae, Microlaimidae, Spirinidae, Cyatholaimidae and Comesomatidae.

Superfamily 2: Choanolaimoidea

The cuticle is ornamented with heterogeneous punctuations and pores. The cephalic region is blunt and not distinctly set off. The cephalic sensilla are in the form of short cones or setae. The amphids are circular, spiral or multispiral. The cheilostome is composed of flaps or rugae. In some taxa there are jaws or movable teeth. The oesophagostome is sometimes divided into two sections with immobile teeth between the sections. The preanal male supplements may be large and cuticularized or simply a series of papillae.

There are three families: Choanolaimidae, Selachinematidae and Ethmolaimidae.

Order 2: Desmoscolecida

This order is characterized by the presence of conspicuous cuticular annulations on the body surface. The annuli may be covered with concretion rings or the cuticle may be ornamented with scales, warts or bristles. The cephalic sensilla reportedly are reduced in number. The internal whorl is absent, papillae of the second and third whorls are papilliform and setiform, respectively. The amphids (vesiculate type) are oval to circular and occupy much of the cephalic region. The somatic setae are tubular and the open distal ends are often elaborate. The posterior oesophagus is only slightly expanded. Pigment spots or ocelli, if present, are just posterior to the oesophagus. Females are amphidelphic with usually outstretched ovaries. The male spicules are generally accompanied by a gubernaculum. Both sexes have three caudal glands.

The order contains two families: Desmoscolecidae and Greeffiellidae.

Order 3: Desmodorida

The order is characterized by the presence of the cephalic capsule or helmet and the conspicuous somatic annuli. The cephalic sensilla are generally in three whorls, the second and third whorls may be combined. The shape of amphids may vary from reniform to elongate loops to simple or multiple spirals. The stoma may or may not be equipped with a dorsal tooth, which may or may not be opposed by subventral denticles.

In this order there are two superfamilies: Desmodoroidea and Draconematoidea.

Superfamily 1: Desmodoroidea

The cephalic sensilla are generally in three whorls, in some the two outermost whorls are combined. The shape of amphids vary from an elongate hook to a simple spiral or multiple spiral. External cuticle is ornamented with fine to coarse annuli. Usually the oesophagus ends in a valved bulb. Numerous setae are present oftenly in the cervical region. Males may have one or two spicules.

This superfamily contains three families: Desmodoridae, Ceramonematidae and Monoposthiidae.

Superfamily 2: Draconematoidea

The body annulations may be ornamented with spines, ridges, granules or internal inflations. The labia range from obscure to well developed. In most genera the presence of numerous anteriorly placed cervical setae obscures the symmetry of the outer circlets of cephalic sensilla. The shape of amphids varies from a simple elongate loop to one that intersects itself and appears as a single spiral. Cephalic adhesion tubes may be present. Elongate setae are scattered over the body. Adhesion tubes are generally present posteriorly. The females are amphidelphic with reflexed ovaries. In males the spicules are accompanied by a gubernaculum.

There are three families: Draconematidae, Epsilonematidae and Prochaetosomatidae.

Order 4: Monhysterida

The cephalic sensilla are in three whorls, the second and third whorls are usually combined but in some taxa the third whorl of four distinct setae is separate. The cuticle may be smooth or may have annuli or ornamentaion. When the annuli are distinct, the somatic setae may be long and in four to eight longitudinal rows. In general, the stoma is funnel-shaped and lightly cuticularized; however, in some taxa the stoma is spacious, heavily cuticularized and equipped with protrusible teeth. The amphids are simple spirals to circular. The normal pattern of distribution is often disrupted by numerous cervical setae. The near-cylindrical oesophagus is sometimes swollen posteriorly. The females have outstreched mono or didelphic ovaries.

There are seven families: Linhomoeidae, Siphonolaimidae, Monhysteridae, Scaptrellidae, Sphaerolaimidae, Xyalinidae and Meyliidae.

Order 5: Araeolaimida

The cephalic sensilla are often in three whorls, the second and third whorls are rarely combined. The first whorl has papilliform, the second coniform and the third has usually setiform papillae. The amphids are simple spirals that appear as elongate loops, shepherd's crooks, question marks, or circular. Body annulation is simple. The stoma is anteriorly funnel-shaped and posteriorly tubular; rarely it is armed. Usually the oesophagus ends in a bulb that may be valved. Females are didelphic. Male preanal supplements are generally tubular, rarely papilloid.

The order contains two suborders: Araeolaimina and Tripylina.

Suborder 1: Araeolaimina

Cephalic sensilla are in three whorls. The first whorl contains papilliform and the third has four long or short setae. The stoma is most often a simple tube, slightly funnel-shaped anteriorly. The diverse amphids are always derivable from simple spirals, sometimes appearing circular. The oesophagus has tuboid endings on the oesophageal radii of the anterior corpus. The basal oesophageal bulb may be valved.

The suborder contains three unassigned families, viz., Axonolaimidae, Camacolaimidae and Tripyloididae, and two superfamilies: Araeolaimoidea and Plectoidea.

Superfamily 1: Araeolaimoidea

The cephalic sensilla are in three whorls; generally the four sensilla of the third whorl are setiform. The amphids are generally simple spirals, elongate loops, or hooks. In some taxa the spiral nature of the amphid is obscure and it appears to be circular. The stoma is not well-developed; anteriorly it is thinly cuticularized and the oesophagostome is tubular or funnel-shaped. The oesophagus is cylindrical or externally tripartite (corpus, isthmus and a swollen valveless terminal bulb). The females are didelphic with outstretched ovaries. The males lack preanal tubular supplements.

There are four families in this superfamily: Araeolaimidae, Cylindrolaimidae, Diplopeltidae and Rhabdolaimidae.

Superfamily 2: Plectoidea

Cephalic sensilla are in three whorls which are usually papilloid. The four sensilla of the outer circlet may be setiform or elaborated into winglike lamellae. The amphids are simple circles, ovals, or shepherd's crook. The stoma is anteriorly tubular or expanded, and posteriorly tubular. The oesophagus is cylindrical anteriorly (with or without an isthmus) and a posterior expanded muscular bulb with a bellows-type valve (family Bastianidae lacks valve). Females are didelphic with reflexed ovaries. Male preanal supplements vary from none to several and tuboid.

The superfamily contains four families: Plectidae, Leptolaimidae, Haliplectidae and Bastianidae.

CLASS 2: SECERNENTEA

The body cuticle varies from four layers to two layers and is generally transversely striated. Laterally, along most of the body length, there is a lateral wing area called lateral field which is marked by longitudinal striae or ridges. In some parasitic forms the lateral alae may extend out from the body contour for a distance equal to the body diameter. The cephalic sensilla are located in the labial region and are porelike or papilliform. In some parasitic groups the number of sensilla are reduced from 16 to 4. The amphids are located dorsolaterally on the lateral labia or anterior extremity and their aperture is commonly pore-like. In some instances they are oval or cleft like or located postlabially (generally in immatures). At the level of the nerve ring there are generally a pair of sensilla called deirids and a pair of sensilla called phasmids in the caudal region. The cells of the hypodermis may be multinucleate.

The oesophagus generally has three glands. The dorsal gland always opens anteriorly, either in the procorpus or in the anterior metacorpus or posterior corpus. The two subventral glands open posterior to the metacorpus. The excretory system opens through a ventromedian cuticularized duct. Collecting tubules extend from the excretory cell either on both sides of the body or on one side. Females are devoid of somatic setae or papillae. First moulting occurs inside the egg, and the second-stage juvenile, with a few exceptions among parasitic forms, hatches out. Caudal papillae present in males, preanal supplements are paired; sometimes there is a ventromedial preanal papilloid supplement. Commonly males possess laterally caudal alae or bursa around the cloaca.

There are three subclasses: Rhabditia, Spiruria and Diplogasteria.

Subclass A: Rhabditia

One of the most distinguishing characteristics of the subclass is the tripartite oesophagus (corpus, isthmus and posterior bulb) at least in immatures. In adult parasitic forms the oesophagus may be clavate or cylindrical; however, the second stage juveniles have the rhabditoid valved bulb or a form reminiscent of it. The valve within the posterior bulb acts as a rolling valve, not as a bellows as in Plectidae. The bipartite stoma generally lacks teeth/denticles; the stoma may be further subdivided into two or more sections. Males generally have well-developed caudal alae that may be supported by rays either papilliform or cuticularized.

Within the subclass the "family" Myenchidae that contains parasites of frogs and leeches is considered *incertae sedis* so as to position and *familia dubium* as to ranking.

The subclass contains two orders: Rhabditida and Strongylida.

Order 1: Rhabditida

The characteristic feature is that the tripartite oesophagus always ends in a muscular bulb that is invariably valved. The number of labia varies from 6-2 or is absent. The tubular stoma may be composed of five or more sections called rhabdions. The excretory tube is cuticularly lined and paired lateral collecting tubes generally run posteriorly from the excretory cell; some taxa have anterior tubules also. Females are mono-prodidelphic or didelphic. The cells of the intestine may have 1-3 nuclei. The hypodermal cells may also be multinucleate. Caudal alae, when present, contain papillae.

The order contains two suborders: Rhabditina and Cephalobina.

Suborder 1: Rhabditina

In most taxa the labia are distinct and bear the papilloid cephalic sensilla and pore-like amphids. The stoma is commonly cylindrical and devoid of distinct rhabdions. The stoma is generally two or more times as long as it is wide. The oesophagus is tripartite and ends in a valved muscular bulb. Females are mono or didelphic. Males usually have paired spicules, a gubernaculum and caudal alae.

There are five superfamilies: Rhabditoidea, Bunonematoidea, Cosmocercoidea, Oxyuroidea and Heterakoidea. The suborder also contains one family *insertae sedis*: Alloionematidae.

Superfamily 1: Rhabditoidea

The superfamily is characterized by the presence of a well-developed cylindrical stoma. The number of labia varies from two to six. The oesophagus, at least in the immature stages, has a muscular posterior bulb with a roller-type valve. The caudal alae of males are supported by five to nine papilloid rays.

The superfamily contains 12 families: Rhabditidae, Rhabditonematidae, Odontorhabditidae, Steinernematidae, Rhabdiasidae, Angiostomatidae, Agfidae, Strongyloididae, Syrphonematidae, Heterorhabditidae, Carabonematidae and Pseudodiplogasteroididae.

Superfamily 2: Bunonematoidea

The superfamily is characterized by the asymmetrical bodies in terms of both the cephalic sensilla and labial distribution. Furthermore, the asymmetry extends to the elaborate ornamentation on the right side of the body; the left side of the body appears near-normal. The male caudal alae are often asymmetrical. Asymmetry is limited to the external covering of the body. The stoma and oesophagus are rhabditoid type. Females are didelphic.

There are two families in this superfamily: Bunonematidae and Pterygorhabditidae.

Superfamily 3: Cosmocercoidea

The number of labia is three or six on which ventrolateral papillae are always present. The stoma is weakly developed and for the most part is surrounded by oesophageal tissue. The posterior bulb of the tripartite oesophagus is always valved. Oesophageal glands are always uninucleate and no caeca are associated with the oesophagus or intestine. A precloacal sucker when present in males, the spicules are equal and in some taxa are functionally replaced by the gubernaculum.

The superfamily contains two families: Cosmocercidae and Atractidae.

Superfamily 4: Oxyuroidea

The taxa are distinguished by the absence of ventrolateral papillae and presence of eight or fused to four cephalic sensilla in a circle. The labia are reduced or absent. The stoma is primarily composed of the cheilostome. A valve is always discernible in the posterior bulb of tripartite oesophagus. Precloacal suckers may be present on males and the spicules may be greatly reduced.

There are three families in the superfamily: Oxyuridae, Thelastomatidae and Rhigonematidae.

Superfamily 5: Heterakoidea

The superfamily is characterized by having ventrolateral papillae and eight labial sensilla are paired in a whorl of four. The labia are well-defined but small. The stoma is primarily composed of the oesophagostome. The oesophagus is tripartite and the posterior bulb is valved. The subventral oesophageal glands are duplicated. Males have a precloacal sucker surrounded by a cuticular ring. The spicules are always paired.

The superfamily contains two families: Heterakidae and Ascaridiidae.

Suborder 2: Cephalobina

This suborder is characterized by having thin-walled expanded vestibular stoma followed by a heavy-walled chamber of about the same width. The funnel-shaped or tubular posterior stoma (oesophagostome) is always narrower than the cheilostome and composed of four separate rhabdions. The oesophagus is tripartite, rhabditoid type and the posterior bulb is always valved. Females are mostly mono-prodelphic with several flexures. Caudal alae are absent on the male tail; however, the tail does bear papillae. The male spicules are paired and a gubernaculum is present.

The superfamily contains four unassigned families: Cephalobidae, Robertiidae, Chambersiellidae, Elaphonematidae and one superfamily: Panagrolaimoidea.

Superfamily 1: Panagrolaimoidea

The cheilostome is thick-walled and as long as it is broad or only slightly longer than broad. The posterior stoma is funnel-shaped, short and lined with small rhabdions. The oesophagus may have a distinctly muscular metacorpus. Females are mono-opisthodelphic. The posterior uterus is vestigial and functions as a seminal receptacle.

The superfamily contains three families: Panagrolaimidae, Alirhabditidae and Brevibuccidae.

Order 2: Strongylida

The cephalic region may have three or six labia or they may be replaced by a corona radiata. The variable stoma may be well-developed or rudimentary but is never collapsed or inconspicuous. The capacious stoma consists primarily of cheilostome; only the base of the cup is surrounded by oesophageal tissue. The stoma, even when reduced, is represented by the cheilostomal vestibule. The oesophagus in larval forms is rhabdiform and the posterior bulb is valved, whereas in adults it is cylindrical to clavate. In the fourth juvenile stage and the adult, the excretory system consists of paired lateral collecting tubules, an excretory cell and paired subventral glands. Other immatures have a single excretory cell. The females have one or two gonads and the uterus is heavily muscularized. The males have unusual caudal alae; in this group it is referred to as a bursa copulatrix. The bursa differs from other caudal alae, i.e., it contains muscles and rays. Males have paired and equal spicules.

This order contains three unassigned families: Diaphanocephalidae, Metastrongylidae, Maupasinidae; and three superfamilies: Strongyloidea, Ancylostomatoidea and Trichostrongyloidea.

Superfamily 1: Strongyloidea

The stoma is always well-developed and spacious. It may be hexagonal in cross-section, globular, cylindrical, or infundibuliform; it is never equipped with teeth or cutting plates. The oral opening may be surrounded by six small labia or a corona radiata. In newly hatched larvae the oesophagus is rabditiform but the adult oesophagus is clavate or near-cylindrical. Males have well-developed bursae and the bursal rays are never fused.

The superfamily contains three families: Strongylidae, Cloacinidae and Syngamidae.

Superfamily 2: Ancylostomatoidea

This group is distinguished by the characteristic capacious, thick-walled, globose stoma that may or may not be equipped with teeth or cutting plates anteriorly. The branches of the dorsal bursal rays in the male bursa are greatly reduced.

The superfamily contains three families: Ancylostomatidae, Uncinariidae and Globocephalidae.

Superfamily 3: Trichostrongyloidea

Inconspicuous labia (three-six) surround the oral opening, in some taxa they may be absent. The stoma is reduced or collapsed and there is no evidence of a corona radiata. The inflated cuticle of the cephalic region may be unusual among strongyles. The thick somatic cuticle often has longitudinal ridges causing these worms to be wiry rather than limp. In second-stage juveniles, the oesophagus is rhabditoid but the posterior bulb may be devoid of valve. In the bursa the lateral rays are well-developed even though the dorsal rays may be atrophied.

The superfamily contains six families: Trichostrongylidae, Amidostomatidae, Strongylacanthidae, Heligmosomatidae, Ollulanidae and Dictyocaulidae.

Subclass B: Spiruria

This is the only subclass in Nematoda that is entirely parasitic and the only group for which no free-living counterparts are identified. The oral opening is surrounded by three or six apical lobes, often called pseudolabia, or the labia are modified into two lateral lobes or "labia". The complement of cephalic sensilla is variable throughout the group and ventrolateral papillae are present only among some taxa in the Ascarida. The stoma may be globose or long and cylindrical; when armed, the armature is cheilostomal in origin. There are taxa where the stoma is not distinguishable; in these taxa the oesophageal tissue extends to the much reduced cheilostome. The oesophagus is variable, it may be clavate, nearly cylindrical or divided into a narrow anterior region and an elongate swollen glandular region (bottle-shaped). The oesophagus is never in any stage rhabditoid in character or valved posteriorly. The collecting tubules of the excretory system may be in the form of an "H" or an inverted tuning fork; in a few instances the system is limited to one side of the body, usually the left. The vagina may be greatly elongate and tortuous.

The generally paired male spicules may be greatly disproportionate in length. Caudal alae with embedded papillae may occur. Eggshells are often highly ornamented.

The subclass contains three orders: Ascaridida, Cammallanida and Spirurida.

Order 1: Ascaridida

The oral opening is generally surrounded by three or six labia, but in some taxa labia are absent. The cephalic sensilla are always evident. Usually there are eight cephalic or labial sensilla; the submedians may be fused then only four sensilla are seen. Phasmids are sometimes large and pocketlike. The stoma varies from being completely reduced to spacious or globose. The oesophagus may vary from club-shaped to nearly cylindrical, never rhabditoid. There may be posterior oesophageal or anterior intestinal cecae. The collecting tubules of the excretory system may extend posteriorly and anteriorly. Males generally have two spicules, in some taxa there may be none or only one. The gubernaculum may be present or absent. Females are generally didelphic, some times polydelphic. The number of uteri may be two, three, four or six.

The order contains five superfamilies: Ascaridoidea, Seuratoidea, Camallanoidea, Dioctophymatoidea and Muspiceoidea *incertae sedis*.

Superfamily 1: Ascaridoidea

Taxa are much bigger in size with the body length of 1-40 cm and cuticle is marked by superficial annuli. The oral opening is generally surrounded by three well-developed lips: one dorsal and two subventral that bear the porelike amphids. The stoma is primarily oesophagostome followed by a cylindrical to clavate oesophagus. Cecae may extend anteriorly or posteriorly from the oesophagointestinal junction. Females are didelphic or polydelphic (multiples of 3, 4 or 6). Males have paired spicules rarely accompanied by a gubernaculum.

The superfamily contains seven families: Ascarididae, Toxocaridae, Anisakidae, Acanthocheilidae, Goeziidae, Crossophoridae and Heterocheilidae.

Superfamily 2: Seuratoidea

Lip region and lips are greatly reduced or absent. Ventrolateral sensilla are present or absent, and the submedian sensilla are doubled or single.

The stoma may be small, and may be poorly or strongly developed and equipped with teeth posteriorly. The oesophagus is cylindrical or slightly clavate. Intestinal cecae are rarely present. Males may or may not have a precloacal sucker, spicules are paired and often accompanied by a well-developed gubernaculum. Caudal alae, if present, are very narrow.

The superfamily contains five families: Seuratidae, Schneidernematidae, Quimperiidae, Subuluridae and Cucullanidae.

Superfamily 3: Camallanoidea

The stoma is well-developed and the main chamber of cheilostome may be either globose or transversely rectangular and internally supported by numerous longitudinal or oblique ridges. The oesophagostome forms the basal stomatal plates. The internal whorl of cephalic sensilla are minute and the external whorl of eight sensilla is partly fused. The oesophagus is short and generally clavate.

This superfamily contains two families: Camallanidae and Anguillicolidae.

Superfamily 4: Dioctophymatoidea

This superfamily was placed in Adenophorea prior to 1981 when it was moved to the Ascaridida within Secernentea/Rhabditia. It was retained with Ascaridida but moved to Secernentea/Spiruria.

The cuticle, with a well-developed oblique fibre layer, lacks the endocuticular layer. Externally the cuticle may be annulated or ornamented with distinct hooklike spines. When well-developed, the stoma is primarily oesophastomal. The oesophagus is externally cylindrical but internally divisible into the corpus and the postcorpus with three ramifying glands. Both males and females have a single gonad. The male tail is expanded into a thickened cuplike bursa copulatrix. Males have a single, elongate spicule.

The superfamily contains two families: Dioctophymatidae and Soboliphymidae.

Superfamily 5: Muspiceoidea *incertae sedis*

These nematodes are arbitrarily relegated not only to the superfamily level, but to inclusion among the Ascaridida. The known taxa are parasites of vertebrates where they cause tumorous and cancerlike damage.

The cephalic sensory organs are greatly reduced. Amphids or cephalic papillae have been reported in only one species. Males are unknown. In females the digestive tract is reduced. The females are reportedly amphidelphic, diovarial, and viviparous. However, the vulva may or may not be present; when absent, larvae are released by the rupturing of the female body; thus reminiscent of *Dioctophyma*.

There are three families all *incertae sedis*: Muspiceidae, Robertdollfusidae and Phlyctainophoridae.

Order 2: Spirurida

The oral opening is surrounded by two lateral labia or pseudolabia, some times by four or more lips. Lips are rarely absent. Due to the variability in lip number, the shape of the oral opening exhibits corresponding variation and may or may not be surrounded by teeth. The amphids are lateral, in some taxa they may be immediately posterior to the labia or pseudolabia. The stoma may be cylindrical and elongate or rudimentary. The oesophagus is generally bipartite (anterior muscular portion and an elongate swollen posterior glandular portion). Multinucleate glands are present in posterior oesophagus. Eclosion larvae are usually provided with a cephalic spine or hook and the presence of a porelike phasmid on the tail.

All known taxa utilize an invertebrate in their life cycle; the definitive hosts are mammals, birds, reptiles and rarely amphibians.

The order contains six superfamilies: Spiruroidea, Drilonematoidea, Physalopteroidea, Dracunculoidea, Diplotriaenoidea and Filarioidea.

Superfamily 1: Spiruroidea

The lateral lips are well-developed. In some taxa lips are absent. There are either four or eight cephalic sensilla in the external whorl. In some taxa the sensilla are doubled or fused. The cephalic and cervical regions may manifest cordons, collarettes or cuticular rings. The oral opening may be circular, hexagonal or dorsoventrally elongated. The stoma is cheilostome, it is always well-developed and anteriorly equipped with teeth. The vulva is generally at mid-body and in only a few taxa it is located posteriorly or anteriorly near the oesophagus.

The superfamily contains five families: Spiruridae, Thelaziidae, Acuariidae, Hedruridae and Tetrameridae.

Superfamily 2: Drilonematoidea

The stoma is greatly reduced and may be surrounded by rudimentary lips (often not visible). The amphids may be postlabial and enlarged. The oesophagus in mature adults may be tripartite (corpus, isthmus and glandular bulb) without a valve. Females have a single elongate ovary. Male spicules, if present, are paired with or without a gubernaculum. The phasmids, sometimes called caudal suckers, are greatly enlarged and may occupy the full width of the tail. The placement of these nematodes in the hierarchy is uncertain; however, their biology and known morphology are characteristic of Spiruria and not Rhabditia.

The superfamily contains six families: Drilonematidae, Ungellidae, Scolecophilidae, Creagrocericidae, Mesidionematidae and Homungellidae.

Superfamily 3: Physalopteroidea

The inner surface of the two large lateral labia (pseudolabia) are generally provided with teeth; interlabia are not present between the pseudolabia. Of the cephalic sensilla, the inner whorl (circumoral) is reduced or absent and the outer whorl consists of four fused sensilla. The anterior cuticle in some taxa is reflexed and forms a collar that may partly cover the pseudolabia. The cephalic region may be ornamented with a very prominent spined bulbous structure; however, there are no cordons, collarettes or rings. In males the caudal alae are well developed when the caudal papillae are pedunculate; when the caudal papillae are non-pedunculate, the caudal alae are absent. In females the ovaries may be four or more, and the vulva may be either anterior or posterior to the mid-body.

The superfamily contains three families: Physalopteridae, Megalobatrachonematidae and Gnathostomatidae.

Superfamily 4: Dracunculoidea

The stoma is generally very reduced and most often is only vestibular. The cephalic sensilla of internal whorl (six) and external whorl (eight) are well developed. In immature females the vulva is at mid-body; however, in gravid females it is atrophied as is the female's posterior intestine. If caudal alae are present on the male tail, they are small and postcloacal.

The superfamily contains three families: Dracunculidae, Philometridae and Micropleuridae.

Superfamily 5: Diplotriaenoidea

The pattern of sensilla surrounding the dorsoventrally elongate oral opening is variable. There may be four small sensilla in the first whorl around the oral opening, followed by the second whorl of six and third whorl of four sensilla. There may be two whorls of four sensilla each, or these may join as paired or fused sensilla. Within the anterior stomatal region are protrusible cuticular structures called tridents located on either side of the anterior oesophagus. The oesophagus is divided into a narrow anterior portion and a wide, elongate, glandular portion. The vulva is anterior and females are oviparous or ovoviviparous. Males have unequal spicules and lack caudal alae.

The superfamily contains two families: Diplotriaenidae and Oswaldofilariidae.

Superfamily 6: Filarioidea

The internal whorl of sensilla may be absent or consist of two or four papillae surrounding the circular or oval oral opening. The external whorl consists of eight sensilla. The stoma is rudimentary. The oesophagus shows little difference between the corpus and the postcorpus with its multinucleate glands. The vulva is generally located in the anterior half of the body. In males the spicules are either equal or unequal and caudal alae may be present or absent. The gubernaculum is always absent.

The superfamily contains five families: Filariidae, Aproctidae, Setariidae, Desmidocercidae and Onchocercidae.

Subclass C: Diplogasteria

Taxa are small to medium sized nematodes, rarely exceeding 3-4 mm. The cuticle may be punctated and has annuli that are sometimes traversed by longitudinal striae. The labia may be well-developed and hexaradiate in symmetry. The full complement of cephalic sensilla is often present, especially in males. In derived taxa the inner whorl of six sensilla may be lacking. In the external whorl of ten sensilla the externolaterals are ventrolateral on the lateral labia. The amphids, when labia are evident, are located dorsolaterally on the lateral lips and the apertures are porelike, small ovals, slits, or clefts. In some larval forms, particularly dauer larvae, the amphids may be postlabial. The variable stoma is primarily cheilostome and is most often armed. Except for rare instances, movable stomatal armature in the form of large teeth, opposable fossores or axial spears is limited to this subclass. As in other nematodes the movable armature is controlled by the three anterior-most muscles of the oesophagus. The ferrule (shaft and apodemes) of the axial spear or fossores, is the product of the oesophagostome. The oesophagus has a muscular corpus divided into an almost cylindrical procorpus and a muscular, almost always valved metacorpus followed by an isthmus and a glandular postcorpus. A valve is never present in the postcorporal bulb. Females have one or two ovaries and males have one testis with the ductus ejaculatoris almost devoid of musculature. Males have paired spicules with or without a gubernaculum or caudal alae.

The subclass is divided into two orders: Diplogasterida and Tylenchida.

Order 1: Diplogasterida

Hexaradiate symmetry is distinct in lips, but they are not well developed. The external circle of labial sensilla may appear setose but they are always short, never long and hairlike. The stoma may be slender, elongate or spacious, or any gradation between these two. The stoma may be equipped with movable teeth, fossores, or a pseudostylet or may not be equipped. The corpus is always muscled and distinct from the postcorpus that is divisible into an isthmus and glandular posterior bulb. The metacorpus is almost always valved. The females are mono or didelphic. Males may or may not have caudal alae; however, a gubernaculum is always present. The male tail most commonly has nine pairs of caudal papillae, three are preanal and six are caudal papillae.

The order contains four families: Diplogasteridae, Odontopharyngidae, Diplogasteroididae and Cylindrocorporidae *incertae sedis*.

Order 2: Tylenchida

The order includes taxa which are algal, fungal and moss feeders and parasites of lower and higher plants but not parasites of animals. The labial region may be distinctly set off or smoothly rounded and well-developed; the hexaradiate symmetry is distinct with a few exceptions. The amphids are mostly lateral in position, and are porelike, oval, slitlike, or clefts located on the lips. The internal whorl of six papillae may be reduced to a visible four or some may be doubled. The hollow stylet is the product of the cheilostome (conus, guiding apparatus, and framework) and the oesophagostome (shaft and knobs). Throughout the order and its suborders the stylet may be present or absent and may or may not bear knobs. The oesophagus is oftenly tripartite (corpus, isthmus and glandular posterior bulb) but shows considerable variation. The corpus is further divisible into the procorpus and metacorpus. The metacorpus is generally valved but may be absent in some females and males. The orifice of the dorsal oesophageal gland opens either into the anterior procorpus or just anterior to the metacorporal valve. The excretory system is asymmetrical. Females are monoprodelphic or didelphic. Except for sex-reversed males, there is only one gonad; spicules are always paired with or without a gubernaculum. Males may have phasmids, caudal papillae are absent. Bursa is usually present with great variation, but it lacks papillary ribs or rays.

The order contains four suborders: Tylenchina (two infra-orders), Hoplolaimina, Criconematina and Hexatylina.

Suborder 1: Tylenchina [Chitwood in Chitwood & Chitwood, 1950]

Cuticle smooth or distinctly annulated, sometimes marked with longitudinal striae or grooves, but annules never retrorse or with scales, spines, appendages or a double cuticle. Cephalic region smooth or annulated; framework with light, rarely heavy sclerotization. Labial plate with six inner labial sensilla in the form of papillae or pits around a pore like round or oval oral opening; prominent labial disc rarely present, six outer labial sensilla not on surface; four cephalic sensilla on surface usually present; amphidial apertures labial, sometimes just postlabial. Stylet generally small, consists of conus, shaft and knobs (small and rounded). Procorpus cylindroid or fusiform; median bulb usually muscular with refractive thickenings but not amalgamated with procorpus. Isthmus elongate, slender. Oesophageal glands forming a basal bulb or rarely extending over intestine. Cardia is three-celled, reduced. Nerve ring circum-oesophageal. Excretory pore in oesophageal region, renette cell post-oesophageal; excretory duct not vesiculate distally. Rectum and anus normally distinct. Tails similar between sexes, elongate-tapering, filiform or short and different between sexes. Females and males have single and outstretched gonads. Ovary with usually a postvulval uterine sac. Crustaformeria (glandular part of uterus) normally with less than 20 cells. Ovary normally outstreched. Oesophagus and stylet normally developed in male. Testis single, outstretched; spermatogenesis continues in adult life. Spicules paired, cephalated, arcuate, distally round to pointed with sensory pore ventrally subterminal; never setaceous or U-shaped. Gubernaculum generally simple, fixed. Bursa simple or lobed, may be highly reduced or rarely absent. Phasmids absent. Prophasmids (papillalike sensilla near vulva or at about the same position on male body) located just dorsal to lateral field present. Cloacal lips seldom forming a penial tube (tubus); hypoptygma on posterior lip, if present, double.

This suborder contains two infraorders: Tylenchata, and Anguinata.

Key to Infraorders of Tylenchina

1. Amphidial apertures minute oblique slits on raised areas on labial plate; cardia absent; fungal feeders and/or parasites of above ground plant parts .... **Anguinata**
2. Amphidial apertures not minute oblique slits on raised areas on labial plate but variously modified from pore to large sinuate slits; cardia present; algal and root feeders, not fungal feeders or parasites of above ground plant parts .... **Tylenchata**

Infraorder 1: Tylenchata

The infraorder includes algal feeders and parasites of underground parts, and not of above ground parts. Taxa are usually smaller in size of less than 1 mm length but may reach a length of 2 mm. Lateral field is distinct with 1-6 incisures, rarely obscure. Cephalic region may be from low flat to high, continuous to offset; framework poorly developed, usually with high arches, rarely sclerotized. Labial sensilla in the form of papillae or pits; oral opening pore like round or oval; prominent labial disc rarely present and four submedian cephalic sensilla usually on surface. Amphidial apertures pore or slit like, usually extending along the lateral sides of cephalic region. Deirids present near the level of excretory pore. Phasmids are absent, but prophasmids present. Orifice of dorsal oesophageal gland generally 1-4 µm behind stylet base. Corpus elongate, slender; procorpus elongate cylindroid, lacking musculature. Metacorpus or median bulb present or absent, muscular or non-muscular, with or without inner refractive thickenings, smaller than basal bulb, and not occupying entire body width cavity. Isthmus elongate-slender. Basal or terminal oesophageal bulb enclosing oesophageal glands present. Cardia present, prominent, tricellular. Rectum and anus distinct. Tails similar between sexes, elongate-tapering, usually filiform. Bursa rarely absent, simple or lobed, not supported by phasmidial pseudoribs. Vulva a transverse slit, lips usually not modified, absent, epiptygma usually absent. Females are mono-prodelphic with ovary outstretched. Postvulval uterine sac shorter than body width or in some genera absent. Males are monorchic and anteriorly outstretched; spicules paired, similar, cephalated, ventrally arcuate, rarely straight accompanied by gubernaculum which is trough shaped, fixed, not protrusible without or with capitulum.

The infraorder contains one superfamily: Tylenchoidea.

Superfamily 1: Tylenchoidea [Orley, 1880 (Chitwood & Chitwood, 1937), Skarbilovich, 1959 (syn. Atlenchoidea) (Golden, 1971)]

Characters of superfamily resemble the infraorder Tylenchata. The superfamily contains four families: Tylenchidae, Atylenchidae, Ecphyadophoridae and Tylodoridae.

Key to Families of Tylenchoidea

1. Cephalic setae present .... Atylenchidae
   Cephalic setae absent 2
2. Stylet over 24 µm long, if shorter then about as long as procorpus, with tubular protractors .... Tylodoridae

Stylet under 24 µm (generally 8-16 µm) long, not as long as procorpus, generally with divergent protractor 3
3. Body extremely attenuated (a = 50-150), appearing glass fibre-like; bursa often lobed .... Ecphyadophoridae
Body not extremely attenuated, not appearing glass fibre-like; bursa not lobed .... Tylenchidae

Family: Tylenchidae [Orley, 1880 (= Boleodoridae) Khan, 1964 (Brzeski & Sauer, 1983)]

Members of this family are algal, lichen, moss feeders and root surface feeders. Taxa are slender and vermiform. Lateral lines 1-6, rarely absent. The transverse lines may be transected by longitudinal ridges. Labial region is usually elevated, rounded and annulated. Labial framework generally weakly developed; stylet usually small and delicate. Amphids variable: small oblique slits to long, sinuous, longitudinal clefts. Deirids present or absent, phasmid-like structures present or absent. Oesophagus divided into slender procorpus, elliptical metacorpus most often valved, long slender isthmus followed by symmetrical pyriform glandular region. Females are generally mono-prodelphic; 12-celled spermatheca often offset; columned uterus with four rows of cells; postuterine sac length less than vulval body width. Male caudal alae leptoderan. Sperm cells with little cytoplasm. Tails elongate-conoid, generally narrowing to filiform outline.

There are five subfamilies: Tylenchinae, Boleodorinae, Duosulciinae, Thadinae and Tanzaniinae.

Key to Subfamilies of Tylenchidae

1. Corpus short and broad, about as long as basal bulb .... Tanzaniinae
Corpus elongate slender, longer than basal bulb 2
2. Amphidial apertures prominent, posterior to level of cephalic papillae, partially eq. covered by a cuticular flap .... Boleodorinae
Amphidial apertures rarely prominent, anterior to cephalic papillae, not covered by a cuticular flap 3
3. Lateral field narrow, with a single ridge .... Duosulciinae
Lateral field broad, with two or three ridges 4
4. Spermatheca axial; tails short-conoid to subcylindroid .... Thadinae
Spermatheca offset; tails elongate, usually filiform .... Tylenchinae

Family: Ecphyadophoridae [Skarbilovich, 1959]

Members of this family are extremely attenuated and have cylindroid corpus, small stylet and lobed bursa. The taxa are small (under 1 mm), extremely slender as to appear fibre-like (width 8-10 µm) and are easily overlooked (unless they are searched for). Cuticle thin, smooth or moderately thick and deeply annulated. Lateral fields have 2-4 incisures, may be obscure or absent. Cephalic region circular, flattened or continuous. Four cephalic papillae on surface, but not projecting as setae. Amphidial apertures pore-like, near oral opening or longitudinal clefts. Prophasmids postmedian, dorso-sublateral, in females just anterior to vulva. Stylet short, attenuated, with needle like solid appearing tip of conus and small round basal knobs. Corpus cylindroid, non-muscular, lacking a postcorporal bulb. Basal bulb present but dorsal gland may extend as lobe over intestine. Vulva transverse, posterior, may be covered with anterior lip flap. Postvulval uterine sac present. Tails very long, filiform. Bursa adanal, lobed projecting outward and backward, rarely simple flap like. Spicules needle like, setose or tylenchoid. Gubernaculum present, fixed, may be obscure. Hypoptygma absent.

The family contains two subfamilies: Ecphyadophorinae and Ecphyadophoroidinae.

Key to Subfamilies of Ecphyadophoridae

1. Amphidial apertures long, slit-like, extending to middle or more of cephalic region; vulva directed outward; not overhung by its anterior lip; spicules arcuate, tip pointing ventrally .... Ecphyadophoroidinae
2. Amphidial apertures indistinct, pore-like vulva directed posteriorly, overhung by its anterior lip; spicules almost straight, tip pointing posteriorly .... Ecphyadophorinae

Family: Atylenchidae [Skarbilovich, 1959]

Unique characteristic of this family is the presence of four cephalic setae. The taxa are slender (upto 1 mm long) with longitudinal ridges all around; annulations formed by transverse grooves crossing longitudinal ridges, as a result minute blocks are formed on the body surface. Lateral fields with 1-3 incisures. Cephalic region small, continuous or set-off, annulated or smooth, with cephalic setae. Stylet moderately developed, slender, about as long as procorpus. Median bulb muscular. Basal bulb enclosing glands; cardia present. Vulva transverse in posterior region. Vagina directed inward. Postvulval uterine sac present. Gonads are single and anteriorly outstretched. Tails are long, filiform. Bursa is conspicuous, adanal, lobed, (absent in Atylenchus). Spicules paired, cephalated, arcuate, pointed. Gubernaculum simple, fixed. Cloacal lips not forming a penial tube.

The family contains two subfamilies: Atylenchinae and Eutylenchinae.

Key to Subfamilies of Atylenchidae

1. Lateral field with two or more ridges, cephalic region broad, annulated, continuous, bursa absent .... Atylenchinae
2. Lateral field with one ridge, cephalic region narrow, smooth, offset, bursa present .... Eutylenchinae

Family: Tylodoridae [Paramonov, 1967 (Siddiqi, 1976)]

Members of this family are deep root feeders. Taxa are medium to large in size (0.6-2 mm), slender; cuticle prominently annulated or smooth. Lateral field with 3-6 incisures at mid-body, rarely obscure. Cephalic region usually small, continuous or offset by constriction; cephalic setae absent; a conspicuous labial disc present. Amphidial apertures longitudinally slitlike near labial plate/disc or cleft-like. Deirids present. Prophasmids dorso-sublateral, postmedian. Stylet well-developed and is about as long as procorpus or longer (upto 100 µm). Conus equal to or longer than the shaft; basal knobs prominent. Protractor muscles tubular around stylet. Orifice of dorsal oesophageal gland at base of stylet. Procorpus usually as long as the stylet; median bulb well-developed with prominent refractive cuticular thickenings; isthmus slender, longer than procorpus. Oesophageal glands enclosed in basal bulb. Vulva with or without lateral membranes. Postvulval uterine sac present, prominent or sometimes absent. Vagina directed inward or forward. Spermatheca large, axial, oblong to more elongate. Gonads are stretched. Crustaformeria (glandular part of uterus) usually a quadricolumella with five to six cells in each row. Tails are elongate, filiform, and cloacal lips are non-tubular. Bursa adanal, simple or lobed. Spicules moderately robust, gubernaculum linear and fixed.

The family contains three subfamilies: Tylodorinae, Epicharinematinae, and Pleurotylenchinae.

Key to Subfamilies of Tylodoridae

1. Stylet over 75 µm long; labial disc conspicuous; spicules of heavy build, robust .... Tylodorinae

Stylet under 35 μm long; labial disc inconspicuous; spicules not of heavy build 2

2. Cuticle thick, strongly annulated, with or without longitudinal ridges; postvulval uterine sac prominent; bursa simple .... Pleurotylenchinae

Cuticle thin, not strongly annulated; postvulval uterine sac absent; bursa lobed .... Epicharinematinae

Infraorder 2: Anguinata (Infraord. n.)

Members are fungal feeders or parasites of lower (mosses, seaweeds) and higher plants, attacking stems, leaves, floral parts and seeds, almost always inciting galls; root parasitism known only for *Subanguina radicicola*, which incites and inhabits galls; also associates of insects (Sychnotylenchidae) but not parasitic in insects or other animals.

Taxa of this infraorder are small to large in size (0.4-3.5 mm), in some genera adults may be obese and sedentary. Cuticle smooth or with fine transverse striations. Lateral field with or without incisions (4, 6 or more). Deirids usually present. Phasmids absent; prophasmids present in postmedian region outside lateral field in female near vulva. Cephalic region is low, cap like, smooth and generally continuous with body contour; framework hexaradiate, with or without faint sclerotization. Oral opening a small round pore surrounded by six labial papillae; amphids indistinct, oval, slit like, slightly dorso-sublateral on raised areas on labial plate. Stylet small (under 15 μm), with small rounded knobs. Orifice of dorsal gland close to stylet base. Median oesophageal bulb present or absent, with or without refractive thickenings. Oesophageal glands tend to be enlarged, forming a basal bulb, or rarely, the dorsal gland extends over intestine dorsally and laterally; without stem like extension at base of basal bulb. Cardia absent; two anteriormost intestinal cells often acting as a valve. Gonads single, anteriorly outstretched, may be reflexed or coiled in swollen adults. Vulva a large transverse slit, posteriorly located; lateral vulval membranes distinct only in *Pterotylenchus*. Spermatheca axial, elongated, sac like (except rounded on *Pseudhalenchus*). A postvulval sac often longer than body width (absent in *Diptenchus*) and may serve as storage for sperm. Sperm round, large, with a prominent translucent cytoplasmic vesicle around the nucleus. Spicules robust, anteriorly expanded, separate or fused medially, tip often truncate or broadly rounded. Gubernaculum simple, trough-like, not protrusible, rarely absent (as in *Nothanguina*). Bursa moderately large, usually subterminal, but may extend to terminus (Sychnotylenchidae) or be adanal. Tails similar between sexes (except when bursa is terminal), usually elongate-conoid, may be cylindroid or filiform; juvenile tails often elongate-conoid to filiform.

The infraorder contains one superfamily: Anguinoidea.

Superfamily 1: Anguinoidea [Nicoll, 1935 (1926) syn. Anguillulinoidea Baylis & Daubney, 1926 (Nicoll, 1935); syn. Nothotylenchoidea Thorne, 1941 (Jairajpuri & Siddiqi, 1969)]

The superfamily possesses characteristics of the infraorder Anguinata, and contains two families: Anguinidae and Sychnotylenchidae.

Key to Families of Anguinoidea

1. Female tail cylindroid or subcylindroid, dissimilar to that of male; bursa enveloping tail terminus; associates of insects .... Sychnotylenchidae
2. Female tail conoid to filiform, rarely subcylindroid, similar to that of male; bursa not enveloping tail terminus; very rarely associates of insects .... Anguinidae

Family: Anguinidae [Nicoll, 1935 (1926) syn. Anguillulinidae, Baylis & Daubney, 1926 syn. Anguinidae, Paramonov, 1962 (Siddiqi, 1971) syn. Ditylenchidae, Golden, 1971 (Fotedar & Handoo, 1978) syn. Nothotylenchidae, Thorne, 1941 (Jairajpuri & Siddiqi, 1969)]

These nematodes are fungus feeders or parasites of stems, leaves, flower parts and seeds, where they usually incite galls, not root parasites (except *Subanguina radicicola* which incites and inhabits root-galls). Adult forms are moderately long (0.4-3 mm), slender or obese. Cephalic region low, smooth; framework hexaradiate. Median bulb with valvular aperture may be present or absent; basal bulb small or large, offset from intestine, or the dorsal gland may become enlarged and extend over intestine as a lobe. Vulva posterior and generally at less than 85% of body length. Postvulval uterine sac present, or rarely absent (*Diptenchus*). Ovary outstretched or reflexed; oocytes may be arranged about a rachis in obese females. Tails similar between sexes, female tail rarely subcylindrical, never cylindrical or hooked. Bursa variable from adanal to subterminal, never enclosing tail tip.

The family contains two subfamilies: Anguininae and Halenchinae.

Key to Subfamilies of Anguinidae

1. Excretory duct widened, sclerotized; exclusively marine, parasitic on sea algae forming galls .... Halenchinae
   Excretory duct not widened or sclerotized; not marine; not parasitic on sea algae .... Anguininae

Family: Sychnotylenchidae [Paramonov, 1967 syn. Neoditylenchidae Kakuliya & Devdariani, 1975]

The taxa are associates of insects mostly bark beetles, but do not parasitize them. Adult forms are smaller in size (0.5-2.5 µm), slender. Cuticle is finely annulated, annules may be distinct. Cephalic region low, smooth; framework hexaradiate, often moderately sclerotized. Lateral sectors of oesophagus may be offset from procorpus by a constriction. Isthmus usually short. Basal bulb generally large, saccate, offset from intestine, lacking a stem like extension at base. Vulva posteriorly located at 88-94% of body length, lacking lateral cuticular flaps. Postvulval uterine sac present. Quadricolumella with six or more cells in each row. Female ovary is outstretched or reflexed, oocytes not arranged about a rachis and tail is small, cylindroid or subcylindroid, occasionally conoid but with rounded tip. Male tail is conoid and completely enveloped by a bursa. Juveniles with generally elongate-conoid tails; third or fourth stage as dauer larvae.

The family contains one subfamily: Sychnotylenchinae.

Suborder 2: Hoplolaimina [Chizhov & Berezina, 1988 syn. Heteroderata Skarbilovich, 1959]

Nematodes from this suborder are obligate parasites of plant roots; none is myceliophagus or entomophagus. The taxa are small to moderate size (0.5-2 mm). Cuticle with distinct outer and inner layers, often strongly annulated, annules never retrorse. Lateral fields with 1-6 incisures, occasionally reduced or absent (*Basirolaimus*). Cephalic framework well-developed, usually with high arches and strongly sclerotized and refractive. Labial disc may or may not be distinct; six labial sensilla in the form of papillae or pits present around a pore-like round or oval oral opening; cephalic sensilla usually not on surface. Amphidial apertures pore or slit like, just below labial disc, rarely postlabial (*Psilenchus, Macrotrophurus*). Deirids are generally absent, or may be present (Psilenchidae, Merliniinae). Phasmids present (not detectable in *Aphasmatylenchus*) in or near tail region (except for migratory scutella of some Hoplolaimidae), small, with porelike apertures, or large, scutellum like, always in lateral position. Prophasmids absent. Stylet usually well-developed; protractors tubular around stylet; basal knobs prominent (absent in Psilenchidae). Orifice of dorsal oesophageal gland is located close to or at some distance from stylet base. Oesophageal glands free in body cavity or enclosed in a basal bulb. Median bulb is well-developed, muscular, with refractive inner thickenings (except in some males with degenerate oesophagus in Hoplolaimoidea). Intestinal cell walls and lumen usually indistinct; rectum and anus distinct. Females are didelphic, amphidelphic (reflexed or coiled in obese forms) and the posterior branch of ovary may be reduced. Vulva a transverse slit, lips usually not modified, median or submedian, in swollen females may be located subterminally or terminally; epiptygma present or absent. Glandular part of uterus tri or rarely quadricolumellate. Spermatheca generally axial. Tails dissimilar between sexes (except Psilenchidae and some Pratylenchidae). Female tail generally short (less than two anal body widths) but may vary to become elongate-conoid or absent in some obese females. Hypoptygma double. Bursa usually enveloping tail, subterminal, adanal (Psilenchidae) or rarely absent; with or without phasmidial pseudoribs. Testis monorchic (may be diorchic in abnormal *Meloidogyne*), anteriorly outstretched. Spicules paired, similar or rarely dissimilar, cephalated, straight to arcuate, with or without distal flanges, independently protrusible. Gubernaculum simple trough shaped or modified rod-like, fixed or protrusible, with or without terminal titillae; capitulum present in several genera.

The suborder contains two superfamilies: Hoplolaimoidea and Dolichodoroidea.

Key to Superfamilies of Hoplolaimina

1. Subventral oesophageal glands enlarged, usually extending past the dorsal gland; distinct sexual dimorphism is present in anterior region .... Hoplolaimoidea
2. Subventral oesophageal glands not enlarged, not extending past the dorsal gland; sexual dimorphism in anterior region is not distinct .... Dolichodoroidea

Superfamily 1: Hoplolaimoidea [Filipjev, 1934 (Paramonov, 1967) syn. Heteroderoidea Filipjev & Schuurmans Stekhoven, 1941 (Golden, 1971)]

Hoplolaimoidi (=sub superfamily proposed by Paramonov, 1967 as Hoplolaimini)

Taxa are small to moderately large nematodes (about 0.5-2 mm). Sexual dimorphism in cephalic region present, and may also be in stylet, oesophagus and body shape, indistinct in some Pratylenchidae. Cuticle with distinct outer and inner layers, strongly annulated; longitudinal striae may be present but longitudinal ridges outside lateral fields absent. Cephalic framework strongly sclerotized and refractive, less developed in males. Labial disc generally offset, distinct; six labial sensilla usually not on the surface. First cephalic annule generally divided into six sectors which may be modified. Deirids absent (except *Pratylenchoides*). Phasmids small, with porelike apertures, or large, scutellum-like, always in lateral position, (absent in *Aphasmatylenchus*). Stylet well-developed, two to five times lip region width; protractors tubular around stylet; conus about as long as shaft, knobs prominent. Oesophageal glands lobed, overlapping intestine (except *Pararotylenchus* and some *Pratylenchoides* spp. in which they form a pseudobulb). Subventral glands enlarged, equal to or usually larger than dorsal gland; nuclei of one or both subventral glands lying posterior to that of the dorsal gland. Median bulb always well-developed (except in some males with degenerated oesophagus), muscular, with refractive thickenings. A cellular cardia absent, but oesophago-intestinal junction provided with a small cuticular valve. Intestinal cell walls and lumen usually indistinct; rectum and anus distinct. Females are didelphic, amphidelphic; posterior branch may be reduced or represented by a uterine sac (rudimentary). Vulva a transverse slit, in swollen females may be located subterminally or terminally; epiptygma present or absent; lateral membranes are inconspicuous or absent. Glandular part of uterus tricolumellate. Spermatheca thick-walled, round, usually axial. Female tail generally short (less than two anal body widths) but may vary to become elongate-conoid, absent in some swollen females. Male tail usually short and with a distinct hyaline terminal portion. Bursa enclosing all or most of tail, absent in forms with tail less than one anal body width long (Meloidogynidae and Heteroderidae). Tails of both sexes are similar in some Pratylenchidae. Spicules paired, similar or dissimilar, cephalated, straight to arcuate, with or without distal flanges, independently protrusible. Gubernaculum fixed or protrusible, with or without terminal titillae; capitulum present in several genera.

The superfamily contains five families: Hoplolaimidae, Heteroderidae, Meloidogynidae, Pratylenchidae, and Rotylenchulidae.

Key to Families of Hoplolaimoidea

1. Mature female round, pear or lemon-shaped behind neck, with anus terminal or nearly so; male with stylet larger than that of female and tail very short or absent, non-bursate (except *Bursadera*). Male develops by metamorphosis (except *Meloidodera*); third and fourth-stage juveniles often swollen 2

   Mature female not round, pear or lemon-shaped, with anus not terminal; male with stylet equal to or smaller than that of female and tail bursate, not very short (except *Verutus*), does not develop

by metamorphosis; third and fourth-stage juveniles normally not swollen 3
2. Excretory pore in mature female opposite or anterior to median bulb; female labial disc dorsoventrally elongated; male with large cap and large transverse slit like amphidial apertures; gall inciting .... Meloidogynidae
Excretory pore in mature female behind median bulb; female labial disc rounded; male with small lip cap and with small oval to round amphidial apertures; not gall inciting .... Heteroderidae
3. Juveniles and females with low arched cephalic framework, generally endoparasites of roots .... Pratylenchidae
Juveniles and females with high arched cephalic framework, generally ectoparasites of roots 4
4. Mature female swollen, sedentary .... Rotylenchulidae
Mature female not swollen, migratory .... Hoplolaimidae

Family: Hoplolaimidae [Filipjev, 1934 (Wieser, 1953) syn. Nemonchidae Skarbilovich, 1959; Aphasmatylenchidae Sher, 1965 (Fotedar & Handoo, 1978); Pararotylenchidae Baldwin & Bell, 1981 (Eroshenko, 1984); Interrotylenchidae Eroshenko, 1984]

This is a large family of the order Tylenchida, and contains numerous economically important migratory ectoparasitic plant nematodes. Taxa are small to moderately large in size (0.6-1.5 mm) and vermiform. Sexual dimorphism is present in cephalic region. Lateral fields have four incisures, rarely reduced or absent. Deirids absent. Phasmids either small, with pore like apertures near or a little anterior to anus or large scutellum like, near anus or much anterior to it anywhere on body behind oesophageal region; absent in *Aphasmatylenchus*. Cephalic region elevated, high arched, frame-work strongly sclerotized. Labial disc is usually distinct and labial sensilla not located on the surface (except *Pararotylenchus*). Oesophageal glands overlap intestine (except *Pararotylenchus*); subventral glands symmetrical in size and location (except *Antarctylus*). Females are didelphic or rarely monodelphic and tail is short, rounded (about one body width or less long), or conical (over two anal body widths, e.g., *Antarctylus*). Male tail short (except *Aphasmatylenchus*). Bursa large, enveloping tail; flaps crenate, sometimes indented at tip, usually lacking phasmidial pseudoribs. Spicules robust or slender, straight to arcuate, with distal flanges which may be reduced. Gubernaculum large, protrusible or fixed.

The family contains three subfamilies: Hoplolaiminae, Aphasmatylenchinae, and Rotylenchoidinae.

Key to Subfamilies of Hoplolaimidae

1. Phasmids present 2
   Phasmids absent .... Aphasmatylenchinae
2. Phasmids large, scutellum-like .... Hoplolaiminae
   Phasmids small, porelike .... Rotylenchoidinae

Family: Rotylenchulidae [Husain & Khan, 1967 (Husain, 1976)]

Taxa are small in size (1 mm or less) with marked sexual dimorphism in adult body shape and in anterior region. Mature female swollen, often sausage or kidney-shaped behind a less swollen neck; immature female vermiform in type genus. Male vermiform, with reduced cephalic sclerotization, stylet and oesophagus. Cuticle annulated and lateral field obliterated in mature female. Lateral field with four incisures, not areolated. Deirids absent. Phasmids pore like, on tail near anus. Cephalic region in female high, rounded to truncate, continuous or rarely offset, lacking longitudinal indentations; sclerotization strong in females and juveniles. Excretory pore in oesophageal region. Stylet in female and juveniles well-developed, two to three cephalic region widths long, conus length upto shaft length, knobs prominent, rounded; male stylet weaker than the female, orifice of dorsal oesophageal gland well separated from base of stylet. Median bulb with large refractive thickenings. Oesophageal glands extending over intestine mostly ventrally or, as in Acontylinae, dorsally. Vulva postmedian or more posterior with round lips; lateral membranes absent. Ovaries reflexed, coiled or straight. Tail in mature female present, or rudimentary but a distinct postvulval region of body always present. Male tail conoid or reduced (*Verutus*); bursa present or absent. Testis single, outstretched. Spicules cephalated, ventrally arcuate, with narrow distal tip and lacking distinct flanges. Gubernaculum large, trough-shaped, fixed. Juveniles vermiform (except later stages of *Verutus*), with distinct or indistinct hyaline terminal portion of tail, straight, ventrally arcuate or curved when relaxed. Eggs laid singly in gelatinous matrix, not retained in body in large numbers. Adult female a sedentary ectoparasite of roots.

The family contains three subfamilies; Rotylenchulinae, Acontylinae, and Verutinae.

Key to Subfamilies of Rotylenchulidae

1. Female and male with distinct tails, young vermiform migratory female stage present; bursa present 2

Female and male with very reduced tails, young vermiform migratory female stage absent; bursa absent .... Verutinae
2. Oesophageal glands mostly dorsal to intestine; one functional ovary; male head inflated; juveniles almost straight on death ....
.... Acontylinae
Oesophageal glands mostly ventral to intestine; two functional ovaries; male head not inflated; juveniles curved on death ....
.... Rotylenchulinae

Family: Pratylenchidae [Thorne, 1949 (Siddiqi, 1963) syn. Nacobbidae Chitwood in Chitwood & Chitwood, 1950 (Golden, 1971); Radopholidae Allen & Sher, 1967 (Khan & Nanjappa, 1972)]

Nematodes of this family are migratory endoparasites of roots. Taxa are vermiform (except *Nacobbus*). Cuticle is prominently annulated and lateral fields have 4-6 incisures, very rarely areolated behind oesophagus. Deirids absent (except *Pratylenchoides*). Phasmids porelike and located well behind anus. Female: Amphids pore like, indistinct, near oral opening which is surrounded by six labial pits. Cephalic region low, anteriorly flattened to broadly rounded, annulated; framework strongly sclerotized; labial disc inconspicuous, dumb-bell shaped (SEM) in *Pratylenchus, Radopholus* and *Achlysiella,* but indistinguishable in *Hirschmanniella*. Stylet strong, length not exceeding three cephalic region widths (except in *Hirschmanniella*); conus about as long as posterior part; knobs large, rounded, usually closely applied to shaft. Orifice of dorsal gland close (usually 2-3 µm) to stylet base. Procorpus slender. Postcorpus strongly muscular, with prominent refractive thickenings. Isthmus short. Oesophago-intestinal junction indistinct, with a refringent valvula. Oesophageal glands extending over intestine; subventral glands asymmetrical, extending past the dorsal gland; three gland nuclei usually lying in tandem. Vulva a transverse slit, submedian to more posterior but not subterminal; lips not modified; lateral membranes absent. Vagina directed inward. Didelphic, amphidelphic or pseudomonoprodelphic. Ovaries outstretched; posterior ovary degenerate in pseudomonodelphic forms. Spermatheca large, rounded, with small round or rod-like sperm when impregnated. Tail conoid, subcylindrical to elongate-conoid, about twice or more anal body width long, with round to pointed tip which may bear a mucro (*Hirschmanniella*), generally with inconspicuous hyaline terminal portion. Male: Stylet and oesophagus similar to those in female or reduced as in Radopholinae. Tail elongate-conoid, bursa terminal or subterminal. Testis single, outstretched, spermatocytes in one or two rows. Spicules similar, cephalated, arcuate, pointed with subterminal opening on dorsal or ventral side. Gubernaculum simple, fixed, or complex with telamon or titillae, protrusible. Hypoptygma present or absent. Juveniles resemble females in having similar anterior region and tail.

The family contains four subfamilies: Pratylenchinae, Hirschmanniellinae, Nacobbinae and Radopholinae.

Key to Subfamilies of Pratylenchidae

1. Oesophageal glands extending over intestine mostly ventrally and ventrolaterally; no marked sexual dimorphism in anterior region 2
   Oesophageal glands extending over intestine mostly dorsally and dorsolaterally; with marked sexual dimorphism in anterior region 3
2. Tails similar between sexes; phasmids near terminus .... Hirschmanniellinae
   Tails dissimilar between sexes; phasmids not near terminus .... Pratylenchinae
3. Mature female spindle-shaped or batatiform, with numerous eggs within body, gall- inciting .... Nacobbinae
   Mature female not spindle-shaped or batatiform, not with numerous eggs within body, not gall-inciting .... Radopholinae

Family: Meloidogynidae [Skarbilovich, 1959 (Wouts, 1973)]

Taxa of this family cause root gall or knots and the female feeding incites multinucleate nurse (giant) cells. Marked sexual dimorphism present. Cuticle striated. Lateral fields with 4 or 5 incisures. Cephalic region low, with one to four annules. In SEM, female labial disc dorsoventrally elongated, dumb-bell-shaped with oral opening a small round pore surrounded by six inner labial pits (sensilla). Framework moderately sclerotized, hexaradiate; lateral sectors equal to or wider than submedians. Stylet moderately strong, male stylet longer and more robust than the female. Orifice of dorsal oesophageal gland closely behind stylet base. Median bulb oval or round, with large refractive thickenings. Oesophageal glands elongated, extending over intestine mostly ventrally, but also laterally. Excretory pore in female opposite or anterior to median bulb, in male usually behind median bulb. Female preadult vermiform stage does not exist (except in *Meloinema*). Mature female: Swollen, sedentary, round, oval to pear-shaped with a protruding neck. Cuticle moderately thick, striated, generally forming typical, fingerprint like perineal pattern terminally. No cyst stage. Vulva subterminal or terminal. Anus near vulval lip; tail rudimentary or absent. Stylet under 25 µm long but in *Meloinema* 30-35 µm long. Median bulb oval or rounded, usually offset, with large refractive thickenings. Didelphic-prodelphic coiled ovaries. Eggs are laid in gelatinous matrix which are secreted by large rectal glands. Male: Vermiform, migratory, nonparasitic, over 1 mm long, posterior end twisted through 90-180°, develops by metamorphosis within a saccate juvenile. Cephalic region rather low and continuous; amphids large transverse slits; labial cap large, prominent; framework moderately sclerotized, lateral sectors wider than submedians. Stylet strong, usually over 20 µm long, knobs prominent. Tail short, lacking a bursa or absent (except *Bursadera*). Spicules large (25-64 µm), pointed distally. Gubernaculum linear to trough-shaped, not protrusible. Cloacal lips non-tuboid, generally with hypoptygma. Juveniles: Second stage hatches, is migratory and infective. Third and fourth-stage juveniles swollen, without stylet in genus, *Meloidogyne*. In *Meloinema* third and fourth stage juveniles vermiform. Cephalic region low, anterioly flattened or rounded. Lateral sectors wider than submedians. Stylet weak to moderately developed, less than 20 µm long in *Meloidogyne,* strongly developed in *Meloinema*. Tail elongate-conical, with minutely rounded tip and conspicuous terminal hyaline portion. Phasmids dot-like on tail, usually anterior to middle.

The family contains two subfamilies: Meloidogyninae and Nacobboderinae.

Key to Subfamilies of Meloidogynidae

1. Preadult male and female juveniles of third and fourth stage saccate; mature female with anus at base of dorsal lip of vulva; male tail hemispherical .... Meloidogyninae
2. Preadult male and female juveniles of third and fourth stage vermiform; mature female with anus at some distance from base of dorsal lip of vulva; male tail conoid- rounded ....
.... Nacobboderinae

Family: Heteroderidae [Filipjev & Schuurmans Stekhoven, 1941 (Skarbilovich, 1947) syn. Heteroderidae Thorne, 1949; Ataloderidae Wouts, 1973 (Krall & Krall, 1978); Meloidoderidae Golden, 1971 (Krall & Krall, 1978)]

Members of this family are sedentary endoparasites of underground parts and exhibit marked sexual dimorphism. Cuticle is strongly annulated, annules usually modified to form a lace-like pattern on swollen females and cysts. Lateral fields have 2-4 incisures. Cephalic region large, annulated; labial disc distinct or indistinct, rounded. Cephalic framework is strongly sclerotized, secondarily reduced in females; lateral sectors equal to or narrower than submedians. Stylet robust, usually over 20 µm long, with conus about half of its total length and prominent basal knobs. Orifice of dorsal oesophageal gland close to stylet base. Median bulb ovoid to rounded, with prominent refractive thickenings. Oesophageal glands elongated, extending over intestine mostly ventrally, but also laterally; subventral glands asymmetrical, longer than dorsal gland. Female: Sedentary, swollen, elongate-oval, lemon-shaped or spheroidal, with a short neck, may or may not form cyst. Female preadult vermiform stage does not exist. Cuticle abnormally thickened, with annulation or surface variously patterned; perineum lacking fingerprint like pattern. Vulva equatorial to usually terminal. Tail absent; anus dorsally subterminal or terminal. Stylet well-developed with prominent knobs. Excretory pore in anterior region, and also opposite or usually posterior to valve of median bulb. Male: Vermiform, develops through metamorphosis within a swollen juvenile (except in *Meloidodera*). Cuticle thick, annulated. Lateral fields have 3-4 incisures, outer bands may be areolated. Cephalic region regularly annulated, offset or continuous, labial disc distinct or indistinct, rounded; framework strongly sclerotized. Stylet robust, 20-46 µm long, with prominent knobs. Oesophagus well-developed. Testis single, anteriorly outstretched. Tail less than one anal body width long, bluntly rounded, hemispherical, or absent; tail end twisted. Bursa absent. Phasmids near cloacal aperture. Second stage juveniles: Infective stage emerges out of egg. Body slender, vermiform, straight to arcuate on death. Stylet robust, over 17 µm long; tail conical, with pronounced hyaline terminal portion; phasmids anterior to middle of tail. Third and fourth stage juveniles swollen, with robust stylet.

The family contains three subfamilies; Heteroderinae, Meloidoderinae and Ataloderinae.

Key to Subfamilies of Heteroderidae

1. Vulva equatorial; male development not through metamorphosis .... Meloidoderinae

   Vulva terminal or subterminal; male development through metamorphosis 2

2. Female turns into a hard walled cyst .... Heteroderinae

   Female does not turn into a hard walled cyst .... Ataloderinae

Superfamily 2: Dolichodoroidea [Chitwood in Chitwood & Chitwood, 1950 (Siddiqi, 1986)]

Taxa are Migratory ectoparasites of roots. Marked sexual dimorphism is absent; juveniles and adults vermiform and similar; mature female not obese. Cuticle prominently annulated, not showing distinct outer and inner layers (except Merliniinae). Lateral fields have 1-6 incisures (one incisure only in *Belonolaimus*). Deirids present only in Merliniinae and Psilenchidae. Amphidial apertures pore like or oval slits, usually near labial sensilla or near cephalic sensilla. Phasmids small with pore like aperture on tail or just preanal as in females of some species of *Neodolichodorus* which have short hemispherical tails, in males extending into bursa, forming a pair of pseudoribs. Cephalic framework mostly with high arches and conspicuous extensions projecting posteriorly, with light or heavy sclerotization; annules generally distinct, basal annule mostly not indented. Labial disc prominent in Dolichodoridae and in some members of the subfamily, Telotylenchinae. Oral aperture small, round or oval, surrounded by six labial sensilla. Stylet short (about 10-12 µm) to long (over 100 µm), with distinct basal knobs (knobs absent in Psilenchidae). Orifice of dorsal oesophageal gland near stylet base. Procorpus is cylindrical; metacorpus is muscular round to oval with refractive cuticular thickenings and isthmus is slender. Basal bulb enclosing oesophageal glands present, or only the dorsal gland enlarging and extending over anterior end of intestine, while subventral glands remaining small and anterior to the dorsal gland and may or may not overlap intestine. Cardia is three-celled, and well-developed (reduced in forms with overlapping glands). Females have outstretched didelphic, amphidelphic ovaries, secondarily becoming pseudomono-prodelphic by the reduction of the posterior branch in *Trophurus*. Vulva a transverse slit, rarely round or oval, median or postmedian, with or without epiptygma; lateral vulval membranes absent. Vagina at right angles to body axis, sclerotized in the subfamily, Dolichodorinae. Crustaformeria (glandular distal part of uterus) is tricolumellate. Spermathecae axial, round (in most groups), lobed (Merliniinae) or sac-like (Psilenchidae). Female tail rarely less than two anal body widths long, and are conoid, cylindroid, subclavate or elongate-filiform. Male monorchic with outstretched testis. Bursa enveloping entire tail or, in Psilenchidae, adanal. Spicules symmetrical, cephalated, ventrally arcuate, with distal flanges and pointed tip bearing subterminal pore or cylindroid with tip broadly rounded and notched (Merliniinae). Gubernaculum simple or modified, fixed or protrusible, without titillae and capitulum. Cloacal lips not modified into a penial tube. Hypoptygma (cloacal lip papillae) usually not visible (except in Merliniinae).

The superfamily contains four families: Dolichodoridae, Belonolaimidae, Psilenchidae and Telotylenchidae.

Key to Families of Dolichodoroidea

1. Tails similar between sexes, generally filiform; bursa adanal .... Psilenchidae
   Tails dissimilar between sexes, not filiform (except terminally only in some Dolichodoridae); bursa enveloping entire tail 2
2. Cephalic region four lobed with reduced lateral lip areas; basal plate strongly sclerotized; vaginal wall usually sclerotized; stylet conus often much longer than shaft 3
   Cephalic region rounded, not four lobed, lateral lip areas not reduced; basal plate or vaginal wall not strongly sclerotized; stylet conus usually about as long as shaft (except Macrotrophurinae) .... Telotylenchidae
3. Oesophageal glands not overlapping intestine; oesophago-intestinal junction more than two body width from median bulb; bursa trilobed .... Dolichodoridae
   Oesophageal glands overlapping intestine; oesophago-intestinal junction about one body width from median bulb; bursa simple, not trilobed .... Belonolaimidae

Family: Dolichodoridae [Chitwood in Chitwood & Chitwood, 1950 (Skarbilovich, 1959)]

Members of this family are generally called as awl nematodes. They are migratory ectoparasites and feed on deep root tissue. Taxa are small to moderately large in size (generally about 1.0-1.5 mm), straight, usually arcuate, or more curved. Cuticle strongly annulated. Lateral fields have 3-4 incisures. Cephalic region usually four lobed; labial disc distinct. Basal plate and arches of the cephalic framework strong sclerotized. Amphidial apertures longitudinal or dorsoventral slitlike. Stylet well developed, usually very long. Oesophageal glands enclosed in a basal bulb. Oesophago-intestinal junction more than one, corresponding to body width from median bulb. Females are didelphic-amphidelphic. Vulva median or submedian, transverse slit like. Vaginal wall sclerotized. Postrectal intestinal sac present. Tails dissimilar between sexes: female tail elongate, usually filiform with spicate tip, or hemispherical to mamillate (*Neodolichodorus*). Male tail short, conoid, completely enveloped by a trilobed bursa. Spicules well-developed, similar, pointed, with or without distal flanges. Gubernaculum large, protrusible; hypoptygma absent.

The family contains two subfamilies: Dolichodorinae and Brachydorinae.

Key to Subfamilies of Dolichodoridae

1. Cephalic region four lobed; labial disc distinct; stylet 50-170 μm long .... Dolichodorinae
2. Cephalic region not four lobed; labial disc indistinct; stylet under 40 μm long .... Brachydorinae

Family: Belonolaimidae [Whitehead, 1960 (Siddiqi, 1970)]

Members of this family are migratory ectoparasites, feed on roots (generally Poaceae) and inhabit soil. Taxa are moderate in size (1-3 mm). Lateral fields have 1, 2 or 4 incisures, areolated. Deirids absent. Phasmids on tail. Amphidial apertures obscure, probably longitudinal cleft like. Cephalic region offset by a constriction, rarely continuous, finely but distinctly annulated (6-10 annules), usually four lobed, lateral sectors much smaller than submedians; labial disc prominent, round to lemon shaped; six pits around oral aperture; framework sclerotized, its inner margins extend as elongated, tubular vestibulum extension, posteriorly giving support for the attachment of most of the protractor myofibrils. Stylet very long (60-150 μm), conus 55-80% of its total length; knobs rounded. Procorpus with convoluted lumen, postcorpus very muscular, oval to rounded, isthmus short and oesophageal glands extending over intestine latero-dorsally, laterally or latero-ventrally; dorsal gland larger and longer than subventrals. Oesophago-intestinal junction indistinct; less than one corresponding body width from median bulb. Intestinal fasciculi present. Vulva with epiptygma. Vagina in a few species sclerotized. Females: Ovaries are didelphic, outstretched; tail cylindroid, subcylindroid or conoid, over one and a half anal body widths long. Male tail elongate-conoid; bursa simple, enclosing tail tip. Testis are monorchic and outstretched. Spicules well-developed. Gubernaculum large, modified, usually with titillae and recurved distal end.

The family contains one subfamily: Belonolaiminae.

Family: Telotylenchidae [Siddiqi, 1960. syn. Tylenchorhynchidae Eliava, 1964; Merliniidae Siddiqi, 1971 (Ryss, 1993); Meiodoridae Siddiqi, 1976 (Ryss, 1993)]

Nematodes of this family are ectoparasites of roots and are small to medium sized (0.5-1.1 mm). Cuticle prominently annulated; longitudinal striae and grooves may be present. Lateral fields have 3-6 incisures. Deirids absent except in Merliniinae. Amphids are labial and apertures pore like. Cephalic region annulated; labial disc indistinct and not marked off from cephalic annules (except *Sauertylenchus*). Oral opening is oval and surrounded by six labial sensilla; framework with light to moderate sclerotization. Stylet under 45 μm long, conus about as long as shaft except *Macrotrophurus*. Oesophageal glands enclosed in a basal bulb or extending over intestine laterally or latero-dorsally. Vulva is transverse and vagina not sclerotized. Females are didelphic and tail generally two to four anal body widths long, terminus variable. Male tail elongate-conoid, enveloped by a simple bursa or trilobed in Meiodorinae. Spicules arcuate with large distal flanges, tip narrowly pointed or indented accompanied by gubernaculum which is rod-like, protrusible or fixed.

The family contains four subfamilies: Telotylenchinae, Macrotrophurinae, Meiodorinae, and Merliniinae.

Key to Subfamilies of Telotylenchidae

1. Deirids present (except in *Scutylenchus*), lateral field with six incisures; male with hypoptygma; spicules cylindrical, not flanged .... Merliniinae
   Deirids absent, lateral field with two to five incisures; male without hypoptygma; spicules flanged 2
2. Amphidial apertures conspicuous, postlabial; stylet over 80 µm long .... Macrotrophurinae
   Amphidial apertures inconspicuous, labial; stylet under 50 µm long 3
3. Male tail conspicuously shorter than that of female; bursa trilobed .... Meiodorinae
   Male tail not conspicuously shorter than that of females; bursa simple .... Telotylenchinae

Family: Psilenchidae [Paramonov, 1967 (Khan, 1969)]

Members of the family Psilenchidae possess weak stylet and elongate tail. They are associates of lower plants, and weak plant parasites feeding on root hairs and epidermal cells. The taxa are small to medium sized (0.5-1.8 mm). Cuticle distinctly annulated. Lateral fields have 4 incisures, inner ones rarely obscure. Amphidial apertures indistinct, porelike, near oral opening, or distinct slit like at base of lip areas. Deirids present. Phasmids distinct, pore like, on tail. Cephalic region generally continuous, finely striated; 6 inner labial and 4 cephalic papillae on surface. Stylet slender, conus much shorter than the shaft, basal knobs present or absent. Median bulb muscular, valvate; basal bulb offset from intestine; cardia prominent. Vulva median or submedian with or without lateral membranes. Spermathecae axial (Psilenchinae) or lobed (Antarctenchinae). Ovaries paired, outstretched, tails filiform or elongate-conoid, similar between sexes. Bursa simple adanal. Spicules tylenchoid, cephalated, pointed distally accompanied by gubernaculum which is simple, trough like, and fixed.

The family contains two subfamilies: Psilenchinae and Antarctenchinae.

Key to Subfamilies of Psilenchinae

1. Stylet knobbed; vulva with lateral membranes and epiptygma .... Antarctenchinae
2. Stylet not knobbed; vulva without lateral membranes and epiptygma .... Psilenchinae

Suborder 3: Criconematina [Siddiqi, 1980]

Members of this suborder are ecto and semiendoparasites of roots. Taxa are small in size (0.1-1 mm) and show marked sexual dimorphism in anterior region. Cuticle either thin and finely annulated, or thick and coarsely annulated; latter may have retrorse annules, scales, spines or an extra cuticular body sheath. Lateral fields present or absent. Deirids absent except in juveniles of some Tylenchuloidea. Phasmids absent. Female: Vermiform, sausage-shaped, or obese only in Tylenchuloidea. Cephalic region smooth or usually with 1-3 coarse annules; framework hexaradiate, with light to heavy sclerotization. Oral aperture dorsoventrally oval or slit like, often appearing I-shaped due to the presence of two lateral liplets. Six lip areas, may be fused to form a labial disc. Submedian lip areas with lobe-like outgrowths in several genera. No sensilla on labial surface; inner labial sensilla open in prestoma. Stylet long or short, but shaft always about 8-10 µm long. Basal knobs well-developed; large knobs may characteristically be anchor shaped. Oesophagus criconematoid: corpus enormously developed, broad-cylindroid with muscular postcorpus amalgamated with procorpus; isthmus either slender and offset from basal bulb (Tylenchuloidea), or broad and amalgamated with it (Criconematoidea, Hemicycliophoroidea); basal bulb small, containing three oesophageal glands (except *Tumiota*). Cardia small, usually indistinct. Excretory pore oesophageal or post-oesophageal; excretory system may produce gelatinous matrix in which eggs are deposited (e.g., *Tylenchulus*). Females monoprodelphic; vulva transversely oval or slit like, located 85% of body length posteriorly. Vagina anteriorly directed. Postvulval uterine sac absent. Spermatheca small, offset, ventral or ventro-lateral to the axis of the gonoduct. Crustaformeria may be very thick-walled, and in *Meloidoderita* hypertrophied to form a cystoid body. Male: Vermiform. Oesophagus degenerated. Stylet also degenerated or absent. Monorchic; gonoduct usually filled by minute round or amoeboid sperm, which are often produced at one stage near the final moult; testis usually obliterated in adult. Bursa weakly developed (except Hemicycliophoroidea), rarely enveloping tail tip (*Tylenchocriconema*), absent in several groups (Tylenchulidae, Sphaeronematidae, most Paratylenchidae). Spicules setaceous, often very long, straight, arcuate, U- or hook-shaped. Gubernaculum simple, linear or crescent like in lateral view, fixed. Cloacal lips narrow, sometimes drawn out as a penial tube. Hypoptygma single, rarely double (*Tylenchocriconema*), or absent (*Tylenchulus*).

The suborder contains three superfamilies: Criconematoidea, Hemicycliophoroidea, and Tylenchuloidea.

Key to Superfamilies of Criconematina

1. Females and juveniles with thick cuticle, bearing coarse, round or retrorse annules; oesophagus with isthmus broad and amalgamated with basal bulb 2

   Females and juveniles with thin cuticle bearing fine round annules; oesophagus with isthmus not amalgamated with basal bulb .... Tylenchuloidea

2. Females and juveniles elongate-vermiform, usually over 0.6 mm long; annules round, lacking cuticular outgrowths; male with elongate tail and high bursa, does not develop by metamorphosis .... Hemicycliophoroidea

   Females and juveniles spindle- or sausage-shaped, usually under 0.5 mm long; annules retrorse (except secondarily becoming rounded in females of most *Hemicriconemoides* spp.), may have cuticular outgrowths; male with short tail and low or no bursa, develops by metamorphosis .... Criconematoidea

Superfamily 1: Criconematoidea [Taylor, 1936 (1914) (Geraert, 1966) syn. Criconematoidi Paramonov, 1967 (sub-superfamily name amended from Criconematini Paramonov, 1967)]

Taxa of this family are ectoparasites of roots. They are small-sized (usually under 0.8 mm) and show marked sexual dimorphism. Females and juveniles sausage to spindle-shaped with thick cuticle and coarse retrorse annules (annules secondarily rounded in females of Hemicriconemoidinae and some Criconematinae), with or without scales, spines and other configurations; males vermiform with not so thick cuticle and annules are always rounded and a degenerate oesophagus hardly showing any structure. Incisures in lateral fields are present in males and absent in juveniles and females. Cephalic region of juveniles and females with one or two annules, an indistinct labial disc bearing I-shaped oral aperture surrounded by six pseudolip areas, with or without submedian lobes; no sensory papillae or pits on surface; in males cephalic region usually continuous, rounded and striated, and framework not appearing in lateral view as 'spectacle mark'. Stylet well developed in juveniles (exceptionally absent in some stages) and females, conus markedly longer than shaft, latter usually about 10-12 µm long, basal knobs prominent, appearing anchor shaped. Orifice of dorsal oesophageal gland 3-6 µm behind stylet base; stylet absent in males. Juveniles and females with well-developed criconematoid oesophagi, procorpus broad, posteriorly expanded and continued into a slightly broader, very muscular, metacorpus having large, elongated refractive thickenings; the two parts forming a broad cylindrical muscular corpus filling the body width. Isthmus short, broad and amalgamated with a small reduced basal bulb offset from intestine and containing the three oesophageal glands. A small non-cellular cardia may be present. Excretory pore at, or behind the base of oesophagus. Intestine syncytial, lacking lumen. Vulva far posterior, ovary anteriorly outstretched. Postvulval uterine sac absent. Male develops by metamorphosis within a sausage shaped juvenile. Testis degenerate in adult; gonoduct packed with numerous, very small, round sperms produced at one stage of development. Spicules elongate-setose, almost straight to arcuate, proximally cephalated and distally pointed; gubernaculum simple, fixed. Male tail short; bursa low, adanal, subterminal or terminal, occasionally absent. Juveniles lack a body sheath.

The superfamily contains one family: Criconematidae.

Family: Criconematidae [Taylor, 1936 (1914) (Thorne, 1949) syn. Ogmidae Southern, 1914; Macroposthoniidae Skarbilovich, 1959: Madinematidae Khan, Chawla & Saha, 1976]

Members are commonly called as ring nematodes. The family Criconematidae have characteristics of the superfamily, as described above. The family contains three subfamilies: Criconematinae, Hemicriconemoidinae, and Macroposthoniinae.

Key to Subfamilies of Criconematidae

1. Female with a cuticular body sheath of round annules; juveniles with scales or spines usually arranged irregularly or in alternating rows .... Hemicriconemoidinae
   Female without a cuticular body sheath; scales or spines in juveniles, if present, almost always arranged in longitudinal rows 2
   Females and juveniles with annules ornamented with scales, spines or other appendages (except females of *Criconema*) .... Criconematinae
   Females and juveniles with smooth or crenate annules, and lacking scales, spines or other appendages .... Macroposthoniinae

Superfamily 2: Hemicycliophoroidea [Skarbilovich, 1959 (Siddiqi, 1980)]

Taxa are small to moderately large in size (0.6-2.0 mm), vermiform, straight to arcuate upon relaxation, with thick cuticle and coarse round annules numbering over 200. Juveniles and females have extra cuticle as a protective sheath, except in *Caloosia*. Lateral fields present except in females of *Caloosia* and of a few other genera. Female cephalic region with 1-3 annules, with a labial disc but lacking submedian lobes. Stylet in juveniles and females elongated (over 50 µm long in females), basal knobs rounded, usually posteriorly sloping. Oesophagus criconematoid, with a short broad isthmus amalgamated with basal bulb. Intestine syncytial, vacuolated, usually extending anteriorly over basal bulb, subterminal; lips variously modified. Females mono prodelphic and tail usually elongate-conoid to filiform but may be cylindroid or rarely hemispherical. Males with degenerated oesophagus, no stylet, an elongated tail and prominent bursa that rarely covers more than half of the tail. Spicules setaceous, long, usually strongly curved to become semicircular, U- or hook-shaped (hemicycliophoridae), but may be arcuate (*Loofia*) or straight (Caloosiidae). Gubernaculum fixed. Cloacal lips usually elongated to form a penial tube. Hypoptygma single, seta-like, at posterior lip of cloaca.

The superfamily contains two families: Hemicycliophoridae and Caloosiidae.

Key to Families of Hemicycliophoroidea

1. Females and juveniles with or without a membranous body sheath; cephalic annules separated; vulva flush with body contour, overhung by its anterior lip; spicules straight ..... Caloosiidae
2. Female and juveniles with a thick body sheath; cephalic annules not separated; vulva rarely flush with body contour but then not overhung by its anterior lip; spicules arcuate, semicircular, U- or hook-shaped .... Hemicycliophoridae

Family: Hemicycliophoridae [Skarbilovich, 1959 (Geraert, 1966)]

Members of this family are commonly called as sheath nematodes for the presence of a characteristic body sheath as a second cuticle. The sheathed juveniles are not found in any other tylenchid group. The body sheath is present in females and juveniles with lateral fields. Cephalic annules not modified or separated. Vulva over half body width long, usually marked by recessed body contour behind it; lips modified and projecting (except in *Loofia*). Vagina straight, curved, not sigmoid. Males with smooth, offset cephalic region, framework in lateral view appearing as 'spectacle mark' and tail longer than that of female. Spicules arcuate (*Loofia*), semicircular (*Hemicycliophora*) or U- and hook-shaped (*Aulosphora*). Cloacal lips elongated to form a penial tube. Bursa is small.

The family contains one subfamily: Hemicycliophorinae.

Family: Caloosiidae [Siddiqi, 1980]

Taxa of this family lack body sheath, and are large sized with elongate filiform tails, male tail shorter than that of female, with a large bursa covering more than one-third of tail. Females and juveniles either lack a body sheath or have a membranous sheath being much thinner than the body cuticle. Lateral fields are present only in the form which have a sheath. Cephalic annules separated and usually modified in juveniles and females. Male cephalic region continuous with body and lack lateral 'spectacle mark'. Vulva transversely oval, less than half body width long, partially overhung by its dorsal lip, thus appearing flush with body contour. Vagina sigmoid. Spicules straight; cloacal lips not elongated and a penial tube is absent.

The family contains one subfamily: Caloosiinae.

Superfamily 3: Tylenchuloidea [Skarbilovich, 1947 (Raski & Siddiqi, 1975) syn. Tylenchulidoidea Raski & Siddiqi, 1975 (= incorrect spelling) Tylenchocriconematoidea Raski & Siddiqi, 1975]

Taxa are small in size, usually under 0.5 mm long. Sexual dimorphism present in several genera (obese females). Cuticle thin, finely annulated (secondarily thickened and without discrenible annulation in obese females). Lateral field with 2-4 incisures are usually present. Female: Lip region smooth lacking prominent annules, continuous with body, with or without submedian lobes. Metacorpus large, muscular, amalgamated with broad procorpus, which may be slender in short-stylet forms. Isthmus slender, not amalgamated with basal bulb which is usually small and rounded; oesophageal glands free in body cavity only in *Tumiota*. Vulva a large transverse slit. Ovary outstretched or coiled in obese females. Uterine wall thick in obese females, may form a cystoid body in *Meloidoderita*. Excretory cell may be abnormally enlarged to produce a gelatinous matrix in which eggs are deposited (e.g., *Tylenchulus*). Male: Oesophagus and stylet degenerate or the latter is absent. Bursa usually absent, otherwise low, adanal to subterminal (enveloping entire tail in *Tylenchocriconema*). Spicules setose, arcuate, with pointed tip. Hypoptygma single, usually absent. Gonoduct packed with minute sperm, testis in adult degenerates. Juveniles: Similar to female in most details.

The superfamily contains three families: Tylenchulidae, Paratylenchidae, and Sphaeronematidae.

Key to Families of Tylenchuloidea

1. Stylet long (usually over 20 μm and with conus abnormally elongated); procorpus broad; female usually vermiform, if saccate, then body elongate obese, enlarging on all sides; males and juveniles usually strongly curving ventrally upon relaxation, with short tails .... Paratylenchidae
   Stylet short (usually under 15 μm and with conus not abnormally elongated); enlarging mostly dorsally; males and juveniles usually not curving upon relaxation, with elongate tails 2
2. Adult female spherical or subspherical, lacking a postvulval region; excretory pore in oesophageal region ...... Sphaeronematidae
3. Adult female not spherical or subspherical, with a distinct postvulval region; excretory pore much behind oesophageal region (except *Boomerangia*) .... Tylenchulidae

Family: Tylenchulidae [Skarbilovich, 1947 (Kirjanova, 1955) syn. Tylenchulidae Raski, 1957]

The taxa small in size (under 0.5 mm). Marked sexual dimorphism in body form in anterior region is present; females are obese, with well developed stylet and oesophagus. Males are slender, and have degenerated stylet and oesophagus. Cuticle thin (except in obese females), finely annulated and lateral fields are present (obscure in obese females). Female: Elongate-obese, ventrally curved, with a distinct postvulval region. Cephalic region rounded, with or without perioral elevation. Stylet about 15 μm or shorter; knobs rounded. Basal bulb enclosing gland, offset from isthmus and intestine. Excretory pore generally well behind oesophagus; enormously developed and produces gelatinous matrix in which eggs are deposited. Vulva a large transverse slit. Postvulval uterine sac absent. Uterus with thick walls, and only one mature egg at a time. No uterine cyst formed. Ovary coiled in obese females. Eggs laid in gelatinous matrix. Anus generally rudimentary or absent; tail is short. Male: Body almost straight when relaxed. Cephalic region rounded, continuous. Stylet reduced or absent. Testis outstretched, degenerate in adult; sperm rounded. Spicules small, slender, arcuate, cephalated. Juveniles: Slender, straight to arcuate when relaxed. Deirids often present. Stylet well developed, 15 μm or shorter. Procorpus and isthmus elongate-slender. Excretory pore generally much posterior to oesophagus. Tail elongate-conoid, minutely rounded.

The family contains one subfamily: Tylenchulinae.

Family: Sphaeronematidae [Raski & Sher, 1952 (Geraert, 1966) syn. Meloidoderitidae Kirjanova & Poghossian, 1973]

Members of this family have sedentary ectoparasitic females. In *Goodeyella* and *Sphaeronema*, the females are sedentary endoparasites of roots, the only known instances of endoparasitism among criconematina. The taxa are small in size (0.1-0.7 mm). Marked sexual dimorphism present. Adult female spherical or subspherical with or without neck. Male with degenerated oesophagus without stylet. Juveniles and female with well developed oesophagus and stylet and excretory pore located in oesophageal region. Female: Cuticle thick with indistinct or no annulation. Cephalic region small, elevated, sclerotization delicate. Stylet 15 µm or shorter (rarely upto 27 µm); conus equal to or slightly longer than shaft, knobs rounded. Oesophagus with cylindrical procorpus, spheroidal muscular metacorpus, a slender isthmus and usually a small basal bulb containing glands, or rarely with glands enlarged and extending over intestine. Vulva a transverse slit, terminal, on a protuberance of body or flush with body surface. Uterus swells to form a thick walled chamber which may fill most of the body cavity. Ovary coiled. Tail absent. Male: Gonoduct filled with small round sperms, testis degenerate in adults. Tail conoid. Bursa absent. Spicules slightly arcuate. Cloacal lips conoid, may be slightly elongated. Juveniles: Vermiform. Cephalic region continuous, elevated, sclerotization pyriform. Stylet stout, less than 15 µm long, with round basal knobs. Tail elongate conoid to small rounded terminus.

The family contains two subfamilies: Sphaeronematinae and Meloidoderitinae.

Key to Subfamilies of Sphaeronematidae

1. Adult female with a neck; uterine walls do not form a protective cystoid body for eggs .... Sphaeronematinae
2. Adult female without a neck; uterine walls form a protective cystoid body for eggs .... Meloidoderitinae

Family: Paratylenchidae [Thorne, 1949 (Raski, 1962) syn. Tylenchocriconematidae Raski & Siddiqi, 1975]

Taxa are small in size (under 0.5 mm except *Tylenchocriconema*), vermiform, curving ventrally when relaxed, plump, adult female if swollen remains cylindroid and vermiform. Cuticle thin, finely or moderately annulated. Lateral fields present, with incisures. Female: Stylet well developed, conus abnormally elongated, knobs small, rounded. Oesophagus with a broad corpus having well developed muscular postcorpus. Isthmus slender, basal bulb offset containing oesophageal glands. Excretory pore in oesophageal region; renette cell not abnormally enlarged. Vulva a large transverse slit, posterior, with or without lateral cuticular flaps. Postvulval uterine sac absent. Tail cylindroid, subcylindroid or tapering. Male: Cephalic region symmetrical or asymmetrical. Stylet rudimentary or absent. Oesophagus degenerate, often completely. Testis outstretched. Spicules small, slender, ventrally arcuate, cephalated, pointed. Gubernaculum small, fixed. Cloacal lips often raised, tubular. Bursa present (*Tylenchocriconema, Cacopaurus*) or absent (*Paratylenchus*). Juveniles: Most juveniles resemble female in stylet, oesophagus and body shape, occasionally later stage juveniles may have degenerated oesophagus and no stylet. Body curving upon relaxation.

The family contains two subfamilies: Paratylenchinae and Tylenchocriconematinae.

Key to Subfamilies of Paratylenchidae

1. Female elongate-slender (L=about 0.5 mm or more; a=40-70); male with asymmetrical head and a distinct bursa ....
   .... Tylenchocriconematinae
   Female not elongate-slender (L under 0.5 mm; a<35); male with symmetrical head and indistinct or no bursa .... Paratylenchinae

Suborder 4: Hexatylina [Siddiqi, 1980 syn. Sphaerulariina Maggenti, 1982; Allantonematina Inglis, 1983; Heterotylenchina Inglis, 1983]

Nematodes of this suborder are primarily entomopathogenic, mostly with free-living mycetophagous or plant parasitic (e.g., *Fergusobia*) habits. Female of several genera di, tri or tetra-morphic according to feeding habits. Entomopathogenic generation with only adult female stage is parasitic in insect or mite haemocoel, other stages in host are nonparasitic. Obese adult females occur in arthropod haemocoel and in plant galls (*Fergusobia*), where they absorb food from general body surface; as they usually have microvilli on body surface and canal like formation in body wall. A cuticle in such forms may be lacking. Cuticle smooth or finely annulated. Lateral fields present or absent. Deirids usually present. Phasmids and prophasmids may be present. Cephalic region generally low, smooth or finely striated; with little or no sclerotization; no labial disc or submedian lobes; amphidial apertures dorsosublateral, pore or oblique slit like, at about the level of four cephalic sensilla which are on the surface; six inner labial sensilla around minute pore like oral opening. Stylet generally under 20 μm long (hypertrophied in preadult insect parasitic female), with or without basal knobs. Orifice of dorsal gland close to or at some distance behind stylet. Oesophagus in entomopathogenic forms not divisible into corpus isthmus and basal region. Oesophageal glands three, two in *Sphaerularia* and *Tripius*, contained in a basal bulb or extending over intestine. Nerve ring circum intestinal in insect parasitic forms, circum oesophageal in free living stage. Excretory duct not vesiculate terminally. A cellular cardia absent. Intestine oligocytous, may be syncytial. Rectum may act as a feeding pump (e.g., *Hexatylus*) in free living stage. Anus porelike, atrophied in saccate females. Tails of juveniles and adults over one anal body width long. Female: Monoprodelphic. Free living female with a short slender stylet, oesophagus with corpus, isthmus and basal region and nerve ring encircling isthmus or subcylindroid with cardia anterior to nerve ring. Preadult entomoparasitic, females with hypertrophied stylet (=pseudostylet) and oesophagus, small. Vulva is elongated and uterus serves as storage for sperms. Adult females often have several eggs and/or juveniles. Crustaformeria in the form of a quadricolumella or with more than four rows of cells, never a tricolumella. Oviduct with two consecutive rings of four cells each (*Hexatylus*) or with two rows of more than three cells each. Ovary single, outstretched, reflexed at tip or coiled; a rachis may be present in forms inhabiting insect haemocoel. Vagina may be tuboid and strongly muscular and uterus may prolapse in some insect parasitic forms. Vulva a large transverse slit, oval, or small pore like, located posteriorly at over 85% of body length. Male: With or rarely without stylet. Oesophagus as in free living female, or rarely degenerated. Testis single, outstretched or with tip reflexed, with or without a rachis. Spicules small (usually under 30 μm long), paired, arcuate, cephalated or in Iotonchiidae large, robust and angular; never setaceous. Gubernaculum if present is simple, fixed. Bursa, if present, simple and without phasmidial pseudoribs. Hypoptygma single; caudal papillae absent (except *Fungiotonchium*).

The suborder contains two superfamilies: Neotylenchoidea (=Sphaerularioidea) and Iotonchioidae.

Key to Superfamilies of Hexatylina

1. Two alternating entomopathogenic generations in host present; two or more types of adult found in host's haemocoel; spicules may be angular .... Iotonchioidea
2. One entomopathogenic generation present; one type of adult (heterosexual female) present in host's haemocoel; spicules not angular .... Sphaerularioidea

Superfamily 1: Sphaerularioidea [Lubbock, 1861 (Poinar, 1975) syn. Neotylenchoidea Thorne, 1941 (Jairajpuri & Siddiqi, 1969)]

Nematodes of this superfamily have two types of generation. First are free living, mycoparasites or phytoparasites. The second type have heterosexual females which are parasitic in insect or mite haemocoel. A complete generation cycle does not occur in insect host. Preadult female: Partially free living. Stylet and oesophageal glands hypertrophied, elongated, extending over intestine or forming a cylinder like bulb, subventral glands larger than the dorsal gland. Vulva small, lacking lip flaps. Uterus very long, may extend to middle of body, often packed with hundreds of sperm when impregnated. Ovary underdeveloped. Tail conoid to subcylindroid. Female penetrates host after impregnation. Male: Nonparasitic, occurs in the environment. Essentially similar to free living female. Mature female: In insect haemocoel. Body generally obese and hypertrophied (except in forms with everted uterus). Stylet and oesophagus nonfunctional; feeding through general body surface or everted uterus. Eversion of uterus takes place in Sphaerulariidae. Uterium (hypertrophied everted uterus) round to sausage shaped, 1-20 mm long, capable of absorbing food from haemolymph, independently, ovary goes on producing eggs. Juveniles usually develop in host's haemocoel to third or fourth stage and exit from host's anus or vulva for further development, either as a free living generation or as partially free living adults, which mate and the impregnated female invades the host. Free living generation: Mycetophagous or plant parasitic. No sexual dimorphism in anterior region. Cuticle finely annulated or smooth. Lateral fields and deirids generally present; phasmids absent; amphids labial, indistinct in lateral view. Cephalic region low or elevated, smooth, continuous or offset; framework moderately sclerotized. Stylet generally under 20 µm long, base with or without knobs or thickenings. Oesophagus cylindroid or fusiform, nonmuscular, with gland free in body cavity or divisible into corpus, isthmus, and basal bulb region containing glands and usually with a stem like extension penetrating into intestine. Nerve ring circumoesophageal or circumintestinal. A cellular cardia absent. Excretory pore anywhere between stylet and oesophageal base. Female: Monodelphic, prodelphic. Vulva a long transverse slit. Postvulval uterine sac present or absent. Ovary outstretched, tip may be reflexed due to excessive growth. Crustaformeria usually in the form of quadricolumella, with four or more cells in each row. Spermatheca axial, elongate. Male: Testis outstretched with spermatocytes generally serially arranged. Spicules stout, arcuate. Gubernaculum trough like, fixed, or absent. Bursa present or absent. Tails conoid, cylindroid, or filiform; similar or dissimilar between sexes.

The superfamily contains three families: Sphaerulariidae, Allantonematidae and Neotylenchidae. *Familia dubia*, Paurodontidae (Thorne, 1941) is also present.

Key to Families of Sphaerularioidea

1. Entomoparasitic female everts uterus, which often hypertrophies and leads an independent life; oesophageal glands of free living female form a basal bulb or a short overlap over intestine .... Sphaerulariidae
   Entomoparasitic female does not evert uterus; oesophageal glands of free living female form a long overlap over intestine 2
2. Free living fungus feeding or plant parasitic generation present .... Neotylenchidae
3. No complete free-living or plant parasitic generation known .... Allantonematidae

Family: Neotylenchidae [Thorne, 1941 syn. Hexatylidae Skarbilovich, 1952 (Paramonov, 1970); Fergusobiidae Goodey, 1963 (Siddiqi & Goodey, 1964); Gymnotylenchidae Siddiqi, 1980; Phaenopsitylenchidae Blinova & Korentchenko, 1986]

Two types of generations occur. First type involves free living, fungus or plant feeding. The other type involves heterosexual females parasitic in the insect haemocoel (insect host of adult parasitic female not known for the type genus). A generation cycle does not complete in insect host. Sexual dimorphism absent in anterior region. Entomoparasitic forms: Preadult female: Partially free-living, fertilized in external environment. Stylet and oesophagus hypertrophied; ovary immature. Uterus long, packed with minute sperm when impregnated. Mature female: Obese, hypertrophied, elongate tuboid or sausage shaped. Uterus not everted. Food is presumably absorbed through general body surface as the stylet and oesophagus are nonfunctional. Reproductive branch much coiled and filling most of body cavity. Oviparous or ovoviviparous. Free living, myco or phyto-parasitic forms: Cuticle smooth or finely striated. Cephalic framework lightly to moderately sclerotized. Stylet under 20 µm long; basal knobs may be bifid. Orifice of dorsal gland close to stylet base. Oesophagus cylindroid or fusiform, nonmuscular; basal bulb absent, glands free in body cavity, extending over intestine. Nerve ring generally circumintestinal. Excretory pore anterior or posterior to nerve ring. Vulva in posterior region. Postvulval uterine sac almost always absent. Monoprodelphic outstretched ovary and may be reflexed secondarily by excessive growth. Oviduct consisting of two consecutive rings of four cells each in type genus (*Gymnotylenchus*). Tail conoid, subcylindroid or cylindroid. Testis outstretched. Bursa present or absent. Spicules small, paired, similar, arcuate, cephalated, distally pointed. Gubernaculum simple, fixed, may be absent (*Gymnotylenchus, Fergusobia*).

The family contains four subfamilies: Neotylenchinae, Fergusobiinae, Gymnotylenchinae, and Rubzovinematinae.

Key to Subfamilies of Neotylenchidae

1. Juveniles and adults of nonentomoparasitic generation partially obese, parasitic in aerial plant galls .... Fergusobiinae
   Juveniles and adults of nonentomoparasitic generation slender, not parasitic in aerial plant galls 2
2. Bursa and gubernaculum absent; vulva small, less than half body width long; migratory and parasitic in root tissues in nonentomoparasitic phase .... Gymnotylenchinae
   Bursa and gubernaculum present; vulva large, half or more of body width long; free living fungal feeders in nonentomoparasitic phase 3
3. Cephalic framework 8 or 12 sectored; excretory pore at or posterior to nerve ring; parasites of Diptera .... Neotylenchinae
4. Cephalic framework 6 sectored; excretory pore anterior to nerve ring; parasites of Siphanoptera .... Rubzovinematinae

Family: Sphaerulariidae [Lubbock, 1861 (Skarbilovich, 1947) syn. Sphaerulariaceae Lubbock, 1861]

Free-living generation may be present (not known for type genus). Only adult heterosexual female or its uterium are parasitic in insect hosts; male not found in host. Parasitic female in host haemocoel everts uterus which normally hypertrophies into a large sac (=uterium) containing oviduct, ovary, eggs and juveniles, leading an independent life. Free living forms: Free living generation present, suspected to be present, or absent. Free living generation forms as described in the superfamily. Partially free living forms: Sexual dimorphism in anterior region is absent. Female: Stylet well-developed, conus with distinct lumen, base plain, tripartite or knobbed. Orifice of dorsal gland just at base of stylet, not detectable in type genus. Oesophageal glands forming a broad cylindroid or bulboid structure, often with a stem like extension at base; glands not enlarged, may form short lobes over intestine. Vulva inconspicuous, small; vagina short. Postvulval uterine sac present or absent. Uterus very long in fertilized females, packed with round sperm. Ovary immature. Tail subcylindroid, with a rounded tip. Male: Similar to female. Oesophageal glands enclosed in a basal bulb, not extending over anterior end of intestine. Testis outstretched, or reflexed. Spicules cephalated, arcuate, accompanied by gubernaculum. Bursa present, subterminal or enveloping tail tip, or absent (*Tripius*).

The family contains one subfamily: Sphaerulariinae.

Family: Paurodontidae [Thorne, 1941 (Massey, 1967) (=*familia dubia*)]

Fungus feeding generation present; entomopathogenic forms not known, but may exist. The taxa are small to large (0.3-2.9 mm). Sexual dimorphism in anterior region is absent. Cephalic region low, rounded or flattened, often distinctly striated; framework six or eight sectored. Stylet under 20 µm long, often knobbed. Oesophagus divisible into corpus, isthmus and basal bulb. Corpus cylindrical or with a nonmuscular fusiform swelling at base. Basal bulb with a stem like extension penetrating into intestine. Vulva posterior. Ovary outstretched; tip may be reflexed. Bursa adanal to enveloping tail. Spicules cephalated. Gubernaculum simple. Tails variable in shape, generally similar between sexes.

The family contains one subfamily: Paurodontinae.

Family: Allantonematidae [Pereira, 1931 (Chitwood & Chitwood, 1937) syn. Contortylenchidae Ruhm, 1956]

Free-living generation does not exist. Single heterosexual generation cycle (hermaphrodites in *Anandranema*), with adult female parasitic in insect or mite haemocoel. Only one type of adult, i.e., heterosexual female present in host (males and secondary females may occur within maternal body in *Scatonema*). Entomoparasitic female are obese, round, oval, spindle-shaped, or elongate sac-like, ventrally or dorsally curved, its body cavity mostly filled with reproductive organs; vulva may be cleft-like (*Contortylenchus*); uterus not everted. Partially free-living forms:

Sexual dimorphism may or may not be present in anterior region. Female: Preadult with immature ovary and generally within juvenile cuticle. Stylet upto 15 µm long, with or without knobs. Oesophageal glands elongated, lobe like, subventral glands extending past the dorsal one. Vulva inconspicuous, small; vagina short; postvulval uterine sac short or absent. Uterus elongated, distended when packed with sperm. Tail conoid to subcylindroid. Male: Essentially similar to female but oesophageal glands not enlarged, oesophagus may be degenerate. Testis outstretched, producing minute, rounded sperm. Spicules arcuate, pointed, usually under 25 µm long. Gubernaculum fixed, rarely absent. Bursa present or absent. Impregnated female invades host.

The family contains two subfamilies: Allantonematinae, and Contortylenchinae.

Key to Subfamilies of Allantonematidae

1. Entomopathogenic female contorted, dorsally curved; vulva deeply cleft .... Contortylenchinae
2. Entomopathogenic female not contorted, usually not dorsally curved; vulva not deeply cleft .... Allantonematinae

Superfamily 2: Iotonchioidea [Goodey, 1953 (Siddiqi, 1986)]

Taxa are known or suspected parasites of insect haemocoel, with partially free living stages; free living feeding generation present. In host, two alternating parasitic generations present. The first generation comprises primarily heterosexual females, which invade the host. The second generation is heterosexual or parthenogenetic nonfeeding females in the host. Primary heterosexual female is the largest in size. Sexual dimorphism in anterior region of partially free living forms present or rarely absent. Cuticle finely striated, with lateral fields; deirids near excretory pore or more anterior; phasmids absent. Female cephalic region low or elevated, hexaradiate, male cephalic region may be tri-or tetralobed and asymmetrical. Female stylet small or large, with or without basal knobs, male stylet generally reduced or absent. Oesophagus cylindroid or fusiform, nondivisible. Oesophageal glands elongated extending over intestine mostly dorsally. Nerve ring apparently circumintestinal. Vulva posterior, may be over two vulval body widths anterior to anus. Postvulval utering sac absent. Spicules large or small, L-shaped or arcuate. Gubernaculum present or absent.

The superfamily contains two families: Iotonchiidae and Parasitylenchidae.

Key to Families of Iotonchioidea

1. Marked sexual dimorphism in anterior region present; vulva more than two body width from anus; spicules robust, angular .... Iotonchiidae
2. Marked sexual dimorphism in anterior region absent; vulva less than two body widths from anus; spicules neither robust nor angular .... Parasitylenchidae

Family: Iotonchiidae [Goodey, 1953 (Skarbilovich, 1959)]

Two types of female occur in the cavity of host body. The first type is heterosexual female curving ventrally and spirally and having vulva at more than two body width from anus. The second type is parthenogenetic female. Alternation of heterosexual and parthenogenetic generations present. Partially free-living forms show marked sexual dimorphism in anterior region, and in males the oesophagus is degenerate and stylet degenerate or absent. Male cephalic region tri or tetralobed, usually asymmetrical. Female stylet generally over 18 µm long, usually without basal knobs, indistinct basal knobs or thickenings may be present. Excretory pore opposite or behind nerve ring. Vulva more than two body widths in front of anus, with or without anterior lip flap. A ventromedian body pore behind vulva present or absent. Vagina strongly muscular; postvulval uterine sac is absent. Tails generally elongate-conoid or filiform. Spicules robust, angular, L-shaped or of an aberrant form, in two parts, proximal part broad and cephalated, distal part slender with rounded or spined tip. Large postanal genital papillae may be present (*Fungiotonchium*). Bursa small, adanal or large, completely enveloping tail.

The family contains one subfamily: Iotonchiinae.

Family: Parasitylenchidae [Siddiqi, 1986]

Two or three types of adults present in host body cavity, primary heterosexual generation alternating with a secondary heterosexual or parthenogenetic generation; rarely nonfeeding parthenogenetic female of the secondary generation occurs in the environment (*Heteromorphotylenchus*), or not known (*Spilotylenchus*). Primary heterosexual female dorsally or ventrally curved, but not curving spirally, when relaxed. Female stylet generally under 18 µm long, with basal thickenings or knobbed. Orifice of dorsal gland close to or further behind stylet base. Excretory pore generally located anteriorly. Vulva less than two body width from anus, lips not modified; no ventral body pore near vulva.

Vagina poorly muscular. Postvulval uterine sac absent. Female tail short, subcylindroid or conoid. Male may occur in host's body cavity (*Parasitylenchus, Kurochkinitylenchus*). Spicules slender, simple, ventrally arcuate, about 20 µm or less long (except 27-30 µm long in *Parasitylenchus macrobursatus*). Bursa enveloping tail, or absent (*Heterotylenchus*). Gubernaculum present or absent.

The family contains five subfamilies: Parasitylenchinae, Heterotylenchinae, Heteromorphotylenchinae, Kurochkinitylenchinae and Spilotylenchinae.

Key to Subfamilies of Parasitylenchidae

1. With two types of alternating heterosexual generations; copulation takes place in host's body cavity 2
   Only one type of heterosexual generation; copulation does not take place in host's body cavity 3
2. Parasitic females dorsally curved; parasites of Siphanoptera .... Kurochkinitylenchinae
   Parasitic females ventrally curved; parasites of Diptera and Coleoptera .... Parasitylenchinae
3. Parthenogenetic nonfeeding female and its eggs found in environment .... Heteromorphotylenchinae
   Parthenogenetic parasitic female and its eggs found in host's coelom 4
4. Parasitic females usually dorsally curved; excretory pore generally anterior to nerve ring; parasites of Siphanoptera .... Spilotylenchinae
5. Parasitic females not dorsally curved; excretory pore generally posterior to nerve ring; parasites of Diptera and Coleoptera .... Heterotylenchinae

Order 3: Aphelenchida [Siddiqi, 1980]

Taxa of this order are soil dwelling or insect associates; trophic habit mycophagous, phytoparasitic, predacious or entomophagous. Body is very small to long (0.2-2.5 mm), vermiform, rarely obese except in some insect parasites. Cuticle thin, usually finely annulated; lateral fields have 0-12 or more incisures. Cephalic region low, rounded, continuous or offset and with weak or moderate sclerotization. Four submedian cephalic papillae and six readily visible labial papillae (plus possibly six, more obscure, labial papillae) are present in cephalic region. Amphidial apertures oval, pore like, dorsosublateral on labial region. Stylet always present, usually 10-20 µm long, but exceptionally upto 185 µm; conus usually shorter than the shaft, but much longer in certain insect ectoparasitic forms. Basal knobs weakly developed or absent. Oesophagus is divisible into procorpus (narrow, cylindrical), rectangular median bulb (strongly developed, offset, ovoid to rounded rectangular bulb with crescentic valve plates) and well-developed oesophageal glands forming a dorsally overlapping lobe (except *Paraphelenchus*). Orifice of all three glands are located within the median bulb. Isthmus usually short or absent. Nerve ring circumoesophageal or circumintestinal. Intestine cellular with distinct lumen. Rectum usually distinct. Anus broad transverse slit with an overhanging anterior lip, but degenerated or absent in some insect parasites or associates. Vulva 60-98% posterior, transverse slit (except oval in *Aphelenchus*). Gonads are single and usually outstretched. Spermatheca axial, if present. Postuterine sac usually present and may act as a spermatheca. Male genital system monorchic, outstretched. Sperm large, rounded, arranged in one or two rows in the gonoduct. Spicules typically rosethorn shaped with prominent apex and rostrum, or derived therefrom, but elongate and cephalated in *Aphelenchus* and *Paraphelenchus*. Gubernaculum usually absent, but well developed and elongate in *Aphelenchus* and *Paraphelenchus*. Bursa usually absent, but present in *Aphelenchus* and some other genera. Usually three pairs (two - five) of caudal papillae present.

The order contains one suborder: Aphelenchina.

Suborder 1: Aphelenchina [Geraert, 1966]

Taxa of the suborder possess characteristics of the order Aphelenchida.

The suborder contains two superfamilies: Aphelenchoidea and Aphelenchoidoidea.

Key to Superfamilies of Aphelenchina

1. Spicules slender, cephalated. Gubernaculum well-developed, elongate; V-shaped in cross section. Lateral fields with six or more incisures. Oesophagus with distinct isthmus and nerve ring circumoesophageal .... Aphelenchoidea
2. Spicules robust, rosethorn-shaped or derived therefrom; typically with a dorsal and ventral limb joined by a transverse bar. Gubernaculum absent or, if present, consisting of a small structure near the distal tip of the dorsal limb of the spicule; never elongate or V-shaped in cross section. Lateral fields usually with four or fewer incisures, exceptionally six. Oesophagus lacking a distinct isthmus which, if visible at all, is a short stump less than the distance from the valve plates to the base of the bulb in length; nerve ring circumintestinal .... Aphelenchoidoidea

Superfamily 1: Aphelenchoidea [Fuchs, 1937 (Thorne, 1949)]

Taxa of this superfamily have the cephalic region low, flattened, continuous with body contour. Lateral fields with six or more incisures. Oesophagus with a distinct isthmus; glands either in a dorsally overlapping lobe (Aphelenchidae) or retained within a non-overlapping basal bulb (Paraphelenchidae). Nerve ring circumoesophageal. Vulva a transverse oval pore (Aphelenchidae) or slit (Paraphelenchidae). Female tail short, sub-cylindroid to conoid and with a broadly rounded terminus which may be mucronate. Spicules slender, ventrally arcuate; cephalated. Gubernaculum well developed, elongate; V-shaped in cross section. Bursa if present, well developed, peloderan and supported by four pairs of ribs.

The superfamily contains two families: Aphelenchidae and Paraphelenchidae.

Key to Families of Aphelenchoidea

1. Male with prominent peloderan bursa supported by four pairs of ribs. Female vulval aperture in the form of an oval pore. Oesophageal glands forming a long dorsally overlapping lobe .... Aphelenchidae
2. Male lacking a bursa. Vulva in the form of a transverse slit. Oesophageal glands small, retained within a non-overlapping basal bulb .... Paraphelenchidae

Family: Aphelenchidae [Fuchs, 1937 (Steiner, 1949)]

Taxa are soil dwelling or found in decaying plant material. Lateral fields have more than six lateral lines, usually 10-12. Vulval aperture in the form of a transverse oval pore. Oesophageal glands free, forming a dorsally overlapping lobe. Female tail short, cylindroid with a rounded tip. Male: Bursa well developed, peloderan in form and with four pairs of supporting ribs, one pair adanal and the other three postanal.

The family contains one subfamily: Aphelenchinae.

Family: Paraphelenchidae [T. Goodey, 1951 (J. B. Goodey, 1960)]

The taxa are soil dwelling. Lateral fields have usually 6-8 lateral lines. Oesophageal glands small, enclosed in an abutting basal bulb which is amalgamated with the isthmus. Vulva a transverse slit. Female tail short, subcylindroid and usually with a mucronate terminus. Male tail has 4-5 pairs of caudal papillae and bursa is absent.

The family contains one subfamily: Paraphelenchinae.

Superfamily 2: Aphelenchoidoidea [Skarbilovich, 1947 (Siddiqi, 1980)]

Taxa have the cephalic region usually high and offset from body contour. Lateral fields have upto four incisures (very exceptionally six). Stylet with basal knobs. Isthmus rudimentary or absent. Nerve ring circumoesophageal. Oesophageal glands in a dorsally overlapping lobe. Vulva in the form of a transverse slit. Female tail conoid to a pointed or narrowly rounded terminus which may be mucronate or otherwise adorned. Spicules robust, rosethorn shaped or derived therefrom; usually with a prominent apex and rostrum. Gubernaculum absent or indistinct; if present it is small, located at the distal tip of the dorsal limb of the spicules and is never elongate, linear or V-shaped in cross section. Tail has 2-3 pairs of caudal papillae, occasionally upto 5 pairs. Bursa absent, a small flap is present in the Parasitaphelenchidae.

The superfamily contains six families: Aphelenchoididae, Acugutturidae, Ektaphelenchidae, Entaphelenchidae, Parasitaphelenchidae and Seinuridae.

Key to Families of Aphelenchoidoidea

1. Stylet of both sexes very long (50-180 µm), attenuated; the conus constituting the majority of the stylet length. Ectoparasites of insects .... Acugutturidae
   Stylet usually about 10-20 µm long, never over 35 µm long and never attenuate with the conus constituting the majority of the stylet length 2
2. Mature females with swollen body, endoparasitic in the haemocoel of beetles. Three adult forms: male; immature female; mature parasitic female .... Entaphelenchidae
   Mature females not swollen or endoparasitic. Two adult forms in the life cycle 3
3. Females with functional anus and elongate tails more than four anal body width long, often becoming attenuate or filiform, but may be more cylindroid with a rounded or spathulate terminus. Male tail elongate, conoid with a spicate terminus .... Seinuridae
   Females with short or medium conoid tails usually less than four anal body width long, but if longer and with a filiform or spicate terminus then the female anus is nonfunctional 4

4. Males with small bursa like flap of cuticle at tail tip .... Parasitaphelenchidae
Males lacking such a bursa 5
5. Females lacking a functional anus and rectum, the intestine ending as a blind diverticulum in the tail region. Stylet typically with a wide lumen .... Ektaphelenchidae
Females with a functional anus and rectum, intestine not ending as a blind diverticulum. Stylet robust, with a narrow lumen and basal knobs or swellings .... Aphelenchoididae

Family: Aphelenchoididae [Skarbilovich, 1947 (Paramonov, 1953)]

Nematodes have slender stylet with narrow lumen and with small basal knobs. Post-uterine sac usually present. Female tail is of medium length, conoid, with pointed or rounded, often mucronate terminus. Spicules paired, separate, rosethorn shaped or derived therefrom. Gubernaculum and bursa are absent.

The family contains two subfamilies: Aphelenchoidinae and Anomyctinae.

Key to Subfamilies of Aphelenchoididae

1. Cephalic region low, rounded, lacking an obvious oral disc .... Aphelenchoidinae
2. Cephalic region high, almost spherical, and with a prominent sclerotized oral disc .... Anomyctinae

Family: Seinuridae [Husain and Khan, 1967 (Baranovskaya, 1981)]

Taxa are mostly known to be or suspected of being predacious. Stylet long and slender with a wide lumen. Median bulb exceptionally prominent, elongate oval or rounded-rectangular with the valve plates situated in the posterior half. Spicules with a prominent apex and rostrum. Tail in female long, elongate, with a usually filiform, rarely rounded terminus.

The family contains one subfamily: Seinurinae.

Family: Ektaphelenchidae [Paramonov, 1964]

Taxa are associates of insects, particularly bark beetles (Coleoptera: Scolytidae). Lumen of stylet is usually wide, but narrow in *Cryptaph-*

elenchus. Rectum and anus absent, or indistinct and nonfunctional. Intestine extends into tail as a blind diverticulum.

The family contains one subfamily: Ektaphelenchinae.

Family: Acugutturidae [Hunt, 1980, n. rank]

Taxa are ectoparasites of cockroaches or noctuid moths. All stages are found on insect hosts. Stylet elongate, 50-185 µm long, attenuated. Anus and rectum difficult to discern and probably nonfunctional.

The family contains two subfamilies: Acugutturinae and Noctuidonematinae.

Key to Subfamilies of Acugutturidae

1. Adults and juveniles not markedly swollen, spicules of normal size (<30 µm) and rostrum not tubular, thorn shaped. Parasites of cockroaches .... Acugutturinae
2. Adults and juveniles markedly swollen, spicules very large (38-109 µm) with a tubular rostrum, massive. Parasites of moths .... Noctuidonematinae

Family: Parasitaphelenchidae [Ruhm, 1956 (Siddiqi, 1980)]

Taxa are mostly associates or endoparasites of insects, particularly bark beetles (Coleoptera: Scolytidae). Two insect associates are known to be implicated in serious diseases of palms or pines. Male tail tip enveloped by small, bursa like flap of cuticle. Spicules stout, may be partially fused and with a prominent rostrum.

The family contains two subfamilies: Parasitaphelenchinae and Bursaphelenchinae.

Key to Subfamilies of Parasitaphetenchidae

1. Fourth stage juvenile endoparasitic in the insect haemocoel. Vulva very posterior (85-90%), spicules partially fused, male tail not strongly recurved .... Parasitaphelenchinae
2. Third stage dauer larvae ectophoretic, exceptionally endophoretic. Vulva more anterior at 70-80%. Spicules usually separate, but reported to be partially fused in some species, male tail strongly recurved .... Bursaphelenchinae

Family: Entaphelenchidae [Nickle, 1970]

Taxa are endoparasites in the haemocoel of insects (Coleoptera). There are three adult forms: male, immature female, and mature parasitic female. Mature parasitic female oviparous or ovoviviparous. Stylet 8-22 µm long, with or without knobs. Vulva 80-90% posterior, more anterior in *Roveaphelenchus*. Male spicules are robust, often with a well developed, pointed rostrum. Bursa absent.

The family contains one subfamily: Entaphelenchinae.

5

Biology of Nematodes

Biology of nematodes is simple and consistently uniform. The life cycle consists of six stages viz., egg, four juvenile stages (J_1-J_4), and adult stage. In the class, Adenophorea, J_1 emerges from the egg and first moulting occurs outside, whereas in Secernentea the first moulting takes place inside the egg and the hatching stage is J_2 with the exception of *Anguina*. Biology of nematodes can be best discussed under egg development, moulting, reproduction and survival.

EGG DEVELOPMENT

Embryonic Development

The general embryogenesis in the fully developed eggs, either after the laying or within egg mass or cyst, commences with the first division by formation of a transverse cleavage in the protoplasm to form two cells, anterior and posterior cells (Fig. 36). Another cleavage in both the cells occurs to form a fourcelled embryo. The four cells may be arranged in a row of one or two. Numerous transverse and longitudinal mitotic divisions of daughter cells occur and finally the blastula stage develops in which cells are arranged so as to form a fluid filled sphere. Next is the gastrula stage which is characterized by the early embryo consisting of saclike body with a wall of two layers. The embryo continues to grow into a coiled larva inside the egg membrane. When a stage of cell consistency is reached, cell multiplication ceases in all organs except the reproductive system. Duration of development from egg to differentiated larval stage varies with the nematode species and environmental condition. *Ditylenchus dipsaci* takes 5 days in water to develop to differentiated larval stage, whereas *Radopholus similis* may take 4-11 days.

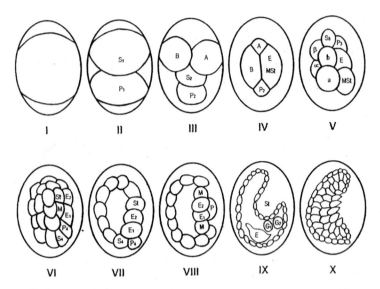

Fig. 36 Embryonic development of nematode (one cell stage to differentiated larvae)

Embryogenesis has been studied in *Caenorhabditis elegans* in detail because of certain characteristics such as short life, easy culture, and few cells composition of juveniles and adults of this nematode. The first cleavage of embryo results in two cells of unequal size. The larger anterior is founder ectodermal cell (AB) and the smaller posterior is germline cell (P_1). The second cleavage results in second somatic daughter cell designated as EMS (ethyl methane sulphonate) and P_2. The EMS cell divides unequally and produces the MS founder cell and E founder cell of the mesentron. The MS cell produces 48 muscle cells, 13 neuronal cells, 9 gland cells and 2 somatic gonad cells, and 8 other cells; 14 cells die and 80 survive.

Shortly after formation of EMS, P_2 divides to produce somatic founder cell C and germline cell P_3. The C cell produces 32 muscle cells, 13 hypodermal cells and 2 neuronal cells; there are 1 dead cell and 47 survivors. The P_3 produces founder cell D that gives rise to 20 muscle cells, and P_4 produces 2 germinal cells, Z_2 and Z_3. At eclosion, the hermaphroditic larva consists of 558 cells. During embryogenesis 671 cells are formed, of these 113 undergo programmed death. It is called programmed death because death of 113 cells is specific, autonomous and invarant during the formation of a differentiated larva.

Post-embryonic Development

The post-embryonic developments include a further growth in the differentiated larva within the egg leading to full development and ready to undergo the first moult and/or hatching.

Hatching

The nematode larva, after the embryonic development, grows further and undergoes the first moult to become second stage juvenile (J_2) before (Secernentea) or after (Adenophorea) the hatching. The tylench J_2 attains a particular growth stage, and if favourable conditions prevail, the larva shows vigorous movement, resulting in bulging of the egg membrane. The larva thrust the stylet vigorously (40-90 thrusts per minute), the frequency further increases till a perforation in the egg shell is achieved through which the larva moves out. In many nematodes, *Globodera*, *Heterodera*, *Pratylenchus*, etc., the larval movement and the entire process are stimulated by chemical stimuli especially through root exudates of susceptible plants. For example, root exudates of potato stimulate hatching, out migration from cyst and movement of *Globodera* larvae in soil to approach the roots. Viability and dormancy of eggs vary considerably. Generally, free eggs as of ectoparasites cannot withstand adverse conditions and hatch quickly, whereas the eggs inside a sac (*Meloidogyne*, *Tylenchulus*) or cyst (*Heterodera*) remain viable in adverse conditions for a longer duration. Hatching of such eggs is initiated by the stimuli from susceptible hosts otherwise they remain dormant.

Moulting

Moulting is a process in which the existing cuticle is usually shed and a new cuticle develops, coupled with some other morphological changes specific to a nematode species or genus. Nematodes undergo four moults to become adult, which is the last developmental stage of the life. The first moulting may occur inside the egg before hatching (Secernentea, e.g., Tylenchida, and Aphelenchida) or outside the egg after hatching (Adenophorea, e.g., Dorylaimida).

During each moulting the entire cuticle, i.e., external cuticle that covers the body surface, and the internal cuticle that invaginates through natural openings and forms the lining of lumen of oesophagus, rectum, anus, vagina and cloaca, is cast-off (shed). In addition, the stylet also disintegrates and disappears. The moulting process initiates with the ceasure of feeding, and the larvae become inactive or sluggish. The body contents may become dense and the existing external cuticle enlarges and separates from the hypodermis, as a result it becomes loose. A new cuticle is secreted by the hypodermis. The older loose cuticle is discarded, and sheds due to abrasion against soil during nematode movement. In some nematodes, a new cuticle develops without shedding of the older external cuticle. This is called as superimposed moulting, e.g., second and third moults of *Meloidogyne*. Concurrent with the loosening of external cuticle, the internal cuticle and stylet also get dissolved. The basal part of the stylet disintegrates first. The new stylet also develops from the base. In some nematodes the stylet does not develop during each moult, e.g., J_3 and J_4 of *Meloidogyne* do not possess stylet.

Duration of a developmental stage as well as the entire life span of nematodes vary greatly. Generally ectoparasites moult quickly, whereas endoparasites, e.g., *Meloidogyne* and *Pratylenchus*, become inactive in soil for several weeks because of low temperature, drought, etc., but once the larvae receive stimuli from the host, they follow a relatively uniform pattern of moulting and development.

REPRODUCTION

The adults feed on undifferentiated (normal host cells) or differentiated cells (nurse cells) to grow further and to mature the gonads for reproduction. The reproduction may be amphimictic or parthenogenetic but never asexual. In amphimictic species, e.g., *Anguina, Globodera, Pratylenchus, Hirschmanniella, Helicotylenchus, Hoplolaimus* and *Rotylenchus*, male and female individuals copulate. In such species males occur in a fairly good number but they are nonparasitic. In parthenogenetic nematodes, e.g., *Meloidogyne, Heterodera, Ditylenchus* and *Radopholus*, males occur in a very low number, but under unfavourable conditions sex reversal in females occurs, as a result male population greatly increases.

Duration of the life cycle from egg to egg varies greatly. Generally it is completed in 2-5 weeks (Table 13). *Seinura celaris*, however, can complete the life cycle in 2-5 days; on the contrary *Xiphinema diversicaudatum* may take one year or more. Frequency of life cycles also varies with the nematodes. *Aphelenchoides* species may complete 10-15 generations, *Meloidogyne* spp. 3-5 generations and *Anguina* spp. one generation within a season of 3-4 months. The reproduction rate in terms of number of eggs laid by an individual (fecundity) also varies. A *Meloidogyne* female may form as many as 2,900 eggs, generally 300-900 eggs, whereas the female of *Hoplolaimus indicus* lays only a few eggs (Table 13).

Reproductive Systems

Nematodes are mostly dioecious having one sex (male or female) in one individual within a species. Usually males are lesser or far lesser than the

Table. 13 Life cycle duration and reproduction rate of some plant parasitic nematodes

Nematode	Duration of Life Cycle (Egg to Egg)	Fecundity (Eggs/Female)
Seinura celaris	2.5-5 days	-
Aphelenchoides spp.	10-15 days	20-40
Radopholus similis	20-25 days	10-32
Rotylenchulus reniformis	25 days	50-75
Hoplolaimus indicus	26-27 days	14-20
Meloidogyne spp.	3-4 weeks	300-900
Heterodera spp.	3-9 weeks	300-600
Tylenchulus semipenetrans	4-8 weeks	40-90
Pratylenchus penetrans	4-11 weeks	30-40

females or may even be completely absent. This scarcity or the absence of males indicates a tendency towards hermaphroditism and parthenogenesis. Hermaphrodite nematodes are rarely known to occur, the best example is *Caenorhabditis elegans* in which male and female gonads are present within an individual. Hermaphroditism is of two types, syngonic and digonic. In Syngonic hermaphroditism the gonad first produces sperms which are stored until the ova mature and fertilization takes place, e.g., *Caenorhabditis briggsae*. In digonic hermaphroditism sperms and ova are produced simultaneously. The progeny that is produced through hermaphroditism has less genetic variability and is not able to successfully withstand environmental fluctuations. In parthenogenesis the eggs develop without fertilization, e.g., *Meloidogyne, Heterodera, Tylenchulus* and *Hemicycliophora*. Males in parthenogenetic species do occur but in a low number, the proportion, however, increases due to sex reversal under conditions of environmental stress. The parthenogenesis is either meiotic or mitotic (*Meloidogyne*).

The reproductive system of nematodes is composed of one or two tubular gonads and the germ cell proliferation may be telogonic or hologonic. In telogonic, the germ cells are proliferated only at the tip of the gonad. The apical end of the telogonic gonad is called the germinal zone which is followed by the growth zone in which the gametogonia enlarge and differentiate. The majority of nematodes including tylenchs are telogonic. In hologonic, the germ cells are produced along the entire length of the gonad, e.g., animal parasitic nematodes belonging to the superfamily, Dioctophymatoidea and Trichuroidea.

Female Reproductive System

The female reproductive system consists of one (monodelphic) or two branches (didelphic) of ovary. The reproductive organ opens outside independently through the vulva. The number and the manner of arrangement of the ovary is considered to be of great taxonomic value. The monodelphic ovary may either be anterior to the vulva (monoprodelphic) or posterior to the vulva (mono-opisthodelphic) (Fig. 37). The position of the vulva varies with the type. It may be posterior to middle, usually far posterior in the monoprodelphic, and anterior to middle or far anterior in the mono-opisthodelphic (Fig. 37). In *Trophurus* the vulva is near mid-body though it is monoprodelphic. In Tylenchida, a large number of families (Tylenchidae, Anguinidae, etc.) possess a single set of reproductive organs, and the arrangement is always monoprodelphic. In didelphic type, amphidelphic arrangement is common in which one branch of the ovary is anterior to the vulva and the other posterior to the vulva (Fig. 37). Position of the vulva is usually in or near the middle of the body. Majority of tylenchs possessing didelphic type of re-

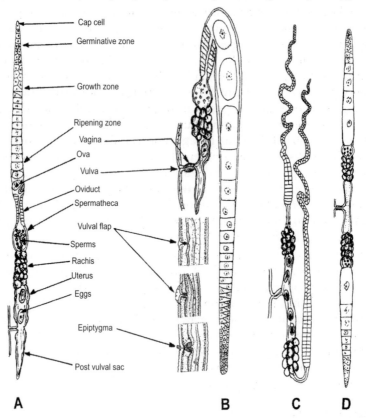

Fig. 37 Diagrammatic presentation of female reproductive organs of a nematode, showing variation in the ovary: monoprodelphic (A), mono-opisthodelphic (B), didelphic prodelphic (C), and didelphic amphidelphic (D).

productive organs have amphidelphic arrangement (Dolichodoridae, Hoplolaimidae, etc.). Another type is didelphic prodelphic in which both branches of reproductive organs are directed anteriorly and the vulva is terminal, e.g., *Meloidogyne, Heterodera*, etc.

The monodelphic condition is considered to have evolved from the didelphic through gradual reduction and eventual loss of one of the gonads (in Tylenchida the posterior gonad). The various stages in the reduction and loss of gonad can be seen in species of *Helicotylenchus* to *Rotylenchoides* and *Varotylus* to *Orientylus*. A distinct post-uterine sac (the lost gonad) is usually present in the monoprodelphic tylenchs, though in some species the rudiments of oviduct and ovary may also be present. In Criconematina, Hexatylina and some genera (*Diptenchus, Dorsalla*) the post uterine sac may be absent. An anterior uterine sac is usually present in mono-opisthodelphic gonad.

The main parts of the female gonad are the ovary containing developing ova, oviduct that provides passage to ova from the ovary, and the uterus where final development and maturation of eggs take place. The uterus joins the vagina which opens through the vulva on the ventral surface of the body. The ovary is a hollow, elongate tube lined with flattened epithelial cells and contains few to numerous oocytes. The apical zone is called germinal or germinative zone and at the tip of this zone there is a cap cell that keeps multiplying to produce ova. In the germinative zone, rapid cell division takes place and the cells that are formed are relatively small. This region is followed by the growth zone which constitutes the greater part of the ovary. The growth zone may be differentiated into an anterior growth zone and a posterior ripening zone. The oocytes accumulate yolk and gradually become bigger and bigger in the growth zone, whereas in the ripening zone the oocytes ripen. The oocytes do not descend into the ripening zone until conditions like temperature, food supply, etc., are suitable for egg ripening. The ovary is straight or outstreched in Tylenchida, exceptions being Anguinidae, Heteroderidae, etc., in which flexures may be present (Fig. 37c). Sometimes the central protoplasmic core called rachis is present, which regulates differentiation of oocytes and keeps them united. In *Anguina* and *Meloidogyne* the oocytes aggregate around the rachis.

The ripe oocytes pass into the oviduct through a tiny canal formed by the oviduct cells (Fig. 37). It is believed that the contraction of the somatic musculature during locomotion helps in oocyte movement, premature advancement of the oocytes into the gonoduct is, however, prevented by sphincters. The oviduct is made up of high columnar epithelial cells and serves as a constriction between the ovary and the uterus. The uterus is the large and most complex part of the reproductive system. It serves the function of fertilization, egg shell formation and ejection of eggs (egg laying). The distal (upper) part of the uterus is differentiated into a spermatheca, sperm storage organ where sperms remain viable. The spermatheca may be axial or non-axial (offset). In bisexual species (hermaphrodite) it is usually full of sperms in mature females, in parthenogenetic females it is small and empty. In the uterus next to the spermatheca, secretory cells are present which form the eggshell or crustaformeria. The vagina is a short, narrow and flattened tube lined with cuticle and provided with well developed musculature (dilator and constrictor). The vagina opens to the exterior through a midventral opening, the vulva. In Tylenchida, the vulva is usually a transverse slit, rarely transversely oval or rounded. The size of the vulva may also be variable. Sometimes lateral vulval membranes, vulval flaps are present (Fig. 37) and serve as useful taxonomic character, eg., *Cephalenchus, Coslenchus, Pterotylenchus,* and *Dolichorhynchus*. In Hoplolaimoidea and some other genera, a cuticular membranous structure called epiptygma located on the vagina or vulva may be present. Specialized musculature is associated with the vulva for dilating during egg laying and closing afterwards. The shape and size of the egg is variable. In Tylenchida, the eggs may be oval, subglobular, elliptical, kidney shaped, etc., its surface is smooth as the outer protein layer is absent. The eggs are usually covered by three distinct membranes, an outer protein layer, middle a chitinous layer, and the inner a lipid layer. A gelatinous matrix (egg sac) is secreted especially by sedentary endo or semiendoparasitic nematodes, e.g., *Meloidogyne, Rotylenchulus,* etc., by rectal cells, vaginal cells, etc., in which eggs are deposited.

The eggs in nematodes are laid outside the body (oviparous) where the embryonic development takes place (exotoky). However, sometimes the eggs develop within the body of the female (endotoky) without being laid, such as *Heterodera* and *Globodera*. The nematodes showing endotoky may either be viviparous, where the egg shell does not form and the juveniles develop inside the uterus, e.g., entomopathogenic nematodes (Fig. 38). In ovoviviparous endotoky, the egg shell is formed but females are not able to lay eggs outside because of obliteration of internal organs, e.g., cyst forming nematodes.

Male Reproductive System

The male reproductive system is tubular consisting of one testis (monorchic) or two testes (diorchic) and opens outside through a common opening, cloaca (Fig. 39). In Tylenchida, the gonad is always monorchic with outstretched testis, pro-diorchic in abnormal males of *Meloidogyne* spp., which develop due to sex reversal of developing fe-

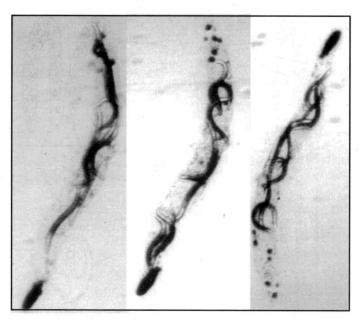

Fig. 38 Females of *Steinernema* sp. showing *endotokia matricida* (progeny develops and feeds inside the mother).

male juveniles under adverse conditions. Major parts of the male gonad are testis, seminal vesicle, vas deferens and spicules. Structurally testis are quite similar to ovary. Proliferation of germ cells is telogonic. The testis are divided into germinative zone and growth zone. The growth zone is further divided into growth zone and ripening zone. At the tip of the germinative zone there is a cap cell which keeps dividing to produce spermatocytes. The young spermatocytes are initially smaller in size but they grow bigger and bigger in the growth zone and mature in the ripening zone. The sperms may be spherical, discoid, conical or elongate and may show amoeboid movements. The testis is covered by a thin epithelial membrane which is continuous with the epithelium of the gonoduct. The spermatocytes are arranged in single or double rows, but may sometimes be in multiple rows. In *Anguina* a rachis similar to its female partner is present. The ripening zone terminates into a seminal vesicle in which the sperms are stored. The next part is a glandular duct, the vas deferens, which stores the developing sperms as they trickle down from the vesicle. In Tylenchida, an ejaculatory duct is also present which helps in the ejection of sperms during mating. The duct tapers gradually and joins the rectum to form the cloaca. Sperms in the vas deferens appear structurally different from those in the female spermatheca. This shows

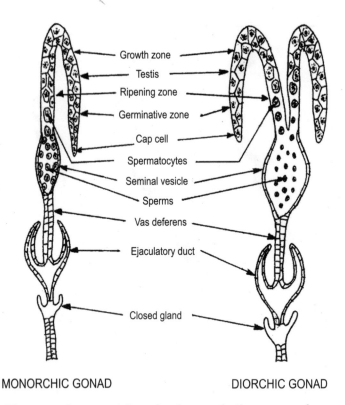

Fig. 39 Diagrammatic presentation of male reproductive organs of a nematode, showing variation also.

that their differentiation continues even after insemination. The cloaca is provided with a male copulatory structure, the spicules and gubernaculum (Fig. 40). The taxonomic value of reproductive structures is of little importance. However, the caudal alae (bursa) which laterally covers the spicules is a useful taxonomic character (detailed under cuticular modifications).

Spicules: The spicules are flat, bladelike or tubular structures and are lodged in the spicular pouches within the rectum (Fig. 40). The specialized cells (spicula primordia) evaginate to form the spicular pouch and invaginate to form the spicule on the dorsal wall of the rectum. The spicules are covered by a sclerotized cuticle and the cuticle is continuous with that of the spicular pouches. The spicules function to open the vulva and transfer sperms during copulation. Structurally the spicules may be differentiated into three regions, capitulum, corpus and lamina (Fig. 41). The capitulum is the proximal end of the spicule. Beyond the capitulum

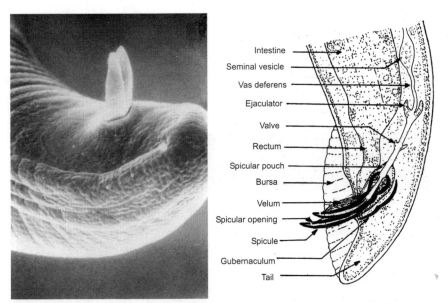

Fig. 40 A SEM picture showing ejaculated spicule of a tylenchid nematode (Courtesy: Anonymous), and a diagrammatic presentation of male reproductive organs of a nematode.

the spicules may narrow in to a section called shaft. The main portion of the spicule is the lamina which tapers to the tip. The lamina may have a longitudinal, wing like membranous extension called the velum, e.g., Tylenchorhynchinae. The velum is a ventrolateral flange that forms a tubular passage for sperm transfer to the vagina when the spicules are protruded. In cross-section each spicule appears as crescentic or tubelike and with a cytoplasmic core in which the sensory nerve is embedded. The Tylenchs possess two spicules, both similar in shape and size. Some other nematodes like *Monoposthia* and *Hydromermis* have only one. Spicules in the pair may or may not be similar in shape and size. The variation in Tylenchida is little (Fig. 41). In Dorylaims, the spicules show considerable variations in shape and size, often within one genus (eg., *Axonchium*).

Gubernaculum: The gubernaculum is a sclerotized plate like structure situated dorsal to the spicules (Fig. 40). It is formed by the dorsal wall of the spicular pouch. It guides the spicule during copulation (spicular protrusion) and prevents it from breaking through the wall of the spicular pouch or cloaca. The simplest gubernaculum is merely a thickened plate or trough called the corpus. In Tylenchida complex gubernaculum is present, which has two grooves to keep the two spicules separate. From the middle of the corpus a plate, coneous arises to form two halves, and the coneous is tipped by a transverse plate called capitulum and two

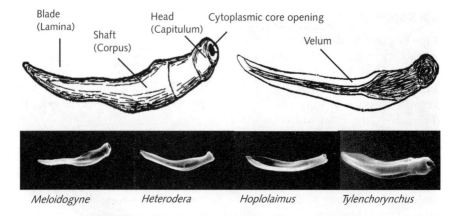

Fig. 41 A diagrammatic presentation of spicules, and its variation in some nematode genera (Courtesy: Rammah and Hirschmann, 1987).

grooves are formed. Toothed lateral extensions (crura) at the bottom of corpus are formed. In some genera (*Hoplolaimus*) titillae are present which are lateral bilobed projections at the distal end of the gubernaculum. In many genera the gubernaculum is protrusible (*Tylenchorhynchus, Dolichorhynchus*) or nonprotrusible (Merliniinae) and may serve as a useful taxonomic characteristic. Generally there are three pairs of muscles that operate gubernaculum. The protractor gubernaculi muscles extend from the ventral body wall anteriorly to the gubernaculum. The second muscle is retractor gubernaculi that extends from the gubernaculum to the dorsal body wall. The seductor gubernaculi muscles extend from the lateral body wall to the gubernaculum.

DIGESTIVE SYSTEM

The inner body tube of nematodes forms the gut or alimentary canal and is distinguished into three parts, foregut (stomodeum), midgut (mesentron), and hindgut (proctodeum) or rectum. The first and third parts of the alimentary canal are lined with cuticle.

Stomodeum

The stomodeum consists of stoma, oesophagus and oesophago-intestinal valve. Although, lips are not strictly a part of the stomodeum, they are discussed here because of their involvement in the feeding.

Lip Region

Lips of plant feeders are immovable. They help the nematode to explore suitability of the host for penetration and are involved in the feeding. The lip region shows great variation. In Secernentea, the lip region is hexaradiate, but in many taxa this is not discernible. Labial papillae, cephalic papillae and amphidial apertures are components of the region. Tylenchids show a wide variation in lip region; full complement of sensilla (6 + 6 + 4 + 2) is rarely present (Fig. 42); most often the second whorl of six labial papillae (surrounding the oral opening) and the outer whorl of four cephalic papillae are visible. In many taxa only the four sensilla of the third whorl are seen. The lip region in tylenchs is often reduced to an unlobed labial plate or there may be two or six lobes with an undivided oral plate (*Meloidodera*). The amphidial apertures are highly variable

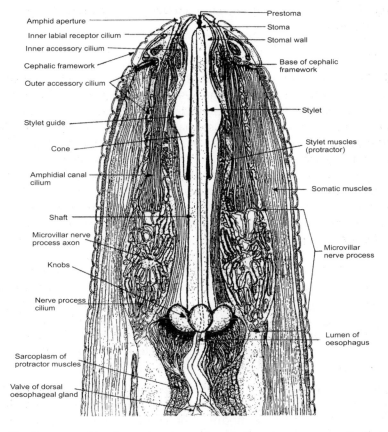

Fig. 42 Structure of feeding apparatus and cephalic sensory organs of a juvenile of plant parasitic nematode (Courtesy: Endo, 1980).

among the Tylenchida. Amphidial apertures are generally round, oval, elliptical, and located laterally. Often the amphids are abutting (touching) the labial disc and in some instances they are on the disc. The obvious difference among these apertures is their relative dorsoventral length. In the family Tylenchidae, the apertures may be elongated, sinuous slits extending posterolaterally, or arc shaped to rounded pits on the labial plate, rarely oblique slit like on the labial plate. In Anguinidae, the aperture is elliptical and directed towards the oral opening.

Among the taxa from Adenophorea, the labial region is most often hexaradiate distinctly or indistinctly, and may bear one to three whorls of cephalic sensilla but never the amphids. The lip region may be smoothly rounded or each sector may be prominent and pyramidal or conical. Amphidial apertures in Adenophorea are highly variable. They are laterally placed and they may be simple ovals, laterally elongate slits, huge dorsoventrally ellipses, large circles, uni- or multiple spirals, or simple pores.

Stoma

It is the mouth cavity or buccal cavity, and forms the feeding apparatus that connects the mouth with the oesophagus (Fig. 42). It forms an extremely important taxonomic character. The nematodes which feed on plants generally possess hollow stoma (except *Trichodorus*) which allows ingestion of cell contents. Stoma of predatory nematodes is relatively larger and is armed with teeth and/or denticles. The nematodes which feed on bacteria and other microorganisms have cylindrical unarmed stoma. Stoma is made up of two parts, the anterior part is cheilostome which is formed by an invagination of the external cuticle and is not surrounded by oesophageal tissue. The posterior part of the stoma is surrounded by anterior oesophagus as it lies embedded in the oesophageal tissue and is called as oesophagostome. Both these parts are, however, heavily cuticularized. Tylenchid nematodes possess hypodermic needlelike feeding apparatus formed by the modification/fusion of stomal (stomatal) wall, hence called as stomatal stylet or spear. In Dorylaims, the stylet is modified into teeth (odontia) and called odontostylet. In the family Trichodoridae, the stylet is solid. Stylet is the most characteristic and easily recognizable feature of plant parasitic nematodes. Stylet has following five parts.

Cephalic framework: The cephalic framework is like an umbrella or inverted basket and usually hexaradiate, forming the internal labial support. It is an anterior modification of the cheilostome that extends over the modified anterior oesophagus. The framework has two functions; it supports the labial region as an internal skeleton, and provides epidermal structure to the protractor muscles of the spear (Fig. 42).

Stylet guide: It is a tubular extension of the stomal cavity or vestibule and is connected at one end with the oral opening, and at the other end with the posterior conical part of spear (Fig. 42).

Stylet cone: It is an anterior conical part of spear, and is also called conus, spear cone, spear tip. It tapers sharply to a pointed tip anteriorly. A fine lumen runs through the centre of the conus and opens in to the exterior through a ventral subterminal opening, the spear aperture (Fig. 42). In Trichodoridae, the lumen is absent as the stylet is solid. Cephalic framework, vestibule and conus are cheilostomal in origin.

Shaft: It is the posterior cylindrical part of the stylet and at the base bears three knobs, one dorsal and two subventral in position (Fig. 42). The knobs provide a point for attachment of spear muscles. A lumen passes through the shaft and connects to the oesophageal lumen.

Stylet muscles: The muscles are attached to the spear, cephalic framework, spear guide at one end, and the body wall near the base of cephalic framework at other end, which help to move the stylet (Fig. 42). These muscles are modified oesophageal muscles of protractor type (muscles serving to extend an arm); the retractor muscles (to bring back the arm) are totally absent in Tylenchids. The shaft and spear muscles are of oesophageal origin.

In resting position, the cone lies inside the stylet guide. During feeding, the forward movement of the stylet is brought about by the contraction of stylet muscles. As a result, the stylet is pushed anteriorly through the stylet guide and mouth into the host cell. The backward movement of the spear is brought about by body pressure and the elasticity of oesophagus. During the process of moulting, the conus sheds along with the body cuticle, while the shaft and knobs disappear by dissolution. After moulting, during the formation of a new spear, the conus reappears first followed by the shaft and knobs.

Modifications in the Feeding Apparatus

All parts of feeding apparatus show great variation in size, shape, etc. These variations provide useful charactersitics for identification at the generic/species level.

Oral opening: The oral opening may be small and round (Tylenchidae), slit like or dorsoventrally oval (Hoplolaimidae), or I-shaped laterally (Criconematidae). Other shapes may also be seen in plant nematodes.

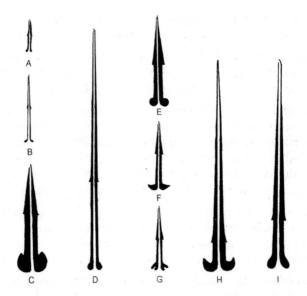

Fig. 43 Major types of variation in stylet of plant parasitic nematodes. Delicate stylet (A), small stylet (B), strong and massive stylet (C), large stylet (D), rounded knobs (E), conoid knobs (F), tulip shaped knobs (C), anchorshaped knobs (H), knobs sloping backward (I), and bifurcated knobs (G).

Size and thickness of spear: The spear or stylet may be delicate (thin and small, *Trichotylenchus*), small (*Duosulcius*), large (*Dolichodorus*), and strong and massive (*Hoplolaimus*) (Fig. 43).

Stylet knobs: The knobs may be rounded (*Aerolaimus*), conoid (*Histotylenchus*), tulip shaped (*Hoplolaimus*), anchor-shaped (*Hemicriconemoides*), sloping backward (*Hemicycliophora*), or bifurcated (*Hexatylus*) (Fig. 43). Intermediate shapes also occur.

Oesophagus or Pharynx

It is the second and the largest part of stomodeum and lies between the stoma and intestine. Oesophagus is primarily a "food transporter" as it pumps food from the host cell through the stoma and transports it to the intestine. Although it shows various diversities, the basic structure remains fairly uniform. Internally the oesophagus is lined with cuticle and externally by a membrane (basal lamella) that separates it from the pseudocoelom. The lumen of oesophagus is generally triradiate, one arm points ventrally and two subdorsally. As a result of three internal sectors, the outer surface of the oesophagus shows three sectors, one dorsal and two subventral. The oesophagus provides an important taxonomic characteristics especially at higher level of classification. The oesophagus

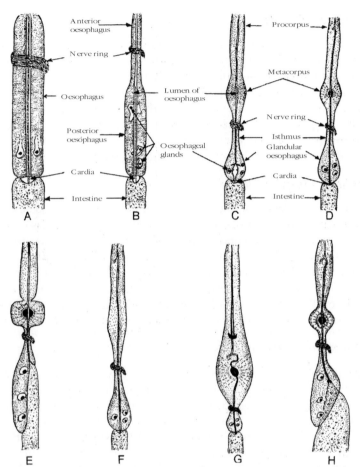

Fig. 44 Oesophagi in the order Mononchida (A), Dorylaimida (B), Rhabditida (C), Tylenchida (D), Aphelenchida (E), and variations within Tylenchida, Neotylenchid type (F), Criconematid type (G), and Hoplolaimid type (H).

is divisible into different parts which vary with the order (Fig. 44). Tylenchid oesophagus has generally four parts: procorpus, metacorpus (median bulb), isthmus, and glandular oesophagus (basal bulb). The metacorpus may have cuticularized valvular apparatus and works as a pump to suck food material from the host through the stoma and transports it to the intestine. Metacorpus is connected to the glandular oesophagus through a short and narrow tubular isthmus. Nerve ring is present around the isthmus. The glandular oesophagus bears oesophageal glands.

Oesophageal Glands

There are three unicellular and uninucleate oesophageal glands located in the glandular oesophagus. One gland is present in the dorsal sector and two glands in the subventral sector of the basal region of the oesophagus. The glands are either enclosed in the basal bulb or hang freely in the body cavity overlapping the intestine. The dorsal gland is usually large and conspicuous with a prominent nucleus. The dorsal gland generally opens near the stylet knobs. The two subventral glands open into median bulb below the valvular apparatus.

Oesophago-intestinal Valve

It is the last part of stomodeum and lies at the junction of the oesophagus and intestine. The oesophago-intestinal valve is also called as cardia. The shape of cardia may be conoid, rounded, flat, disc like, etc. Lumen of the cardia is tri-radiate, made up of three cells and is lined with cuticle. The function of cardia is to regulate the passage of food from oesophagus to intestine and to prevent regurgitation (backward flow of food) particularly when the intestine is full of fluids and has a high turgor pressure. The cardia is more developed in the members of Tylenchida.

Modifications in Oesophagus

The shape of the oesophagus differs greatly with the order and sometimes within an order (Fig. 44). In the order Mononchida, it is simple cylindrical (nonpartite). Bipartite (two parts) or bottle shaped oesophagus with an anterior narrow part and posterior wider part is characteristic of the order Dorylaimida. The oesophagus is further complicated by the presence of another muscular swelling called median bulb. A true median bulb is present in Tylenchida as it contains sclerotized crescentic thickening called valvular apparatus or valves. In Rhabditida, the valve is absent and it is called pseudobulb. In Aphelenchida, the median bulb is rectangular and the glandular oesophagus is bilobed, one is large and other is small. The intestines are connected to the smaller lobe.

Tylenchid oesophagus: Within the order Tylenchida, the oesophagus shows great diversity, and forms an important taxonomic character useful at higher level of identification. There are following three types of tylenchid oesophagi (Fig. 44):

Neotylenchid type: It is a typical tylenchid oesophagus, but valvular plate of the median bulb is completely absent. The median bulb is either present as an inconspicuous swelling or absent (Fig. 44). Neotylenchid type of oesophagus is characteristic of the family Neotylenchidae.

Criconematid type: The procorpus and median bulb are amalgamated (fused) and form the major portion of the oesophagus (Fig. 44). Isthmus and basal bulb are small. Members of the family Criconematidae have this type of oesophagus.

Hoplolaimid type: The median bulb is well-developed and possesses large crescentic valvular apparatus. The basal bulb is lobe-like, and extends over the anterior part of the intestine (Fig. 44).

Mesenteron (Intestine)

Intestine (mid-gut) is the largest part of the alimentary canal of nematodes. It is derived from embryonic endoderm and is the first tissue invaginated during gastrulation. The intestine is a hollow tubular structure made up of a single layer of endodermal epithelial cells. Numerous taxa show three distinct regions in the intestine, designated as anterior ventricular region, mid-gut (mid-intestine) and prerectum. In Dorylaimida, the prerectum is easily recognized by long microvilli and the ventricular region is distinguished by packed cell inclusions and insoluble spherocrystals (Fig. 45). In Tylenchida, recognizable subdivisions of intestine does not occur. Externally the intestine is separated from the pseudocoelom by a basal lamella which plays a vital role in the transport of materials from intestine to pseudocoelom and to other tissues of the body. The plasma membrane that lines the lumen of the intestine is thrown into very fine fingerlike projections called microvilli, which increase the surface area of intestinal cells for absorption (Fig. 45). In all regions of the intestine microvilli are present. Below the microvilli, there may be an area of dense fibrils (terminal web) which is connected by cytoplasmic connections to the rest of the cell cytoplasm. The terminal web may extend into the base of the microvilli. The intestinal cells are both secretory and absorptive in function and are generally rich in mitochondria, golgi complex, endoplasmic reticulum, ribosomes, glycogen, lipids, fatty acids, etc. The fats, fatty acids and proteins occur as globules, while glycogen is in the liquid state. The waste products may be present in the form of crystals. The food moves in the intestine by the ingestion of more food and also by the locomotory activity of the animal. Some nematodes possess intestinal muscles which aid in the passage of food by bringing about peristaltic movement of the intestine.

The intestinal cells throughout the mesenteron may have the same or differing characteristics, and such a condition is designated as homocytous (similar in shape and size) or heterocytous (dissimilar). If all the cells in cross section are of equal height, then the intestine is isocytous; if different in height the condition is anisocytous. Nomenclature is also applied to the total number of cells in the intestine: oligocytous (upto 128 cells), polycytous (256-8,192 cells) and myriocytous

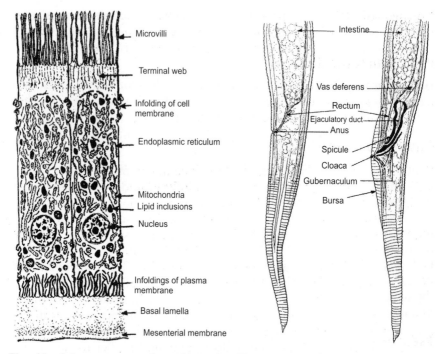

Fig. 45 A transverse section of intestine (Courtesy: Kessel *et al.*, 1981), and posterior ends of a typical female/male nematodes showing intestine, rectum, anal opening and other structures.

(16, 384 cells or more). The cell shape is also affected by number; oligocytous intestinal cells are longitudinally elongate and rectangular to hexagonal, polycytous intestines have cuboidal cells, and myriocytous intestines have tall columnar cells. The shape of the intestinal lumen is also dictated by the number of intestinal cells: oligocytous, cylindrical/ rounded lumen; polycytous, subpolygonal lumen; myriocytous, the lumen is a multifolded or flattened tube. Intestinal cell descriptions are further complicated by being uninucleate to polynucleate. The latter condition is only known among mermithids in Adenophorea, but is common among Secernentea. The intestine in criconematids and many adults of Heteroderidae, Meloidogynidae, etc., is syncytial lacking a definite lumen, though it may be present in the juvenile stages. In some species of Dolichodoridae, the intestine extends into the tail cavity forming a blind sac of varying length. The presence/absence of the sac and sometimes its length may be useful in the identification of species. In some nematodes, particularly Tylenchorhynchinae, lateral canals which are also referred to as intestinal fasciculi or serpentine canals are present. These canals extend into the tail cavity, if intestinal sac is present. By and large, the intestine does not provide any significant taxonomic characteristic.

Proctodeum (Rectum)

Rectum or hind-gut is formed due to invagination of ectoderm during the embryogenesis. The proctodeum is a short, flattened, subtriangular or an irregular tube. Its lumen is lined with cuticle which is in continuity with the external body cuticle and sheds during moultings. An intestino-rectal valve (pylorus) formed by the intestinal cells is usually present, separating the intestine from the rectum. The closure of this valve is brought about by a unicellular annular sphincter muscle. In tylenchid nematodes, rectal glands, three in females and six in males, are present which empty into the rectum, with an exception of six very large unicellular rectal glands present in the females and three in males of *Meloidogyne*. These glands produce a gelatinous matrix in which eggs are laid. The anus is minute pore like in Tylenchida, but in other nematode groups it may be large slit like. In male nematodes the terminal duct of the reproductive system opens into the rectum forming a common opening, the cloaca.

Pseudocoelom (Body Cavity)

The nematodes, irrespective of habitat, possess a well defined body cavity extending along the entire length. The body cavity is devoid of mesodermal lining, hence it is called pseudocoelom instead of coelom. The cavity is filled with pseudocoelomic fluid. The fluid is rich in protein and other dissolved substances. A variety of materials, pseudocoelomocytes, pseudocoelomic membranes, mesentries, etc., are also present within the pseudocoelomic fluid. The fluid has a high osmotic value and is responsible for the turgescent state of the nematode body which is necessary for the functioning of somatic musculature.

EXCRETORY SYSTEM

The excretory system in nematodes is inconsistent and inconspicuous, still it constitutes one of the important taxonomic characters at the highest level of classification. In the class Adenophorea, the excretory cell (renette) opens directly into an excretory pore without a lateral canal. In the class Secernentea, excretory canal is present that opens through the excretory pore.

The cells of the excretory system in nematodes are hypodermal (ectodermal) in origin. In general, the system consists of a glandular organ called excretory cell (also known as sinus cell, renette or ventral gland), a pore or socket cell located in the region of nerve ring, and a collecting tube or tubular organ called excretory canal may be present in the Secernentea. In primitive forms the tubular organ is absent. In Adenophorea, the excretory system is usually absent. If present, it is a primitive form (glandular type) but qualifies an orthodox excretory system, having a discrete cell and pore forming cell. It is composed of a ventral gland with a short or long neck (duct) that terminates in a pouch called ampulla, which empties outside through an excretory pore located on the ventral side in the region of nerve ring (Fig. 46). The duct may or may not be lined with cuticle. In Dorylaimida, usually such a structure is not seen except an excretory pore.

The Secernentea has an advanced excretory system (tubular type). The basic and ancestral secernentean system, which is generally com-

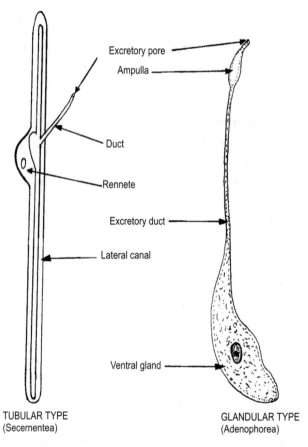

Fig. 46 A diagrammatic presentation of the excretory system of nematodes belonging to the class, Secernentea and Adenophorea.

posed of a pair of longitudinal canals, located one each in the lateral chords, runs almost the entire length of the nematode. The canals unite terminally or in the middle and open in the excretory duct which empties outside through the excretory pore located in the region of nerve ring. The excretory systems in different orders of Secernentea are modified from the ancestral system. In Tylenchida, most oftenly one excretory canal is present in either lateral chord and is called as asymmetrical. The excretory pore connects to renette (excretory cell) through an excretory duct. The duct is easily visible as it is highly cuticularized, e.g., *Halenchus, Allotylenchus,* etc. The excretory duct, which is connected to the lateral canal is also called as collecting tubule as it runs anterior to posterior to collect the excreta. In all tylenchs the lateral canal is not present, e.g., family Criconematidae that lacks the canal. In some nematodes the renette is highly enlarged, like in *Tylenchulus* it occupies nearly 30% of the body cavity and secretes a gelatinous matrix in which eggs are deposited. The excretory pore in Tylenchida is generally located laterally or ventrolaterally, usually in the region of nerve ring.

The exact function of the excretory system in nematodes is not known, because ammonia, carbon dioxide, nitrogenous and other wastes are excreted through the body wall and digestive system. It is likely that the excretory organs are supplementary to ensure proper body turgidity through excretion of liquid materials and/or osmoregulation. If the excretory organs would have been essential, they should have been present in the entire nematode phylum. Many Adenophorea taxa lack excretory system, but they generally have numerous hypodermal glands, caudal glands, tubular gland setae, coelomocytes, etc., which may efficiently perform excretion.

NERVOUS SYSTEM AND SENSORY STRUCTURES

Nervous System

A nervous system is meant to transmit stimuli received externally by way of somatic sensory organs to the central nervous system and then to the internal tissues, where they are translated into a proper response. Hence, the nervous system acts to mediate the nematode's activities through stimulation, coordination and responsive actions. The nervous system of nematodes is poorly developed. It consists of a circumoesophageal commissure, four nerve chords and associated ganglia. The circumoesophageal commissure, commonly called as nerve ring, is considered to be the central nervous system or brain of the nematode. The ring is present around isthmus (between median bulb and basal lobe) in

Tylenchids and Aphelenchids, and around the anterior part of the oesophagus of Dorylaims and Mononchids (Fig. 44). The nerve ring is chiefly composed of nerve fibres and ganglia, which are concentrated in nerve cell bodies and form nerve centres. Considering the complexity of the nervous system, it can be better explained under anterior and posterior nervous system.

Anterior Nervous System

On the anterior side of the nerve ring, six small cephalic papillary ganglia (two lateral, two subdorsal and two subventral) are present, from them six papillary nerves proceed anteriorly to innervate cephalic sensory organs (Fig. 47). The two subdorsal and two subventral nerves trifurcate, sending one nerve to each inner labial papilla, outer labial papilla and cephalic papilla. The two lateral nerves bifurcate giving one nerve to each labial papilla. These nerves pass through the body cavity lying close to the oesophagus. One lateral ganglion is attached to each lateral side of the ring. A sub-division of lateral ganglia gives rise to two amphidial ganglia. A nerve from each amphidial ganglion proceeds to amphids present in the cephalic region on the lateral side. In the oesophagus three nerves are present, which run along the length of the nematode body. These nerves are connected with each other by commissures and also with the nerve ring.

Four large ganglia (two lateral and two ventral) and one small dorsal ganglion are attached to the nerve ring posteriorly. Two small subdorsal ganglia may also be present (Fig. 47). From the lateral ganglia, lateral somatic nerves arise which are mainly of sensory nature and are provided with several ganglia in their course along the lateral chords of the body. In the region of oesophagus, two cervical papillary nerves, i.e., one from each lateral nerve, arise and proceed to deirids or cervical papillae. A mid-dorsal somatic nerve originates from the dorsal ganglion. It runs posteriorly in the dorsal chord, a few ganglia have been observed throughout its length. It connects directly with innervation processes from somatic muscles and is considered chiefly as a motor nerve. Dorsolateral nerves, arising from the two dorsolateral ganglia, may occur between the dorsal and lateral nerves. The two subventral ganglia give rise to two subventral nerves which bifurcate. One branch of each nerve fuses together to form the retrovesicular ganglion in a region behind the excretory pore. The remaining two branches, i.e., one from each ventral nerve, become lateroventral nerves and run upto the anal region. Several commissures running in the hypodermis, connect to the longitudinal (lateral, dorsal and ventral) nerves at regular intervals along the entire length of the body.

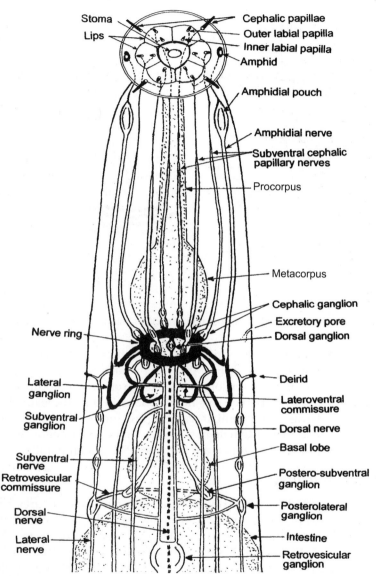

Fig. 47 Schematic diagram of anterior nervous system of a nematode

Posterior Nervous System

Female nematode: In the anal region of female nematode, lateral nerves arise from two lumbar ganglia (Fig. 48). The ventral nerve forms a single or paired anal or pre-anal ganglia from which a pair of connectives extends to the lumbar ganglia. Further posteriorly, the ventral nerve gives two internal branches, which extend around the rectum and unite to form rectal commissure. From the two lumbar ganglia caudal nerves arise which innervate the pair of phasmids.

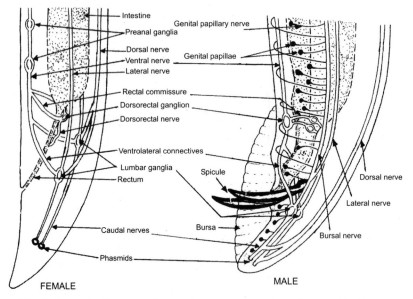

Fig. 48 Schematic diagram of posterior nervous system of a nematode

Male nematode: The posterior nervous system of male nematodes is relatively more complicated (Fig. 48). The pre-anal male genital papillae are formed from bipolar sensory nerve cells from a longitudinal strand which is called bursal nerve. This nerve is located on either side or near the lateral chords. Two lumbar ganglia are formed from the lateral nerves similar to that of female nematodes. Branches of nerves from the enlarged lumbar ganglia innervate the post-anal genital papillae.

Sensory Structures

Since nematodes do not have eyes or ears, the sensory structures mostly present in the cephalic region, keep them aware of the outer environment and help them to approach a susceptible host. The responses to various stimuli indicate that nematodes possess sensitive sensory structures with a high degree of nervous coordination. Sensory structures receive stimuli and transfer them to nerve centres. Initially, stimuli are in the form of electric signals which are converted into chemical signals during hydrolysis of acetylcholine by acetylcholinesterase at synapse. Functionally, the sensory structures of nematodes are chemoreceptors, mechanoreceptors and thermoreceptors. It is assumed that a sensory organ may also perform two or more functions.

The sensory organs commonly called as sensilla (sensillum singular) are made up of cuticle and hypodermis. They are mostly located on the external cuticle, and are known as exteroreceptors. Nematodes also have interoreceptors, the sensory organs located internally on the cuticular lin-

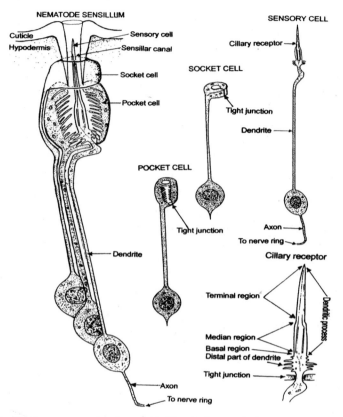

Fig. 49 Basic sensory structure (sensillum) of nematode

ing of oesophagus, vagina, etc. The sensory organs in Tylenchida are relatively poorly developed compared to Dorylaimida or other groups. A sensillum has two basic parts, neuronal and non-neuronal (Fig. 49). The neuronal part consists of a sensory cell (nerve cell or neuron). The anterior extension of the sensory cell is referred to as dendrites. The cell body can also be classified as a bipolar sensory neuron with the distal part extension as dendrites. At the terminal portion of the dendrite an axon emerges which proceeds to the nerve ring.

The non-neuronal part consists of a socket cell and a pocket cell (Fig. 49) which enclose and support the neuron. At the upper end of the neuron, the dendrite enters in the sensillum canal and passes via the sensillar pouch. The pouch is surrounded by a supporting cell (socket cell) and a long-neck sheath cell called pocket cell. The distal end of the pocket cell terminates behind the socket cell and forms a sheath around the expanded sensillar lumen, the entire structure is called sensillar pouch. At the base, the neuron sends an extension (axon) to the circumoesophageal commissure (nerve ring).

For recognition of any change in the external environment, nerve fibres must penetrate through the cuticle and the penetrated nerve must be modified into a sensory region or structure. Specialized sensory structures are found mainly in the anterior and posterior regions of the nematode body. The various types of sensory organs which occur on the external surface (exteroreceptors) of nematode body are: cephalic sensory organ complex (cephalic sensory structures and amphids), deirids, hemizonid, hemizonion, phasmids, cephalids and caudalids (Figs. 50-53).

Cephalic Sensory Structures (Labial and Cephalic Papillae)

The anterior region of nematodes, like other multicellular organisms, possesses greater importance in physical activities such as movement and feeding. To recognize a passage wide enough for movement, availability of susceptible host for feeding, and suitable environment for survival, the head region of nematodes is provided with almost two-thirds of the sensory structures present in the nematode. Plant parasitic nematodes from Tylenchida, e.g., *Ditylenchus dipsaci* possess 16 sensory structures (12 labial papillae and 4 cephalic papillae) in the cephalic region, which can be best viewed at an *enface* view of the head (Fig. 50). Phytonematodes from Adenophorea (Dorylaimida) may also have 16 cephalic sensory structures.

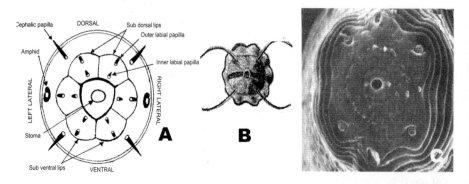

Fig. 50 Schematic diagram of cephalic sensory structures of a nematode (A), and SEM *enface* view showing papilloid cephalic papillae (B), and other structures (C).

Nematodes possess six lips or labia, arranged around the mouth (stoma) in a hexaradiate fashion. Four lips are submedian (2 subdorsal and 2 subventral) and two are lateral in position (one on each lateral side). The twelve labial papillae are arranged in two whorls or circles on the lips. Hence, each lip has one inner labial papilla and one outer labial papilla. The four cephalic papillae are located usually outside the labia in submedian region, two each are subdorsal and subventral (Fig. 50A-B).

Functionally, the outer labial papillae and cephalic papillae are mechanoreceptors, whereas the inner labial papillae perform two functions, chemoreception as well as mechanoreception. Structurally, the labial and cephalic papillae have a number of centrally located dendritic nerve processes, which partly consists of a cilia, i.e., a basal body and a number of peripheral and central ciliary filaments. The cephalic papillae are usually depressed, but may be rarely papilloid or filamentous as in Atylenchidae, e.g., 9-12 mm long papillae in *Atylenchus* and *Eutylenchus* (Fig. 50). The outer labial papillae are blindly beneath the cuticle of the head and they may or may not have a protuberance. The outer labial papillae and cephalic papillae lie very close to each other and the former may be absent. In Tylenchida, the inner labial papillae usually have a pore like opening (pit) either around the stoma or inside an invaginated area called prestoma or vestibule. The position and shape of labial papillae and presence or absence of cephalic papillae may vary with the shape and size of lips. This variation is of immense taxonomic significance. In some groups, e.g., criconematids, the cephalic sensory structures are greatly reduced or lost. This can be correlated with heavy cuticularization of the body surface. Such nematodes possess a long stylet, as a result they may feed without rubbing lips on the host surface to receive chemostimuli.

Amphids

Amphids are found in all groups of nematodes and form an integral part of the anterior sensory organ complex (Fig. 51). They form an important diagnostic characteristic for groups as well as within the group. Amphids are paired and present laterally in the cephalic region. Some nematologists considered them as the two missing lateral cephalic papillae, but structurally as well as functionally they are quite different from the papillae. The amphid has an amphidial aperture situated in either labial or postlabial region and has an opening exteriorly. The opening has a terminal amphidial duct that contains the distal ends (radially arranged cilia) of several dendrites which constitute the lateral nerve bundle (Fig. 51A). Posterior to the duct, the dendrites are enclosed by a sensillar pouch (fuses) which, in turn, is enveloped by the anterior portion of an amphidial gland. Perikaryons of the dendrites are concentrated in ganglia which ensheath the oesophagus anterior to the nerve ring. The amphidial gland consist of several microvilli and rough endoplasmic reticulum. Anteriorly the gland is broadened and contains a nerve bundle. The amphidial gland is connected to the hypodermal chord through a large lateral cell. The shape and size of the amphidial aperture vary greatly which may be oval, slit like, ellipsoidal, horseshoe shaped,

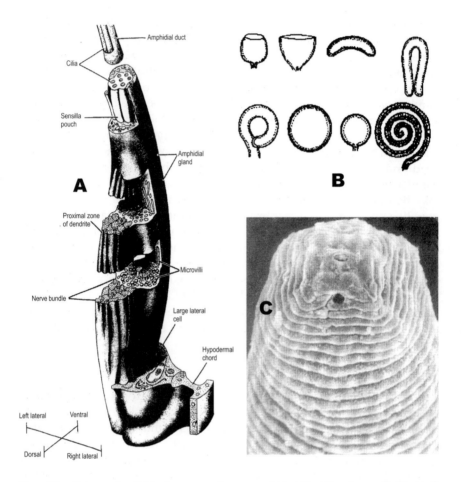

Fig. 51 Schematic 3-D diagram of an amphid (A), (Courtesy: Baldwin & Hirschmann, 1975). Different shapes of amphid (B), and SEM of head region showing amphid and papillae (C) (Courtesy: Siddiqi, 1980).

unispiral, circular, multispiral, etc. (Fig. 51B). In tylenchs, amphids are rather simple type and located close to the oral aperture. Amphids are considered as chemoreceptors for detecting change in the outside environment. They may also perform some secretory and photoreceptive function.

Cephalids

They are two pairs of band like biconvex structures forming a complete ring in the anterior cephalic region on the dorsal and ventral sides, and broken at lateral fields (Fig. 52A). One pair, called anterior cephalids, is

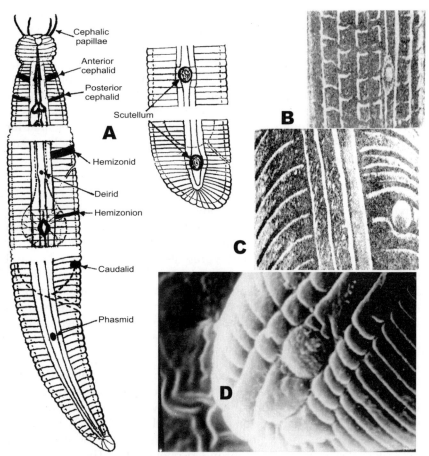

Fig. 52 Schematic presentation of sensory structures present on the nematode body (A). SEM of different regions of the body showing deirid (B), phasmid (C), and scutellum (D). (Courtesy: Sauer, 1985).

just behind the cephalic region and the second pair, posterior cephalids, at some distance behind the previous pair. Anterior cephalids are more prominent than the posterior cephalids. The exact function of cephalids is unknown, but it is assumed that they perform the function of some sensory reception, probably pressure receptor.

Deirids

Deirids can also be considered as a part of the anterior sensory organ complex. They are a pair of small protuberances, one on each lateral side, in the centre of lateral fields in the region of nerve ring or excretory pore (Figs. 52A-B). They are present in the families Tylenchorhynchinae,

Dolichodorinae, Tylenchidae, Neotylenchidae and many criconematids. They are also referred to as cervical papillae and have a structure similar to that of the four cephalic papillae and are considered as two missing cephalic papillae, which have migrated to this lower position. Like phasmids, they are somatic sensilla and function as mechanoreceptors. Their location on the body just before it assumes maximum width may help the nematode to determine the size of the pore space as to whether it can successfully pass through it. However, it is not absolutely essential for terrestrial life as a majority of nematodes lack it. Deirids do not have any exterior opening.

Hemizonid and hemizonion

Hemizonid (= belt or girdle) is a biconvex structure forming a semicircle on the ventral side between lateral fields (Fig. 52A). It is located between the cuticle and hypodermis. It is present only on the ventral side, anterior or posterior to excretory pore and represents ventrolateral or subventral commissure that connects the nerve ring to the ventral nerve chord. Hemizonion is structurally identical to hemizonid but is smaller and located posterior to hemizonid, possibly represents a smaller lateroventral commissure. It is not always found even in the presence of hemizonid.

Phasmids

These are lateral sensory organs, and have immense taxonomic value, e.g., Secernentea have phasmids and are called Phasmidians, and Adenophorea which lack phasmids are known as Aphasmidians. The phasmids occur in a pair, one on each side of the tail (Fig. 53), but sometimes present in pre-anal region (*Helicotylenchus*) or even more anteriorly (*Tylenchus, Hoplolaimus*) or absent in Criconematidae. They are usually located in the centre of the lateral field (*Tylenchus, Cephalenchus*). In *Scutylenchus,* phasmids open exteriorly through a pore. In the members of Hoplolaimidae, phasmids are large plate like or shield like, without an external pore and are called scutella (scutellum singular) (Fig. 52D). Scutella may be located at different levels, e.g., in the region of tail (*Scutellonema*), oesophagus (*Hoplolaimus*) and vulva (*Tylenchus*). They are somatic sensilla and internally possess a pouch containing sensory receptors which are supplied by lateral caudal nerves. Functionally, phasmids are chemoreceptors, but can also serve excretory function because of their association with glands.

Caudalids

These are present in front of the tail. Caudalids represent paired commissures linking the pre-anal ganglion to the lumbar ganglion. They are

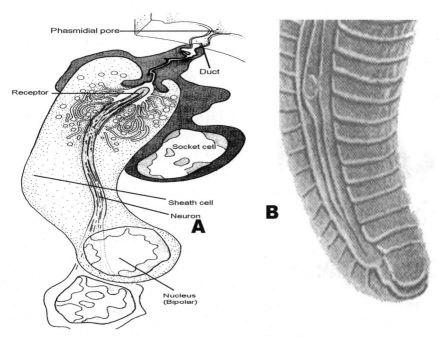

Fig. 53 Schematic diagram (A), and SEM (B), of a phasmid. (Courtesy: Sauer, 1985).

usually present in the family Hoplolaiminae. The function of caudalids is not known, but could be similar to that of cephalids or hemizonids.

General papillae and supplementary organs of males

Male nematodes have various forms of papillae that occur on the ventral side of the posterior end. They may be enclosed in bursa or consist of raised areas of thin cuticle. Supplementary organs are located in the pre-anal region. The shape of these organs may vary from simple papillae to cuticular swelling bearing denticles and provided with glands. Plant parasitic males of Secernentea have simple genital papillae (Fig. 48), whereas the males of Adenophorea (*Xiphinema, Longidorus, Trichodorus*) are provided with well developed pre-anal supplementary organs. Post-anal genital papillae are also present. All these organs are tactile in function and assist males during copulation, hence they are well developed in amphimictic species.

Campaniform organs and ocelli

Campaniform organs are present in some marine nematodes. They consist of cup like pits in which a thin sheath of cuticle projects. On the sheath a sensory cell is attached. Functionally, they are mechanoreceptors for touch and pressure. Ocelli are eye like sensory organs which are found in some marine and fresh water nematodes. These organs are embedded in the oesophageal musculature. They consist of a lens like cuticular body that rests on a chromatic unit. Their function is not properly known.

NEMATODE PHYSIOLOGY

Nematodes are basically aquatic animals but have adopted terrestrial habitat; probably for this reason their body contains more than 90% water. Water content of the nematode, however, varies with the life stage, e.g., hardly 1% water in quiescent stage (w/w), 15-20% in dormant J_1/J_2 larvae, and 70-80% in active J_1/J_2 larvae few hours after hatching. In addition to water, nematodes contain carbohydrates, lipids, fatty acids, nucleic acids, sterols, plant growth regulators and other substances (Karssen, 2002), whereas the egg shell contains carbohydrates, nitrogen, calcium, aluminium, copper, magnesium, selenium, iron, manganese, mucopolysaccharides, glucose, fructose, chitin and other elements (Agrios, 2005). Eggshell is the only part in nematodes that contains chitin. Egg sac matrix contains glycoproteins. Cuticle of adult females of *Heterodera* contains tanned lipoproteins. Unlike other animals, physiology of phytonematodes has not been studied thoroughly. Most of the information available on this aspect of nematodes is based on the studies that were conducted on *Ascaris lumbricoides* or *Caenorhabditis elegans*.

Respiration

Nematodes perform cuticular respiration, and like other animals they are aerobic but also capable of surviving prolonged anaerobic periods. Under oxygen deficient condition, mobility of nematodes decreases but they regain wriggling soon when are transferred to an environment rich in oxygen, such as fresh water. It has been observed that at 10% or less oxygen level, nematodes respire well and maintain metabolic activities at a normal rate. Respiration rates in nematodes (QO_2) are measured in nl O_2/hour. In adult nematodes the QO_2 may range from 1.25 to 8.8 nl O_2/hour at 20°C (Anon., 2006). Oxygen consumption rate decreases with starvation as catabolized reserves shift from carbohydrates to lipids and proteins. Marine nematodes, however, survive better in anaerobic state. Nematode development is affected with available O_2, e.g., development and egg production are retarded at low O_2 level, but hatching and larval movement are not affected, rather plant nematodes move faster under high CO_2 concentration. Carbon dioxide upto 8% has been found generally suitable for nematode survival and other activities. High concentration of CO_2, however, inhibits respiration. Under low level of

O_2, nematodes show lactate dehydrogenase activity for the first 12-16 hours, followed by a shift over to alcohol dehydrogenase to produce ethanol (Dasgupta, 1998). The ethanol production helps nematodes to survive anaerobically, serves as a carbon source, and is safer than lactic acid. Generally high O_2 concentration is injurious to nematodes, because of production of greater amount of H_2O_2 during the respiration which acts as poison for them.

Metabolism

Metabolic activities in nematodes require O_2 and are determined as nl $O_2/\mu g$ fresh weight/hour. The metabolic rates for adult nematodes have been determined as 1.15 nl $O_2/\mu g$ (*Rhabditis cucumeris*) to 4.43 nl $O_2/\mu g$ fresh weight/hour (*Mesorhabditis labiata*) at 20°C (Anon., 2006). These rates, however, decrease with starvation. Plant nematodes draw energy mainly from carbohydrates, lipids and proteins. Metabolism of these food reserves largely resembles that of other animals. However, considerable variation occurs that may be due to lack of adequate research on physiological aspects of plant parasitic nematodes.

Carbohydrate Metabolism

In plant parasitic nematodes, lipids are major carbohydrate reserves which are oxidized to generate energy for various processes and activities (Geraert, 2006). Another carbohydrate which has been found in nematodes is trehalose. It is a nonreducing disaccharide and is also found in insects and fungi, but not in mammals. Nematodes (*Ascaris*) contain 2% trehalose of dry weight. The metabolism of glycogen synthesis in nematode seems to be similar to mammals and trehalose synthesis similar to insects (Anon., 2006). In animal parasitic nematodes, lipids are the chief source only during juvenile stage, whereas the adults and mature stages of free living and plant nematodes depend on glycogen which is 20-30% as compared to 2-9% in juveniles. The glycogen is utilized in aerobic state. Nematodes also use glucose and fructose in respiration. The typical carbohydrate metabolism includes fermentation and glycolysis, TCA cycle and hexose monophosphate shunt, which are summarized here under:

Fermentation and *Glycolysis*

Fermentation is a process where sugars are metabolized to alcohol in the absence of O_2, whereas in glycolysis glucose is converted into pyruvate with the release of energy in the form of ATP and NADH (Nelson and Cox, 2007). In nematodes, for example *Ascaris*, a typical glycolytic cycle similar to other animals operates where glycogen or glucose is anaerobically broken down as phosphoenol pyruvate (PEP), then instead of converting PEP to pyruvate as occurs in mammals, CO_2 fixation in cytoplasm occurs in nematodes, and PEP is converted to oxaloacetate catalyzed by PEP carboxykinase (Anon., 2006). The oxaloacetate is then reduced to malate which is transferred to mitochondrion where it is partly oxidized back to pyruvate and partly reduced to succinate (Fig. 54). These steps are reversal of TCA cycle. Hence, the immediate

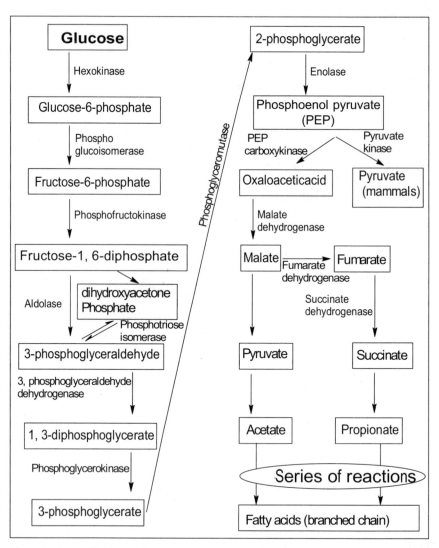

Fig. 54 A modified glycolysis as observed in *Ascaris*

end products are pyruvate and succinate, which are then converted by a series of steps into branched chain fatty acids (2-methylbutyrate). This type of pathway has also been detected in several other nematodes.

Hexose Monophosphate Shunt

It is believed that hexose monophosphate (HMP) shunt also operates in plant parasitic nematodes (Fig. 55). In HMP shunt glucose-6-phosphate is converted to phosphoglucono-8-lactone in the presence of glucose-6-

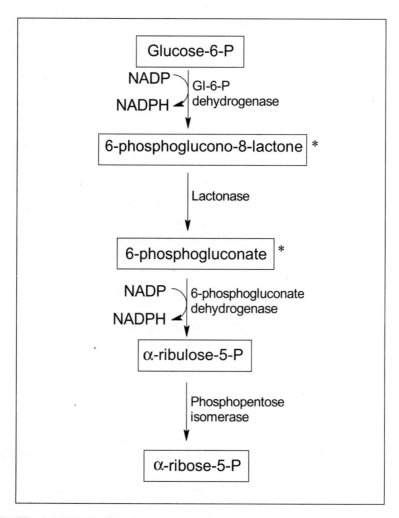

Fig. 55 A typical Hexose Monophosphate Shunt (*some of its steps have been observed in nematodes).

phosphate dehydrogenase (Sinha, 2004). Phosphogluconate is formed which is converted to ribulose-5-phosphate by the enzyme 6-phosphogluconate dehydrogenase, and the final product is α-ribose-5-phosphate. Both the dehydrogenase enzymes have been found in *Ditylenchus* and *Ascaris* species (Fig. 55). The HMP shunt appears to be less important for plant nematodes, as it may take part in the synthesis of nucleic acids, pentose sugars and in the formation of reduced NADP.

Tricarboxylic Acid (TCA) Cycle

The TCA cycle is also known as Kreb's cycle. A typical TCA cycle involves the oxidation of pyruvate to the products, oxalic acid via acetyl coenzyme – A with the release of energy and CO_2 (Goodwin and Mercer, 2001). In nematodes (*Ascaris*), partial reverse TCA cycle operates that involves reduction of fumarate to succinate. However, under aerobic conditions, the electron flow in the cytochrome chain is same as in mammals, i.e., from NADH to oxygen and from succinate to oxygen (Fig. 56). Under anaerobic condition, the electron flows from NADH to fumarate, which is then reduced to succinate. Hence, the electron receptor is oxygen and fumarate under aerobic and anaerobic conditions, respectively (Dasgupta, 1998). Intermediates of tricarboxylic acid, which are also involved in amino acid synthesis, lipid oxidation and other processes have been found in nematodes. In plant parasitic nematodes it is generally agreed that similar metabolic pathways occur. However, α-ketoglutarate dehydrogenase enzyme has not been detected, but other enzymes of the succeeding steps have been analyzed. This indicates for operating of an alternative pathway in nematodes.

Lipid and Sterol Metabolism

Lipid metabolism is a vital pathway because plant nematodes derive major energy from the oxidation of lipids. Lipid is not soluble in water but on hydrolysis it is converted to fatty acid and glycerol (Nelson and Cox, 2007). Lipid, as a stored food, is more efficient in releasing the energy, and yields energy almost double of that obtained from carbohydrates or proteins during oxidation. It is a constituent of plasma membrane and organic membrane largely in the form of phospholipids. Free fatty acids and triglycerides are formed upon hydrolysis of lipid and are oxidized to supply energy during starvation of nematodes (Dasgupta, 1998). Plant parasitic nematodes contain 30-40% lipid per dry weight which is stored in the hypodermis, especially in lateral chords, noncontractile part of muscle cells, intestine, etc. Fatty acids formed due to the hydrolysis of lipids are oxidized to produce acetyl coenzyme–A in mitochondria, utilizing an ATP. In plant nematodes such as *Aphelenchus avenae* and also

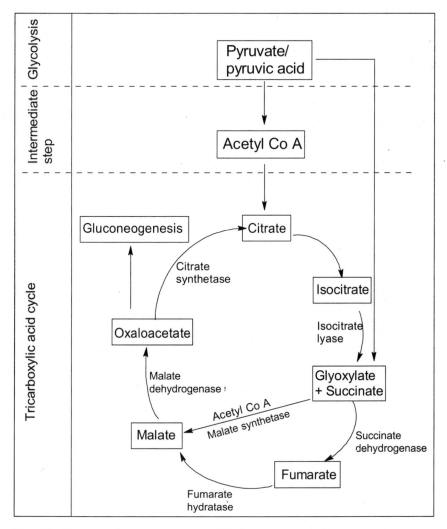

Fig. 56 The typical tricarboxylic acid (TCA) cycle observed in nematodes

Caenorhabditis elegans, lipid hydrolysis occurs under aerobic conditions. In some nematodes glyoxylate cycle has also been detected. *Ascaris* spp. synthesize ascarylose (important for egg production) by utilizing glucose-6-phosphate and NADPH, and the end product, ascarylose dinucleotide phosphate is condensed as glucon to form ascaroside, sterol or hormone. Sterols are chemical messengers with specific structure in one part of an organism that elicit a specific physiological, biochemical or developmental response in another part of an organism. Sterol hormones in nematodes regulate moulting and egg production. Doubts, however, exist regarding the presence of sterol hormone in plant parasitic nematodes, but detection of enzymes like squalene epoxidase and squalene oxide cyclase have provided evidence for sterol biosynthesis in nematodes. These enzymes convert lanosterols to cholesterols. In some nematodes, ecdysteroids similar to ecdysone have been reported to be synthesized (Chitwood, 1987). Similarly, epoxyfarnesoic acid methyl esters, similar to insect juvenile hormones, have been detected in nematodes. The sterol hormones regulate moulting in developmental process of nematodes. Sterol metabolism has been studied in *C. elegans* (Fig. 57). Some of the major steps are C_{24} dealkylation, C_7 desaturation, and C_5 reduction. This nematode also undergoes direct nuclear methyla-

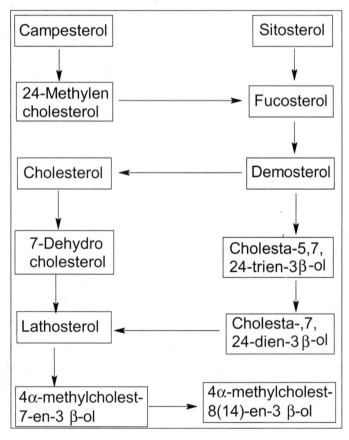

Fig. 57 Synthesis of sterols detected in *Caenorhabditis elegans*

tion to produce 4-methylcholest-8 (14)-enol. The methylation is a unique feature and known to occur only in nematodes.

Protein Metabolism

Free-living nematodes and some plant parasitic nematodes such as *Ditylenchus tricornis* and *Meloidogyne* spp. possess the ability to synthesize all amino acids essential for them, however, other nematodes lack this capability. Protein metabolism has been studied in detail in *Ascaris*. In the muscle cells and ovary of this nematode alanine glutamic acid, aminotransferase and aspartic acid have been detected, the former appears to play a greater role in the transamination. The transamination, however, varies in the intestinal cells and may occur with alanine, valine and methionine. In *Aphelenchoides ritzemabosi*, alanine is synthesized from aspartic acid/glutamic acid and pyruvate with the role of a cofactor (Vitamin B_6, pyridoxal phosphate). Amino acids are oxidized liberating CO_2 and NH_3 with the help of amino acid decarboxylase and vitamin B_6. This process is called as transdeamination which is reversed transamination. Ammonia may also be formed from purine via uric acid pathway in nematodes (Sinha, 2004). Enzymes like adenase, guanase, deaminase, xanthenine oxidase and uricase have been detected in nematodes. Absence of urease in nematodes suggests that NH_3 is not formed from urea. In young *Ascarida galli* (not adults) purine is completely degraded to urea or ammonia, but in other nematodes this metabolism (purine degradation) has not been detected.

Physiology of Reproduction

Animal behaviour is elicited in response to visual, auditory, tactile, olfactory and gustatory sensory information from the surrounding environment. However, in nematodes chemical communication appears to be largely operative. The chemicals involved in the transmission of information to the perceiving individual(s) are referred to as semiochemicals, which are further categorized as pheromones, kairomones, allomones, attractants, deterrents, parapheromones, etc. The semiochemicals involved in the intraspecific communication with regard to sex are known as sex pheromones. The sex pheromones are secreted by exocrine gland (with duct) and elicit behavioural response such as attraction/aggregation for sex, willingness of female for mating, etc. Sex pheromones are synthesized by the adult females to communicate the message of the availability of female for mating to the perceiving male individual. Sex pheromones are believed to be involved in the reproduction of nematodes, amphimictic species in particular. The sex pheromones have been detected in amphimictic species of *Globodera*,

Hoplolaimus, *Rotylenchulus* and also in parthenogenetic species, *Heterodera* and *Meloidogyne*. Only mature males are sensitive to the sex pheromones and respond to these chemicals. Sex pheromones are generally secreted by females, but sometimes males may also secrete.

Physiology of Hatching

Hatching of eggs is a complex phenomenon that involves mechanical and chemical means in the formation of a hatching pore and emergence of the larva. The process of hatching commences with the out diffusion of egg fluid solute from the egg. As a result the egg becomes soft (less turgid). This decreases the pressure on the juvenile. The juvenile, however, becomes turgid and mobile. Water content of the body may increase from 10-20% to 40-50% (*Globodera*, *Heterodera* spp.). The important enzymes that have been detected in nematode eggs are leucine aminopeptidase, pseudocollagenase and lipase. Hatching process in *G. rostochiensis* has been studied in detail. The J_2 moves all around inside the egg to have a general exploration of the internal egg shell. Thereafter, it explores the shell at a few places by pressing the lip region to locate a weaker point to start perforation. The nematode thrusts the stylet at a selected point on the egg shell. Initially, the frequency of stylet thrusting is slower but it becomes faster with the formation of a finer slit. Concurrent to the thrusting, the pharyngeal gland secretions are introduced at the site of thrusting. The sensory organs located in the lip region are responsible for sensing and executing the task of pore formation. In polyphagous and non-host specific nematodes, e.g., *Caenorhabditis xenoplax*, *Tylenchorhynchus maximus*, etc., hatching occurs due to a mechanical action without the aid of enzymes.

In some phytonematodes hatching is stimulated with the root exudates. The root exudates of potato terminate the dormancy of *G. rostochiensis* and stimulate the hatching process. Some chemicals, such as allyl isothiocyanate from root exudates of mustard, neutralize the stimulatory effect of root exudates of potato on *Globodera*. Similarly, root exudates of tomato and tobacco stimulate the hatching in *Meloidogyne* eggs. Three chemicals, viz., glycinolepin A, glycinolepin B, and glycinolepin C, isolated from kidney bean roots have been found to be responsible for stimulation in the hatching of *Heterodera glycines*.

Dormancy and Quiescence

Under unfavourable environmental conditions metabolic activity in nematodes is decreased or suspended for varied durations depending on the nematode species and length of the condition. The dormancy may be metabolic or developmental. However, both are interlinked, when metabolic processes of quiescent stages of *Anguina tritici* (J_2), *Ditylenchus dipsaci* (J_4), and *Aphelenchoides besseyi* (adult) are suspended, development of the nematode also ceases and remains in the same stage during the period of quiescence. Several authors have used the term suspended animation for dormancy and quiescence that occurs under environmental stress. Metabolic dormancy implies to quiescence and includes retardation, suspension or temporary cessation of metabolic activities, development and body movement. The quiescence may be terminated with the termination of prevailing unfavourable environment. Hence, quiescence is induced by environmental extremes. Diapause is a similar term but it is not induced by any factor, rather it is a lalent period that occurs at a fixed stage in the life cycle of every organism, causing it to undergo physiological dormancy (diapause). Hence, it is a process which is related to ageing. The quiescence may be anhydrobiotic, osmobiotic, anoxybiotic, cryptobiotic and thermobiotic.

Anhydrobiosis: Although water is essential for the structural and functional maintenance of nematode body and living processes, at the same time water may cause severe injury to the body at extremely lower and higher temperatures. Some nematodes have the ability to survive lethal low temperatures. They lose body water, suspend metabolic processes and become dehydrated, coiled, shrivelled and clamped together. This state is referred to as quiescent state of anhydrobiosis. *Ditylenchus* spp. represent a typical example of anhydrobiosis. The J_4 larvae lose 90% or more of body water and can survive in this state for upto 2-3 years. Similarly, *Anguina tritici* (J_2) in anhydrobiotic state may survive upto 32 years. Fourth stage larvae of *Aphelenchoides besseyi* and *Aphelenchus avenae* also transform into quiescent state of anhydrobiosis during summer. The cyst wall of *Heterodera* spp. that becomes tanned protects the eggs inside, enabling them to survive upto five years. Some metabolic changes, however, may occur during anhydrobiosis which vary with the nematode. Generally, greater food reserves in the form of carbohydrates and lipids may prolong the survival in quiescence. Higher levels of inositol (greater than glycogen level) have been detected in quiescent *A. tritici* and *D. dipsaci*. These nematodes convert glycerol → glycogen → glucose-1-phosphate → uridine diphosphate glucose (UDPG) via UDP pyrophosphorylase → glucose → glucose-6-phosphate via glyoxylate cycle and glyconeogenesis (Fig. 51). Glycogen and lipids are utilized during induction of anhydrobiosis. In *Aphelenchus avenae* trehalose is synthesized which is reduced to half due to production of UDPG. In the quiescent state glycerol may substitute for water, whereas during revival (rehydration) the amount of lipids is reduced but glycogen is increased. Higher concentration of ATPs has been detected during quiescence.

Osmobiosis: Osmobiosis is an adapted state of nematodes to survive in hypertonic solution. Tolerance to higher osmotic potentials varies with the nematode type. Saprophytic nematodes may tolerate high osmotic potentials, whereas marine nematodes can survive in high salinity. Plant parasitic nematodes are relatively more sensitive but still they may survive in a solution of upto 10 atm, however, the pathogenicity may be affected. *Meloidogyne incognita* causes less damage to plants in saline soils. Some plant nematodes are capable of osmoregulations through the movement of ions. Juveniles and eggs of *Meloidogyne* spp. can tolerate higher osmotic potentials.

Anoxybiosis: Anoxybiosis is the term that refers to survival of nematodes in oxygen tension situations. An oxygen supply of 5-21% is best suited to plant parasitic nematodes, but they can also survive and perform normal life activities at a concentration less than 10%. Under low oxygen levels, lipid and glycogen are oxidatively and fermentatively catabolized, respectively. Sometimes a lower concentration of oxygen becomes essential for certain metabolic activities, e.g., *Aphelenchus avenae* efficiently utilizes glycogen under anaerobic condition, but is hardly able to utilize 8% glycogen under aerobic conditions. A number of nematodes, e.g., *Meloidogyne* spp., *A. avenae* and others, have shown to survive under anoxybiotic conditions.

Cryptobiosis: Some nematodes have tolerance against a wide range of low temperatures (0-5°C). *Meloidogyne javanica* or *Paratrichodorus minor*, are sensitve to low temperature and cannot survive at 5-10°C, whereas *Pratylenchus* and *Tylenchulus* species can survive at this temperature. Eggs are more tolerant than juveniles and can withstand -4°C. During cryptobiosis, metabolic activities are greatly reduced, solute concentration increases and water concentration decreases, as a result the cytoplasm becomes viscous, thus preventing freezing. This helps the nematode in resisting low temperatures.

Thermobiosis: Generally nematodes are more sensitive to higher temperatures than lower temperatures. There are, however, a few nematodes which show thermobiotic quiescence. Thermal sensitivity of nematodes increases further with the increase in soil moisture. *M. javanica* has shown survival in a wide range of low and high temperature (5-45°C). Tolerance to temperature may further increase if the nematode is inside the host tissue or eggs inside the egg mass. *Heterodera avenae* can survive at 45°C. Eggs inside the egg mass or cyst may survive upto 50°C temperature. Prolonged temperature of 45-50°C may cause sex reversal of females to males as occurs in *Meloidogyne*, *Heterodera*, etc.

6

Feeding and Trophic Relationship of Nematodes

FEEDING AND FEEDING BEHAVIOUR

Plant parasitic nematodes are obligate parasites requiring living tissue to feed on. The feeding is an important phase of life cycle of nematodes as it determines the nature and extent of damage to the plant as well as pace of reproduction and consequently the population density. Feeding is a complex phenomenon involving numerous steps such as movement, orientation, exploration, penetration, oesophageal gland secretions, and finally ingestion of cell contents.

Movement

Parasitic nematodes by themselves move very slowly in their free-living stage while away from the host in soil. They hardly move one or two meters in a year. Hence, the importance of larval movement is in the localized activities such as locating and exploring the host surface. Two types of movements have been recorded in nematodes (Fig. 58). Undulatory or serpentine movements (snake-like) in a sinusidal path is commonest and is found in all nematodes except a few in which the larvae show telescopic or earthworm-like movement.

Undulatory movement: The larva forms bending waves which pass in anterio-posterior direction of the body. The spindle-shaped muscles beneath the hypodermis extend only anterio-posteriorly. The bending waves are formed by the coordinated contraction and expansion of the longitudinal muscles on opposite sides of the nematode to move on its sides dorso-ventrally in a sinusidal path. During the movement, one lateral side lies on the surface of the plane whereas the dorsal and ventral sides lie on the left and/or right sides (Fig. 58).

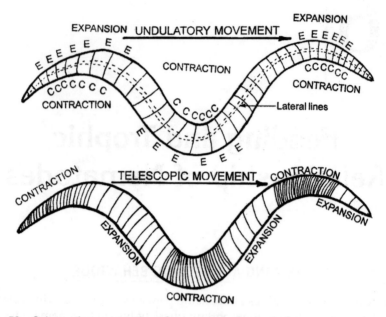

Fig. 58 Schematic presentation of movement of nematodes

Telescopic movement: This movement involves alternate lengthening and shortening or expansion and contraction of the body, just like an earthworm moves (Fig. 58). This type of motion is common in those nematodes which lack lateral lines, e.g., the family, Criconematidae (*Criconema, Ogma*, etc). These nematodes have strong annulations on the cuticle and are sometimes provided with backward directed cuticular outgrowths (annules) (Fig. 30). These annules ensure better grip and prevent backward slipping of the body during forward movement. The telescopic movement differs from undulatory movement in three respects. In the telescopic movement contraction of muscles starts from posterior to anterior end; contraction of longitudinal muscles may occur simultaneously in any section of the body; and length of the body changes during the movement.

Orientation

Soil factors, viz., moisture, texture, temperature, water content and chemical composition, influence the direction of movement of phytonematodes. Generally, nematodes move slowly and randomly, but they show active and oriented movement while in close proximity of roots of susceptible plants. The root exudates form a chemical gradient between the nematode and host which enables the larva to move in the direction of roots (Bird, 2004). Exudates from the roots of potato and tomato attract *Globodera rostochiensis* and *G. pallida,* and *M. incognita* and *M. javanica,* respectively. The larvae of *Pratylenchus penetrans* cover a distance of 15-20 cm in just 5 minutes when they receive chemical stimuli from root exudates of white clover. Phytohormones, e.g., auxin secreted by roots, binds to amphids, phasmids and other sensory structures of nematodes while in close vicinity of roots. This may allow the nematodes to sense and follow an auxin gradient through sensory receptors during oriented movement (Curtis, 2007a).

Exploration

Nematodes first explore the host upon reaching its surface. They move over it, turn their head frequently towards the surface and then probe it with the stylet. The probing tends to increase but locomotion decreases and finally stops. The main objective of head movement is to locate a weaker spot or point which could give better chemical emanation (stimuli) to be detected by the sensory organs. The nematode then presses the lip region against the host surface, apparently to obtain better chemical stimuli to start penetration and feeding. The stylet is thrust, with increasing vigour at 2 or 3 selected points. The stylet is repositioned repeatedly on these points to finally select a most suitable point which is thrusted continuously until a perforation is achieved.

Perforation and Penetration

Nematodes generally make a perforation on the host surface by physical and mechanical pressure exerted through the stylet thrustings (Fig. 59). Orientation of larval head to the host surface is important in providing necessary physical pressure for perforation. The larva arches its body to bring the head and stylet at right angles to the root surface at the time of thrusting. When a nematode strikes the host surface with its stylet, the host cell surface as a normal reaction exerts equal but opposite pressure. The opposite reaction helps the nematode to assess strength and hardiness of the host surface. The nematode then thrusts harder at the same point or repositions the stylet to thrust nearer the last one and keeps piercing continuous. With the movement of stylet, the larva is able to locate the periphery of a perforation if achieved, and thrusts the stylet repeatedly to make the perforation of sufficient width. Thereafter, the tip of the stylet is repeatedly inserted to inject saliva. The head is moved from side to side to give a rasping action (scrap roughly) to the stylet. The nematode then starts deeper thrusts. The frequency of stylet thrust may vary with the nematode species. Members of the genus, *Pratylenchus*

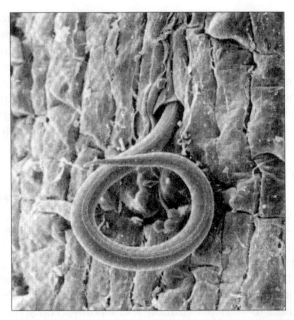

Fig. 59 A juvenile of root knot nematode penetrating the host root (Courtesy: USDA)

or *Tylenchus* make 2-6 thrusts per second, whereas *Paratylenchus* may thrust once in 4-8 seconds.

Salivation (Oesophageal Gland Secretions)

After making the perforation or penetration, nematodes secrete saliva from the oesophageal glands into the host cell. The secretions from oesophageal glands contain several enzymes and other substances which induce dramatic alterations at cellular and subcellular level in the host to satisfy nutritional requirement of the nematode (Vanholme *et al.*, 2004). The saliva passes from the oesophageal bulb to the stylet tip, and from the tip to host cell cytoplasm. This passage of saliva occurs in majority of plant parasitic nematodes except members of the family Trichodoridae. In *Trichodorus, Paratrichodorus*, etc., the saliva passes through a temporary feeding tube formed in the space between the lip region and host surface. The injected saliva has two functions: (i) it performs extracorporeal digestion of cell contents through enzymes, and (ii) dilutes cell contents, which help in easy ingestion of cell contents by the nematode. Enzymes secreted by phytonematodes have been enlisted in Table 14.

Table 14 Some important enzymes secreted by plant nematodes and involved in the parasitism

Enzymes	Plant Parasitic Nematode
Acid phosphatase	*Ditylenchus triformis, Meloidogyne* spp., *Tylenchulus semipenetrans*
Alkaline phosphatase	*Meloidogyne* species
Amylase	*Aphelenchoides sacchari, T. semipenetrans, D. dipsaci, D. destructor, D. triformis, Heterodera schachtii, Globodera rostochiensis, M. javanica*
β-galactosidase	*Globodera rostochiensis*
β-glucosidase	*Aphelenchoides fragariae, P. penetrans, Globodera rostochiensis*
Cellulase	*Aphelenchoides avenae, A. sacchari, D. dipsaci, D. myceliophagus, D. triformis, Radopholus similis, Helicotylenchus nannus, Pratylenchus penetrans, P. zeae, Heterodera trifolii, H. schachtii, M. arenaria, M. incognita, T. semipenetrans*
Chitinase	*D. dipsaci, D. myceliophagus, D. destructor*
Invertase	*A. sacchari, M. arenaria, R. similis, P. penetrans, P. redivivus*
Nonspecific esterases	*M. hapla, M. javanica*
Pectinase	*D. dipsaci, D. destructor, R. similis, P. penetrans, M. arenaria, M. hapla*
Pectinmethylesterase	*D. triformis, D. myceliophagus, D. dipsaci* (alfalfa race)
Polygalacturonase	*A. avenae, P. zeae, D. dipsaci, D. myceliophagus, D. destructor*
Proteinase	*G. rostochiensis, H. schachtii, D. dipsaci, D. destructor, D. allii, D. triformis, A. ritzemabosi, P. redivivus*

Ingestion and Duration of Feeding

The median bulb acts as a pumping machine during food ingestion. The bulb dilates (becomes larger/wider) and contracts several times in a second. Following the saliva injection in host cell, the median bulb of oesophagus pulsates and as a result the cell contents are ingested. In some nematodes, the pulsation of bulb continues during salivation. Phytonematodes generally ingest cytoplasm and cell sap along with fine cell organelles, except intact nucleus. Ectoparasitic nematodes are poorly evolved in parasitism as one female feeds on several hundred cells to produce just a few eggs, whereas sedentary endoparasites feed only on a few cells (nurse cells) and produce hundreds of eggs.

Duration of feeding varies with nematode species and depends on the mode of parasitism. Migratory ectoparasites, e.g., *Tylenchus, Cephalenchus, Tylenchorhynchus, Trichodorus*, etc., may feed for a few seconds at a site. During this short period, the nematode performs all acts of feeding. *Xiphinema* feeds for several minutes to hours at one site on the terminal galls. *Longidorus* feeds for several hours or even days at one site. Migratory endoparasite, *Radopholus similis*, feeds on several cells at one time by causing dissolution of the cell walls. Sedentary ectoparasites (*Cacopaurus pestis* and *Hemicycliophora arenaria*) or sedentary endoparasites (*Meloidogyne, Heterodera, Nacobbus*, etc.) are known to feed at one site throughout their life.

MODE OF PARASITISM

All plant parasitic nematodes, whether root, stem or leaf parasites, have a sharp and pointed stylet at their anterior end, which they use to puncture the epidermis and cell wall or to displace the cells while moving intercellularly. During feeding and causing mechanical damage to plant tissue, nematodes secrete saliva from the oesophageal glands which induce cell wall dissolution and/or extracorporeal digestion. In addition, enzymes induce metabolic changes in the host plant. Most common enzymes in nematodes are cellulase, protease and amylase. Enzyme activity in nematodes is influenced by parasitic habits. For instance, in migratory endoparasites, e.g., *Pratylenchus penetrans*, the cellulase activity has been recorded seven times greater than the sedentary endoparasite, *Heterodera trifolii*. Plant parasitism by nematodes is confined to three orders: Dorylaimida (Class: Adenophorea), Tylenchida, and Aphelenchida (Class: Secernentea).

Parasitism by the Order Dorylaimida

The phytonematodes of this group are migratory ectoparasites of underground parts. They either browse (nibble) the root surface, e.g., *Trichodorus* and *Paratrichodorus*, or feed for longer periods at specific sites from deeper root tissues, e.g., *Longidorus* and *Xiphinema*. In addition to direct damage, these nematodes vector several important plant viruses. For example, *X. index* and *L. elongatus* transmit arabis mosaic virus and tomato black ring virus (nepoviruses), respectively, whereas *Trichodorus* and *Paratrichodorus* spp. vector tobra viruses (tobacco rattle virus, pea early browning virus and pepper ring spot virus). Two families, viz., Trichodoridae and Longidoridae of the order Dorylaimida, contain major plant parasitic nematodes which differ considerably in the parasitism.

Parasitism by the Family Trichodoridae

These nematodes are also known as stubby root nematodes. They usually feed gregariously near the tips of rapidly growing roots, on epidermal

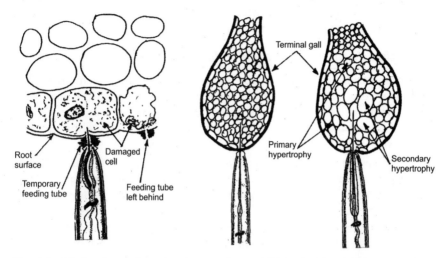

Fig. 60 Mode of parasitism by the members of Trichodoridae and Longidoridae

cells and root hairs, and occasionally on subepidermal cells (Duncan, 2005). Due to massive feeding on epidermal cells the root growth retards, nematodes then move to the apical region of the root and feed there in a similar manner. When the apical region of actively growing roots has lost attraction, nematodes move away in search of new feeding sites and attack emerging lateral roots. As a result the longitudinal growth of lateral roots ceases and they become stubby (short and thick) (Fig. 60).

In all trichodorid nematodes, the stylet is a ventrally curved mural tooth with a solid tip that ensures an effective piercing of cell wall rather than ingesting the cell contents. The nutrients are ingested from the cell through a self-made feeding tube composed of hardened saliva. This feeding tube is left on the host cell after each feeding cycle which takes only a few minutes.

During the initial phase of parasitism the nematode makes a perforation in the cell wall by thrusting the stylet rapidly and repeatedly. When a perforation is achieved, the nematode first releases saliva and inserts the tip of the stylet repeatedly. The saliva between the anterior end of the nematode and host cell surface around the perforation gets hardened to form a feeding tube along the stylet (Hunt *et al.*, 2005). Further injection of saliva into the cell leads to accumulation of cytoplasm at the perforation site. The nucleus is also drawn along with the cytoplasm. When sufficient cytoplasm has been accumulated at the perforation site, the nematode starts making deep thrusts by the stylet. The tonoplast (the inner plasma membrane, bordering the vacuoles) ruptures, and as a result cytoplasm and cell sap get mixed together. This liquefaction of cell contents facilitates ingestion through the temporary feeding tube.

Parasitism by the Family Longidoridae

The longidorids are largest in size among plant parasitic nematodes and possess a very long, hollow and axial stylet measuring 100-300 µm. Thus they can feed deep within the plant root. Parasitism by the genera, *Xiphinema* and *Longidorus* may represent the entire family, although they differ considerably.

Xiphinema index and *X. diversicaudatum* feed preferentially on root tips and may cause terminal galls (Fig. 60). These galls appear to be attractive for the female nematodes as they provide greater nutrients essential for egg production. The galls are formed due to excessive cell division (hyperplasia) of root tip cells. These cells are metabolically more active and seem to be indispensable for reproduction of the nematode. *Xiphinema* species feed ectoparasitically for many hours even for a day on root cells. Species of *Longidorus* attack only root tips and induce terminal galls as by *X. index*. *Longidorus* species insert the stylet nearly fully into the root tip before they start salivation and feeding for several hours. The root tip cells first undergo hypertrophy at the initial stage of gall formation, followed by hyperplasia and then secondary hypertrophy at the actual feeding site. However, the root tip modifications induced by these species are not uniform for the whole genus. Some species induce hyperplasia in the cells of root tips as by *Xiphinema* spp.

Parasitism by the Order Tylenchida

Members of the order Tylenchida show great diversity in parasitic behaviour and adaptation that culminate in sedentary endoparasites such as *Meloidogyne, Heterodera, Globodera, Nacobbus*, etc., of underground parts or *Anguina* and *Ditylenchus* species of aboveground parts (Siddiqi, 2002). In general, tylenchs are most advanced or evolved nematodes, particularly sedentary endoparasites, compared with other nematodes. Parasitism by Tylenchida can be discussed under parasites of underground parts and parasites of aboveground parts.

Parasites of Underground Parts

Ectoparasites

Migratory ectoparasites: The nematodes belonging to this category remain vermiform throughout the life, e.g., *Tylenchus, Paratylenchus, Belonolaimus, Rotylenchus, Tylenchorhynchus, Helicotylenchus,* etc. The genus, *Tylenchus* and some species of *Tylenchorhynchus,* are surface feeders (Fig. 61). They have a short stylet and feed on epidermal cells for a

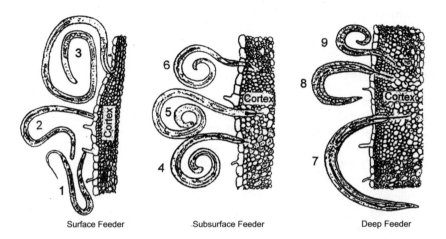

Fig. 61 Diagrammatic representation of ectoparasitism by different tylenchid nematodes: *Cephalenchus* (1), *Tylenchorhynchus* (2), *Belonolaimus* (3), *Rotylenchus* (4), *Hoplolaimus* (5), *Helicotylenchus* (6), *Hemicycliophora* (7), *Macroposthonia* (8), and *Paratylenchus* (9).

few minutes or longer. Whereas, *Paratylenchus* species are subsurface feeders and feed for many hours or even days at the same site on root hairs, epidermal cells or cortical cells (few cell depth) without inflicting any visible damage or symptom. *Belonolaimus* and *Hemicycliophora* species are categorized as deep feeders (Fig. 61). These nematodes possess a long stylet which helps them to withdraw the nutrients from root cortex to several cells depth. *Hemicycliophora* species feed on root tips causing terminal galls.

Sedentary ectoparasites: A few nematode species such as *Cacopaurus pestis* (parasite of walnut), *Hemicycliophora arenaria* and *Paratylenchus* (*Gracilacus*) *epacris* show sedentary ectoparasitism (Siddiqi, 2000). These

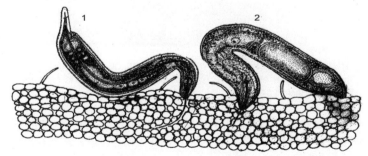

Fig. 62 Sedentary ectoparasitism shown by *Paratylenchus* (*Gracilacus*) *epacris* (1), and *Cacopaurus pestis* (2).

nematodes possess a long stylet and feed from deeper root cells. The adult female becomes slightly swollen and remains attached to the feeding site (Figs. 62-64).

Endoparasites

Migratory endoparasites: The nematodes belonging to this category remain vermiform during the entire life span and may invade roots or leaves at any life stage (Bajaj, 2002). They usually feed on cortical cells. Due to intercellular migration of the nematode in root cortex, the cells are destroyed. This type of parasitism involves a considerable role of enzymes (Table 14) and is common in the members of families Pratylenchidae and Hoplolaimidae (Geraert, 2006). The necrotic lesions on the root caused by *Pratylenchus* spp. develop due to mechanical and chemical damage to cortical cells (Fig. 63). The saliva of *P. penetrans* contains the enzyme β-glucosidase which hydrolyzes amygdlin in root cells and causes the release of benzaldehyde and hydrogen cyanide. These chemicals, especially hydrogen cyanide, cause necrosis to the host cells and tissue.

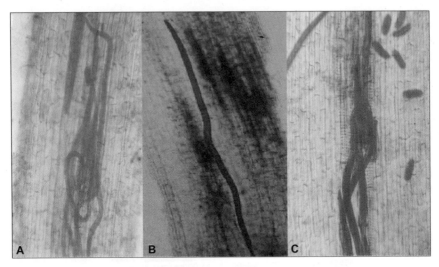

Fig. 63 Pictures showing migratory endoparasitic nematodes inside the host tissue: *Pratylenchus penetrans* (A), *Radopholus similis* (B), and *Hirschmanniella* sp. (C) (Courtesy: IMP, USDA)

The burrowing nematode, *Radopholus similis* penetrates the root epidermis and moves through the cortex (Fig. 63). During this process, the nematode releases cellulase and pectinase enzymes that cause dissolution of the cell wall and middle lamella. The migration and feeding by the nematode cause formation of cavities and tunnels in the cortex which appear in the form of necrosis or decay on the root surface. *Hirschmanniella* spp. also migrate in cortex and cause necrosis in roots. *Hoplolaimus* feeds on phloem parenchyma, and the migration and feeding cause the roots to become flaccid (soft and weak).

Sedentary endoparasites: The nematodes showing sedentary endoparasitism are considered highly evolved root parasites. They invade roots at a definite life stage. For example, second stage juveniles of *Meloidogyne, Heterodera,* and *Nacobbus* fully penetrate the roots. After penetration, the larvae migrate to specific sites where they induce the formation of a highly specialized nurse cell system (Curtis, 2007b) to regulate continuous supply of nutrients until the nematode completes its life cycle. Female nematodes parasitize the host and become saccate, lemon or pear shaped, and remain completely inside the host tissue (Figs. 64, 70, 72).

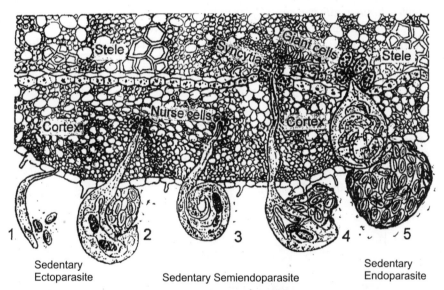

Fig. 64 Sedentary parasitism shown by plant nematodes. 1. *Cacopaurus pestis* (sedentary ectoparasite), 2. *Tylenchulus semipenetrans,* 3. *Trophotylenchus* sp., 4. *Rotylenchulus reniformis* (semiendoparasite), and 5. *Meloidogyne* sp. (sedentary endoparasite).

Second stage juveniles of *Meloidogyne,* after penetrating the root, move intercellularly and establish themselves at the permanent feeding sites, orienting the head towards developing stelar region in the zone of cell elongation with their bodies lying mostly in the cortex. The larvae feed preferentially on primary phloem or adjacent parenchyma. Secretions from the oesophageal glands into these cells lead to the formation of 3-6 giant cells to meet the increased and regular supply of nutrients to the developing nematode larva (Karssen, 2006). The giant cells are formed through repeated endomitosis without cytokinesis (Fig. 71). Concurrent with the establishment of giant cells, root cells around the nematode and its feeding site undergo hyperplasia and hypertrophy, giving rise to root swelling and eventually the gall is formed (Huang, 1985). A similar pattern of elaborate nurse cell system is formed by *Heterodera* and *Globodera* species (Endo, 1986). In place of giant cells, these nematodes produce 3-6 syncytia which are formed due to the dissolution of cell walls and subsequent cytoplasmic fusion of neighbouring cells (Fig. 73). Syncytia, structurally, functionally and chemically are similar to giant cells. Cyst nematodes do not form root galls.

Semiendoparasites: There are some nematode species which partly invade the root and become sedentary after locating a suitable feeding site. The posterior half of the body protrudes outside the root in soil, e.g., *Rotylenchulus reniformis* and *Tylenchulus semipenetrans* (Bajaj, 2002). The J_2 of *T. semipenetrans* feeds on epidermal cells, while the young females (vermiform) partially penetrate deeper in cortex. The posterior half, however, remains outside the root tissue. A trophic nurse cell system is formed around the female head to provide continuous supply of nutrients (Duncan, 2005). At maturation the anterior part of the female becomes irregularly slender and the posterior half in the soil assumes obesity. Similarly, the young female of *R. reniformis* (vermiform) penetrates the root till its head approaches the stelar region, and induces the formation of specialized nurse cells in the pericycle region. The female becomes sedentary and the posterior half outside the host in the soil assumes kidney shape (Fig. 65).

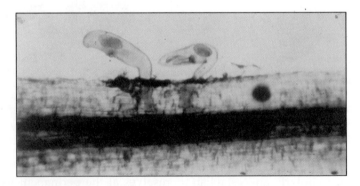

Fig. 65 Adult females of *Rotylenchulus reniformis* showing semiendoparasitism (Courtesy: IMP, USDA).

Parasites of Aboveground Parts

Nematodes which attack stems, leaves, buds and flowers are active under cool and humid condition. A film of water on the plant surface enables the nematodes to move on the surface. Majority of the foliar nematodes belong to the order Aphelenchida. However, *Anguina* and *Ditylenchus* species from the order Tylenchida are classical examples of foliar parasitism. Stem and bulb nematodes, *Ditylenchus* spp. feed endoparasitically on parenchymatous tissue of stems, bulbs or leaves. Due to enzymatic dissolution of middle lamella, cells are macerated leading to softening and swelling of the infected part (Southey, 1993). The unique feature in the life cycle of *Ditylenchus* spp. is the ability of the fourth stage (pre-adult) larvae to transform into a quiescent or resistant state to survive under dry and desiccated environment (Gaur, 2002).

The genus *Anguina* includes the species which are seed borne and are fully migratory ectoparasites during vegetative growth of the host plant, but become endoparasitic during reproductive phase of the plant growth (Agrios, 2005). Second stage juveniles in quiescent state of anhydrobiosis may survive inside seeds for as many as 32 years. Soil moisture terminates the quiescent state and the resulting J_2 feeds ectoparasitically on growing points between the compact leaf sheath. When flower primordia are initiated, the larvae invade them and feed endoparasitically (McDonald and Nicol, 2005). As a result ear cockles or seed galls are formed in place of healthy grains. Inside a gall thousands of nematode larvae in quiescent state may survive under adverse conditions (Fig. 66).

Fig. 66 Second stage larvae of *Anguina tritici* inside a seed gall (cockle)

The leaf and bud nematodes, *Aphelenchoides* spp. are mycophagous in nature, and feed ecto and endoparasitically on leaves and buds (Hunt, 1993). *A. besseyi* is a seed borne ectoparasite, its adult females in quiescent state of anhydrobiosis survive on seeds between glumes and grains (Prasad, 2002). Nematodes are more frequently present on undamaged and healthy seeds. The quiescent state terminates when infested seeds come in contact with water and the nematodes move on the plant surface through a film of water to feed upon leaf tips and margins (Bridge *et al.*, 2005a). At the time of flowering, they migrate to the apical portion and feed on ovaries, stamens, lodicules and embryo. *A. ritzemabosi* and *A. fragariae* are migratory endoparasites but also feed ectoparasitically (Southey, 1993). The nematodes move on the plant surface and enter the leaves through stomata and feed on mesophyll cells or ectoparasitically in bud axil. During the movement and feeding the larvae secrete various enzymes (Table 14). At the advent of adverse conditions, predominently adult females enter into a quiescent state of cryptobiosis to survive under extremely low temperature. Some other important nematodes of the order Aphelenchida are fig nematode (*Schistonchus* spp.), eucalyptus nematode (*Fergusobia* spp.), pine wilt nematode (*Bursaphalenchus* spp.) and red-ring nematode (*Radinaphelenchus* spp.). All these nematodes are transmitted by insect vectors. *S. caprifici* and *F. tumifaciens* incite galls in the leaves/flowers of fig and eucaplyptus with the aid of insect vectors, *Blastophaga psenes* and *Fergusonina* sp., respectively. *B. xylophilus* and *R. cocophilus* after being transmitted by *Monochamus* spp. and *Rhynchophorus palmarum*, feed on stem parenchyma and cause wilt in pine and red-ring in coconut, respectively.

HOST PARASITE RELATIONSHIP

The host parasite relationship may include the events that take place during parasitization of host plant by a nematode and also the changes/alterations which occur in the host and parasite. In a simplest way, host parasite relationship can be defined as the modifications (morphological, anatomical and physiological) that occur in the host plant and nematode as a result of infection by the latter. The modification may be largely morphological, anatomical and/or physiological. Although all plant parasitic nematodes modify the host as a substrate, some nematodes induce conspicuous modifications. This chapter gives detailed information on those nematodes where a modification is pronounced and highly demarkable. Broadly, the host modification may be of two types: morphological and physiological, which are discussed hereunder.

Morphological Modifications

Host Parasite Relationship of Dorylaim Nematodes

Major plant nematodes of the order Dorylaimida belong to two families, viz., Trichodoridae and Longidoridae, which are migratory ectoparasites of annual and perennial crops. Plant nematodes from the family Trichodoridae are commonly known as stubby root nematodes. During the initial phase of parasitism they aggregate near the tips of rapidly growing roots where they feed gregariously on epidermal cells and root hairs followed by the feeding on apical root portion and finally on emerging lateral roots (Karanastasi et al., 2001). As a result, longitudinal growth of the lateral roots ceases and they become stubby in appearance (short and thick, Fig. 67). The stubby roots lack root hairs and become almost nonfunctional with regard to absorption of water and minerals.

Fig. 67 Stubby root growth of citrus caused by *Trichodorus* species (Courtesy: IMP, USDA).

The longidorid nematodes being largest in size among plant parasitic nematodes possess a long and hollow stylet that enables them to feed upto several cells depth in the cortex. *Xiphinema* spp. feed ectoparasitically on root tips for many hours and cause terminal galls (Fig. 60, 68). The galls are formed as a result of hyperplasia (excessive cell division). The galls are metabolically highly active and have excess inflow of nutrients from the adjoining tissue and roots to provide greater and continuous supply of nutrients to the female nematode. Species of *Longidorus* feed on root tips only, and the duration of feeding may last for several hours, some times for days. These nematodes also induce terminal galls as by *Xiphinema* spp. In the process of gall formation, root tip cells first undergo hypertrophy followed by hyperplasia and then secondary hypertrophy at the actual feeding site (Fig. 60, 68).

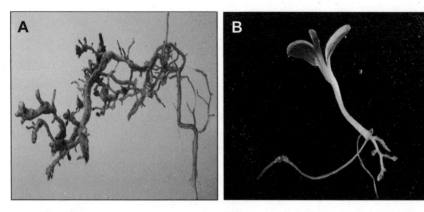

Fig. 68 Terminal galls as a result of parasitism by *Xiphinema diversicaudatum* on rose (A), and *Longidorus africans* on lettuce (B) (Courtesy: www.ucdnema).

Host Parasite Relationship of Tylenchid Nematodes

Generally the migratory ectoparasitic nematodes do not cause any specific change in the host. However, semiendo and endoparasites cause conspicuous alteration in the host.

Semiendoparasitic nematodes

Some nematodes, e.g., *Tylenchulus semipenetrans, Rotylenchulus reniformis, Trophotylenchus, Sphaerotylenchus*, etc., partly invade the roots and become sedentary after locating a suitable feeding site (Bajaj, 2002). The anterior half of the body enters into the cortex whereas the posterior half remains outside the root tissue in soil (Fig. 69). Soon after establishment, a highly sophisticated system of trophic nurse cells is initiated around the female head, consisting of 6-8 cortical cells (*T. semipenetrans*), or 3-8 pericycle cells (*R. reniformis*). These cells are devoid of vacuoles and have thickened walls, enlarged nucleus and nucleolus. At maturity the anterior part of the female becomes irregularly slender and the posterior half in soil becomes obese (swollen) with thick cuticle. The female lays eggs in a gelatinous matrix which is secreted by specialized rectal cells. The infected root part appears bulgy and dirty due to protruding posterior part of the nematode, associated egg mass and the adhering soil particles, organic matter, etc.

Migratory endoparasitic nematodes

These nematodes remain vermiform during the entire life span and may invade roots or leaves at any life stage, e.g., *Pratylenchus, Radopholus,*

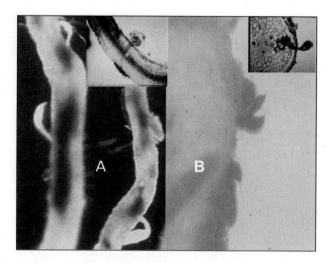

Fig. 69 Adult females of *Rotylenchulus reniformis* (A), and *Tylenchulus semipenetrans* (B), infecting plant roots and inset longitudinal sections of respective root tissues (Courtesy: E.G. McGawley).

Hirschmanniella, etc. They usually feed on cortical cells. Due to intercellular migration and feeding of *P. penetrans* in root cortex, the cells are destroyed and necrotic lesions appear on the root surface (Karssen *et al.,* 2001) (Fig. 70). The necrosis develops due to mechanical and chemical damage to cortical cells. The saliva of *P. penetrans* contains the enzyme β-glucosidase which hydrolyzes amygdlin in root cells and causes the release of benzaldehyde and hydrogen cyanide. These chemicals, especially hydrogen cyanide, cause necrosis of tissue.

Plant lectins may help the burrowing nematode, *R. similis* to locate the host root (Wuyts *et al.,* 2003). The nematode penetrates the root epidermis and moves through the cortex (Fig. 63). During this process, the nematode releases cellulase and pectinase enzymes that cause dissolution of the cell wall and middle lamella. Cavities and tunnels are formed in the cortex due to migration and feeding of the nematode and appear on root surface as necrosis or decay. *Hirschmanniella* spp. also migrate in the cortex and cause root necrosis. Some species of *Hoplolaimus* (*H. indicus*) are endoparasites and feed on the phloem parenchyma, as a result the roots become flaccid (soft and weak).

Sedentary endoparasitic nematodes

The nematodes of this category possess highly evolved parasitism and cause intense modification in the host tissue. Second stage juveniles of *Meloidogyne, Heterodera,* and *Nacobbus* invade roots at a definite life stage.

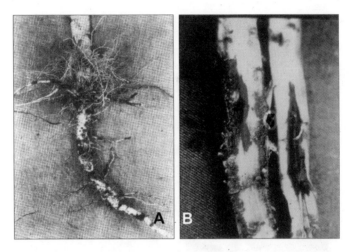

Fig. 70 Root necrosis on tobacco (A), and banana (B), caused by *Pratylenchus penetrans* and *Radopholus similis*, respectively.

After penetration, the larvae migrate intercellularly to reach the stelar tissue where they induce the formation of a highly specialized nurse cell system to regulate continuous supply of nutrients until the nematode completes its life cycle (Endo, 1987). Female nematodes parasitize the host and become saccate, lemon or pear shaped, and remain completely inside the host tissue (Figs. 26, 71, 73). These nematodes develop a specialized and complex relationship with the host. As a perfect example of highly adapted root parasitism, sedentary endoparasitic nematodes enjoy a guaranteed and continuous supply of food and water from the nurse cells and protection within the root tissue (Khan *et al.*, 2005).

The giant cells (nurse cells) are highly specialized cellular adaptations induced and maintained by the developing *Meloidogyne* female in susceptible plants (Castillo *et al.*, 2001). They are formed around the head region of the female. Formation of giant cells is essential for successful parasitism. In case a juvenile fails to induce formation of giant cells in the root, it will either starve or migrate out. The roots containing giant cells shift their metabolism in the direction of increased synthesis of proteins, i.e., towards the infected root parts, and reduce transport of nutrients to shoot (DiVito *et al.*, 2004). The giant cells are formed through repeated endomitosis without cytokinesis (Fig. 71). They are 3-6 (2-12) in number and many times bigger than a normal cell. They are multinucleate, and the size and shape of nuclei vary greatly. Giant cells are rich in cytoplasm, and Golgi apparatus, mitochondria, ribosomes, polysomes and endoplasmic reticulum are abundant. The central vacu-

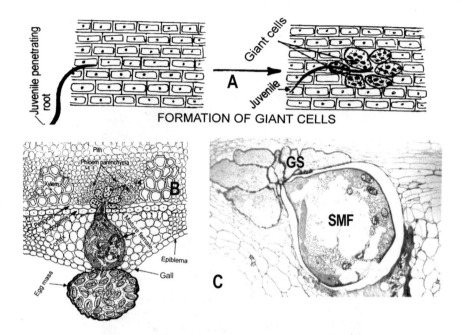

Fig. 71 Schematic presentation of formation of giant cells (A), and host parasite relationship (B), of root-knot nematodes. C- Transverse section of galled tissue showing saccate *Meloidogyne* female (SMF), and giant cells (GS), (C, Courtesy: IMP, USDA).

ole disappears but small vacuoles remain in the cytoplasm. Wall ingrowths traverse the giant cell cytoplasm that greatly increases the surface area of plasmalemma (Huang, 1985). These structures facilitate the flow of carbohydrates from the surrounding tissues to giant cells to meet the nutritional requirements of the female nematode.

Concurrent with the establishment of giant cells, root tissue around the nematode and its feeding site undergoes hyperplasia (repeated mitotic divisions) and hypertrophy (cell enlargement) giving rise to swellings, and eventually root galls are formed (Fig. 72). Stelar parenchyma and pericycle cells undergo hyperplasia while the surrounding cortical cells become hypertrophied (Khan, 1997). Galls usually develop one-two days after juvenile penetration. Galling is induced by the growth regulators released from the subventral glands of the second stage juveniles or when tryptophan is released upon hydrolysis of plant proteins by the juveniles (Hussey, 1987). The tryptophan reacts with endogenous phenolic acids to yield auxins, which incite hyperplasia and hypertrophy in the root tissue (Huang, 1985).

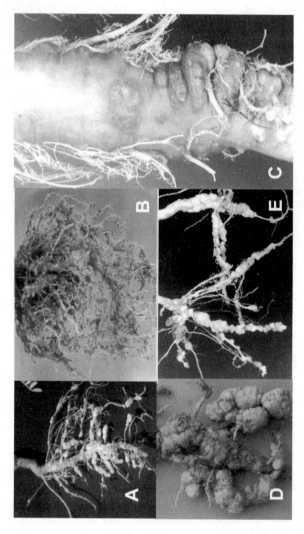

Fig. 72 Galls caused by root-knot nematodes on chickpea (A), tomato (B), radish (C), bottle gourd (D), and pumpkin (E).

A similar pattern of elaborate nurse cell system consisting of 3-6 enlarged cells around the head region is formed by *Heterodera* and *Globodera* species (Endo, 1986). These cells are called syncytia instead of giant cells because they are formed through the dissolution of cell walls and subsequent cytoplasmic fusion of neighbouring cells (Figs. 73 A-C).

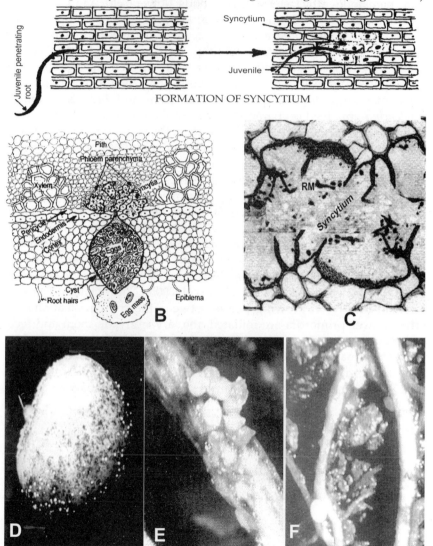

Fig. 73 Schematic presentation of formation of syncytium (A), and modifications in host plant (root) and pathogen (*Heterodera, Globodera*, etc.) (B). Root section showing remnants (RM) of cell wall in a multinucleated syncytium (C). Cysts of *Globodera* species attached to the potato tuber (D) and roots (E, F).

Syncytia, structurally, functionally and chemically are similar to giant cells (Patel and Patel, 2002). Cyst nematodes do not form root galls (Fig. 73 D-F).

Foliar nematodes

Some nematodes attack stems, leaves, buds, flowers, etc., and induce certain modifications in the host tissue to get food and protection. These nematodes are largely migratory endoparasites and belong to two orders, Tylenchida and Aphelenchida (Siddiqi, 2002). Foliar tylenchid nematodes are *Ditylenchus* and *Anguina* spp., which feed ecto and endoparasitically. A film of water is necessary for the movement of both the nematode genera on plant surface. Stem and bulb nematodes, *Ditylenchus* spp. feed on parenchymatous cells of stems, bulbs or leaves and on leaf margins. During the parasitism, the nematode larvae secrete pectic enzymes which cause dissolution of middle lamella (Southey, 1993). As a result, the host cells are macerated leading to softening and swelling of infected parts of stems or leaves. Under environmental extremes, fourth stage larvae of *Ditylenchus* spp. except *D. myceliophagus* transform into a quiescent or resistant state, which is commonly called "eelworm wool" or "nematode wool" to survive in dry and desiccated condition for many years (Jairajpuri, 2002). *Anguina* spp. commonly known as seed gall nematode also enters into a quiescent state of anhydrobiosis (second stage juveniles) and may survive inside seeds for as many as 32 years (McDonald and Nicol, 2005). The juveniles (J_2) feeds ectoparasitically on growing points between the compact leaf sheath. When flower primordia are initiated, the larvae invade them and feed endoparasitically. As a result ear cockles or seed galls are formed in place of normal grains. The galls are small, pronounced, localized swellings which may also develop on leaves or stems in addition to grains (Fig. 74). The gall is a highly organized structure and primarily provides

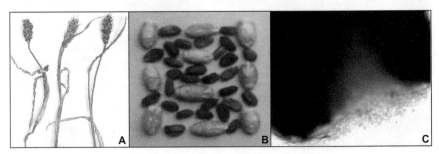

Fig. 74 Earheads of wheat infected with *Anguina tritici* (A), seed galls along with healthy seeds of wheat (B), and the second stage larvae errupted from a ruptured seed gall (C).

Fig. 75 Yellowing and whitening of leaf tip of rice caused by *Aphelenchoides besseyi* (A), leaf browning and necrosis caused by *A. ritzemabosi* on chrysanthemum (B), and *A. fragariae* on strawberry (C).

protection to nematodes against desiccation. A gall may contain thousands of second stage larvae in quiescent state which become active when they come in contact with moisture (Khan and Athar, 1996).

The foliar nematodes belonging to the order Aphelenchida are leaf and bud nematodes, *Aphelenchoides* spp. and also *Aphelenchus* spp. which feed ecto and endoparasitically on leaves and buds. All species need a film of water to move upward on plant surface. *A. besseyi* is strictly a seed borne ectoparasite, its adult females in quiescent state of anhydrobiosis survive on seeds between glume and grain (Bridge et al., 2005b). The nematodes feed on the tips of folded leaves and cause white tip of rice (Fig. 75A). At the time of flowering, they migrate to the apical portion and feed on ovaries, stamens, lodicules and embryo. *A. ritzemabosi* and *A. fragariae* are migratory endoparasites but they also feed ectoparasitically. The nematodes enter the leaves through stomata and feed on mesophyll cells (Southey, 1993). The nematode activity inside the leaf causes characteristic interveinal discolouration or necrosis (Fig. 75 B-C). The larvae also feed ectoparasitically in bud axil. When the condition of the leaf or its part becomes poor, nematodes come out through the stomata and enter into a fresh leaf or its part in the same manner. The nematode also attacks flower buds. *Aphelenchoides* species (predominently adult females) except *A. ritzemabosi* has the ability to enter into the quiescent state of cryptobiosis to survive under extreme low temperature (Brown et al., 1993). *Radinaphelenchus cocophilus* (red-ring nematode) and *Bursaphelenchus xylophilus* (pine wilt or timber nematode) are two most important foliar nematodes, and cause red-ring in coconut and wilt in conifers, respectively (Fig. 76). Both the nematodes need in-

Fig. 76 Red ring of coconut caused by *Radinaphelenchus cocophilus* (A), L.S. of diseased trunk (B), vector insect (C), (Courtesy: www.asd-cr.com/images/redring) and wilt of pine caused by *Bursaphelenchus xylophilus* (D), L.S. of diseased trunk (E), and vector insect (F), (Courtesy: S.C. White).

sect vectors for transmission from diseased plants to healthy plants. The red-ring nematode is vectored by the black palm weevil, *Rhynchophorus palmarum* and sugarcane weevil, *Metamasius* spp. (Griffith et al., 2005), whereas the wilt nematode is transmitted by *Monochamus alternatus* and *M. carolinensis* (Agrios, 2005). These nematodes have affinity to fungi and feed upon them under adverse conditions (Sriwati et al., 2007).

Physiological Modifications

Nematode infection causes a varied range of changes in physiological functioning of the host plant. These changes largely occur in the absorption of water and minerals, photosynthesis and respiration.

Absorption of Water

Majority of plant nematodes feed on fine lateral roots which are actively involved in the absorption of water and minerals from soil. Root injury due to penetration and/or feeding by the nematodes leads to impairment of the efficiency of root system to absorb water. As a result, the nematode affected plants absorb less amount of water and show symptoms of water stress, temporary wilting, etc., especially during active period of transpiration. The morphological changes in the host root, such as galls, stubby roots, sparse roots, etc., induced by the nematodes may greatly retard absorption and upward movement of water and nutrients. *Meloidogyne* spp. and *Heterodera* spp. induce formation of giant cells and syncytia in the stelar tissue and extensively disrupt xylem tissue. The infection may greatly reduce the permeability of roots to water, e.g., *Pratylenchus penetrans* infection in roots causes thickening of walls of cortical cells with pectin material and also accumulation of granules and phenols in the endodermis. These changes contribute towards decreased efficiency of roots for absorption of water and minerals.

Transpiration

Stomata are the main avenues responsible for major water loss from plants. Closing and opening of stomata is sensitive to light, relative humidity and temperature because all these three factors are involved in maintaining turgidity of guard cells which form stomatal pores. Information on transpiration rates in relation to nematode infection is, however, contradictory. It has been found that during the early phase of infection, transpiration rate increases but later it decreases. Infection of tobacco and tomato by *M. incognita* or *M. hapla* have resulted in increased transpiration rate during the first 8 weeks after infection followed by a decrease (Wilcox Lee and Lorea, 1987). This decrease was apparently due to extensive development of galls which may occur in 8 weeks or so due to population build-up by the second/third generation of the nematode (Khan and Khan, 1997). Most of the studies have shown that under severe infection, transpiration rates are lower than the healthy plants. Some studies, however, have revealed that transpiration rates are not influenced as in the case of infection by *G. rostochiensis* and *P. penetrans* in potato (Williamson and Gleason, 2003).

Photosynthesis

Photosynthesis is a most characteristic phenomenon of green plants where light energy is captured by leaf pigments located in chloroplast and is transformed into chemical energy of ATP, reducing power of NADPH (light reaction), etc. This chemical energy is used in the fixation of CO_2 to synthesize carbohydrates and other organic compounds (dark reaction) etc. The carbohydrates formed during the photosynthesis also provide raw material for the synthesis of proteins, fats and other cell constituents. Infection by endoparasitic nematodes, such as root-knot and cyst forming nematodes, can reduce the photosynthetic rate. It is believed that the nematode infection causes partial closure of stomata that may result due to inhibition in the production of cytokinins and gibberellins. These hormones are produced in the roots and translocated upwards. In fact, nematode infection stimulates the production of IAA (Bird and Koltai, 2000). This leads to an imbalance in the ratio of IAA versus cytokinins and gibberellins. As a result, overall production of the phytohormones decreases considerably, and their upward translocation is also affected. Stomatal closure may also occur due to reduced rate of absorption of water by the roots damaged by nematodes. As a result of partial or full closure of stomata, diffusion of CO_2 correspondingly decreases. This leads to decrease in the CO_2 concentration in the leaf cells. Under low CO_2 concentration, in addition to corresponding, lower CO_2 fixation, the main enzyme of C3 cycle, ribulose biphosphate (RuBP) carboxylase acts as RuBP oxygenase leading to photorespiration that involves wasteful release of CO_2 without production of carbohydrates. The photosynthesis may also slow down in C4 plants due to partial closure of stomata, although they have an alternative cycle (C4 cycle) to check photorespiration. Phytohormones such as gibberellins, cytokinins and auxins, influence translocation. A disbalance in the phytohormone production caused by nematodes also leads to disruption in absorption of water and minerals that may affect the stomatal closure and consequently the photosynthesis.

Infection in plant roots by *Meloidogyne, Heterodera, Globodera*, etc., induces formation of nurse cells and regulates greater translocation of photosynthates towards infected root tissue while other parts (foliage) experience shortage (Wyss, 2002). Due to inadequate supply of water, nutrients, photosynthates and energy, growth and development of leaf tissue and its constituents especially chlorophyll pigments are adversely affected (Khan and Khan, 1997). Foliage becomes stunted and mildly chlorotic which is a common symptom of nematode attack. Thus, poor growth of foliage subsequently leads to decreased photosynthetic activity.

Respiration

It is a process of oxidation of carbohydrates synthesized during photosynthesis. Sugars, especially glucose, is oxidized to liberate energy in the form of ATP and NADPH. There are scanty reports on the effect of nematode infection on respiration. A few studies conducted on this aspect have shown that respiration rate increases due to nematode infection, especially at later stage of infection, but during the early stage it remains uninfluenced. It may be noted that in nematode infected plants metabolism shifts towards greater production of proteins, fats, etc., which are diverted to infected root parts. The plant needs ATPs and NADPHs for the production of these secondary photosynthates. Hence, to meet the increased demand of ATPs and NADPHs, the plant has to accelerate the respiration rates.

7

Ecology of Nematodes

Ecology is the study of structure and function of nature. It can also be explained as the study of an organism in relation to all its biotic and abiotic factors. Also, "the study of structure and function of ecosystem". Ecosystems are self-sufficient habitats where living and nonliving components of the environment interact to exchange energy and matter in a continuing cycle. Agro-ecosystems or crop systems are special systems, relatively artificial but stable. In an agro-ecosystem, a number of nematode species may feed on a given crop. Those with small populations are numerous, whereas those comprising high populations are a few and appear as major pests. In a severe form of a disease, population of a particular nematode species becomes high because of its biotic characteristics and ability to reproduce under the prevailing environment. The biological characteristics of a population of a species are its biotic potential, density, birth rate, death rate, age, distribution, growth, etc. All these characteristics are used to measure increase or decrease of a population. The factors responsible for the increase or decrease in a given time are very important, in fact, they are responsible for an outbreak. Ecology of nematodes can be discussed under the headings, population dynamics, factors affecting the population, and interaction of nematodes with other microorganisms.

POPULATION DYNAMICS OF PLANT NEMATODES

A population is a collection of individuals of a species of an organism from a geographical area or location. Population of a living organism is always in a state of dynamism and it exhibits age-specific rates of development, reproduction and mortality. The population has a measurable sex and age composition which may or may not be stable. The term "population dynamics" is used to convey any change in numbers, age-wise distribution, sex ratio (female/male) and behaviour of a population of an organism through time and space. Population of plant nematodes depends on reproductive potential, host plant species and length of the time nematodes remains in an environment favourable for their reproduction. The population level is dependent of the nematode's ability to live successfully in the soil. Hence, the imbalance between birth rate and death rate is responsible for population change, which is determined by the characteristics of nematode and host plant and the environmental influences. A population in which a balance between birth and death rates has been reached and is maintained at a uniform level may be called a state of dynamic equilibrium or a stable population. This seldom occurs in agricultural soils, except in perennial vegetations where the environment is fairly uniform. In agricultural lands wide fluctuations in the structure of nematode population and community may occur within a season and different seasons, particularly when the land is used for a variety of crops (Mennan and Handoo, 2006; Viketoft, 2007). Continuous monoculture tends to narrow the spectrum of nematode community, e.g., a field with continuous monoculture of tobacco may contain *Meloidogyne incognita*, *Pratylenchus penetrans*, and *Tylenchorhychus claytoni*. Fields where tobacco is rotated with other crops contain atleast six to eight or even more species of plant nematodes. In general, the nematode populations and communities are largely governed by three major factors, nematode characteristics (Buena *et al.*, 2006; Jones *et al.*, 2006; Timper *et al.*, 2006), host characteristics (Ramclam and Araya, 2006; Fourie *et al.*, 2007) and environmental conditions (Patrice *et al.*, 2002; Dabirea *et al.*, 2005).

Nematode Characteristics

Nematodes, like other organisms have definable life history strategies. These strategies can be defined in terms of number of life stages (egg, J_1, J_2, J_3, J_4, adult), duration of each stage, fecundity rates (eggs/individual), life expectancy, sex ratio and functionality of males. Males of *Meloidogyne*, *Heterodera*, *Radopholus*, etc., do not have a role in reproduction as the females lay viable eggs without mating (parthenogenesis), whereas the reproduction in *Globodera pallida*, *Hoplolaimus indicus*, *Helicotylenchus multicinctus*, etc., is amphimictic, needing males in fairly good number to mate with females to deposit fertilized eggs. The strategies can also be defined in terms of feeding habits (J_3 and J_4 of *Meloidogyne* do not feed, while of *Heterodera* they feed) and the mode of parasitism (ecto, endo and semiendoparasite, etc.). The following characteristics may greatly influence population densities of nematodes.

Reproduction rate: Nematode life cycle and reproduction rate vary widely, and are important in determining the reproductive potential. *Seinura celeris* reproduces in two and a half days, while *Xiphinema diversicaudatum* reproduces once in a year. Fecundity rates (number of eggs/female individual) also vary widely, e.g., *Heterodera* and *Meloidogyne* species upto 2,900 eggs, *Anguina tritici* upto 2,500 eggs, *Ditylenchus dipsaci* 200-500 eggs, and *Pratylenchus pratensis* 10-35 eggs.

Initial population: Nematode reproduction may be adversely affected if the initial population is too low (underpopulation) or too high (overpopulation). Reproduction rate is usually a straight line relationship until preplant densities are so high that nematodes hinder each other, predators or parasites appear, or nutritional capacity of host becomes limited. With any of these factors the curve would become sinusidal. As a rule of nature, all the individuals would not be able to complete their life cycle successfully. If initial population is too low, there would be further decline in the final population. Similarly, high population density may adversely influence nematode development, and sex ratio may change towards maleness.

Inter or intraspecific competition: Overcrowding of nematodes in or around roots leads to decrease in the population. The decrease results due to intense competition between individuals of different species or within a species for space and food, hence only the healthy individuals survive while the rest die due to injury or starvation, e.g., *G. rostochiensis* male/female ratio increases with increase in the population density and tends to reduce reproductive rate. It has been found that in a field heavily infested with *G. rostochiensis*, only 1 in 40 or 1 in 80 larvae could reach the maturity while the remaining die due to intense intraspecific competition for space and/or root invasion. At a moderate population 1 in about 5 larvae of *G. rostochiensis* attains maturity. Contest competition occurs due to mass invasion of the roots and reduction in tuber production, as a result nutrient supply becomes limited, hence nematodes starve. In addition, scramble competition among *Globodera* larvae occurs that leads to a decrease in the number of eggs per cyst at high population density (Van Den *et al.*, 2006). In *Meloidogyne* species, high population brings about sex reversal towards maleness. Low population also lowers the reproductive rates in both parthenogenetic and amphimictic species.

Quiescence: All nematodes can survive temporary periods of adverse conditions at certain stages of their lives, but some species possess a special mechanism to tolerate extremes of environmental conditions by transforming into a resistant and dormant state which is termed as "quiescent state", and the phenomenon is known as quiescence. *Anguina* *tritici* (II stage) inside cockle can survive dry and hot/cold weather for upto 32 years. *Ditylenchus dipsaci* (IV stage) may survive for many years in the cryptobiotic state. Cryptobiosis refers to survival at low temperature where all reversible life processes are suspended for several years. Similar terms have been coined to refer to tolerance to dehydration (anhydrobiosis), high osmotic pressure (osmobiosis) and low oxygen concentration (anoxybiosis). Nematodes can undergo quiescence in a particular developmental stage. If that stage has passed, the nematode would not be able to transform into quiescent state and will die under the adversity. The second stage larvae of *A. tritici*, fourth stage of *D. dipsaci*, adults of *Aphelenchoides besseyi* (predominently females) and *Hemicycliophora* spp. can undergo quiescence. Eggs are more tolerant to unfavourable conditions than other life stages. Eggs within cysts of *Heterodera* spp. can survive under low and high temperature for 1-8 years or even longer. *M. hapla* may overwinter in developmental stages within the tissue of perennials. Sometimes *M. javanica juveniles* (J_2) lose most of the water content from the body and coil to transform into a state of anhydrobiosis to overcome drought and high temperature; whereas *Aphelenchoides* species may or may not coil (Jagdale and Grewal, 2006) and overwinter in soil, dormant buds and infected dry leaves (Otsubo *et al.*, 2006).

Host Characteristics

The status of host in terms of availability of nutrients, their quantity and quality, and the factors influencing the host vulnerability (liability to be attacked) to nematode infection, are major determinants in the population dynamics. Hence, plant characteristics may be considered from two points of view. Firstly its relative susceptibility to nematode infection, and secondly, the plant's relative suitability as a substrate for the nematode feeding and reproduction. Host status, as an efficient host, is determined by the performance of the parasite (Table 15). An efficient host (good, suitable and congenial) is a plant upon which nematodes can build up greater densities; whereas poor hosts do not support nematode reproduction, e.g., resistant cultivars to *Pratylenchus penetrans* are suitable for the feeding of nematode, but the cultivar is not vulnerable to damage as the invaded tissue suddenly dies (necrosis). The necrotic tissue is not suitable for feeding, hence the nematode is restricted within the lesion. A similar mechanism also operates in resistant cultivars of chrysanthemum against *Aphelenchoides ritzemabosi*.

Nematode population on a host with respect to time increases upto a certain limit and that is referred to as "ceiling level" (Fig. 77). An efficient host has a high ceiling level, whereas a poor host has a low ceiling level.

Table 15 Host status in relation to plant growth and nematode reproduction

Nematode Reproduction	Plant Growth	
	Good	Poor
Good	Tolerant	Susceptible
Poor	Resistant	Hypersusceptible

The nonhosts which do not support nematode reproduction have no ceiling level. *Globodera rostochiensis* and *G. pallida* on potato, *Heterodera avenae*, *Anguina tritici* on wheat, *Meloidogyne incognita* on tomato and *Tylenchorhynchus brassicae* on cabbage have a high ceiling level. Low ceiling level is found with *M. javanica* on pepper, *M. incognita* on wheat, *H. avenae* on maize; whereas *M. incognita* and *M. javanica* on oat and rye, *T. semipenetrans* on papaya, *H. glycines* on potato, etc., have no ceiling level.

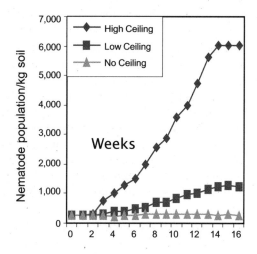

Fig. 77 Nematode population over time showing high, low and no ceiling level

Root exudates of plants may attract or repel nematodes. Root exudates from mustard contain allyl isothiocyanate which neutralizes the stimulatory effects of root exudates from susceptible potato cultivars on *Globodera* spp. Similarly, root exudates from *Tagetes erecta* (α- terthienyl), *T. patula* (α- diethienyl), and *Azadirachta indica* (nimbidin, thionemone, azadirachtin, etc.) cause adverse effects on plant parasitic nematodes. Higher concentration of methoxy substituted benzoxazinoids in plant roots may determine the host status, as these chemicals are nematicidal in nature (Zasada *et al.*, 2007). Sometimes root exudates of plants attract nematodes or even stimulate egg hatching. Root exudates of susceptible cultivars of tomato, tobacco, cotton, etc., stimulate hatching of *Meloidogyne* eggs and help juveniles to move in the direction of roots. Similar effects of root exudates from potato have been observed on the hatching of *Golobodera* eggs.

Environmental Conditions

Environmental factors can greatly influence the population dynamics of nematodes. Environmental conditions are generally considered density independent until they directly affect the nutrient availability to nematodes, e.g., an abrupt change in temperature may make the host plant weak and nutritionally poor. As a result supply of food to the feeding site gets reduced. The change in environmental conditions may also become favourable to predators and parasites of nematodes. Sometimes environmental factors, such as temperature and nutrient status of host, may alter the sex ratio. High temperatures cause sex reversal in females of root knot nematodes leading to greater number of males. Such a condition may also directly influence the population dynamics.

Life history strategies may undergo tactical shifts in response to the impact of density-dependent and density-independent environmental conditions. The tactical shifts are referred to as plasticity (due to independent factors) and adaptability (due to dependent factors) of nematodes, depending on the nature of the condition or factor. Density-independent conditions are those which do not directly affect density of the nematodes, e.g., environmental conditions such as high or low temperatures and unsuitable host. Density dependent conditions can directly influence nematode density, e.g., food limitation may starve the nematode or would alter sex ratio which may promote parthenogenesis, cryptobiosis, etc., as an adaptation.

Soil is the only segment of the environment to which nematodes are directly exposed. All plant parasitic nematodes, whether root parasites or foliar parasites, inhabit the soil for a varying length of time. As such, the soil composition and soil environment would have a tremendous effect on the normal growth and development of nematode which ultimately affects the population. Plant parasitic nematodes, especially endoparasites, once inside the host become independent of direct influence of soil environments but the host plant remains exposed, hence even though the phytonematodes may be inside the host, they are still directly influenced by the surrounding environments. Broadly speaking, the factors which influence nematode populations and consequently their pathogenicity, are placed in two major categories, i.e., biotic and abiotic factors.

Biotic Factors

The biotic factors which influence nematode populations include soil organisms, and the products of living organisms such as organic matter, plant exudates, etc.

Soil Organisms

Soil is an abode for nematodes as well as for various other micro and macroorganisms. Presence of certain fungi, bacteria or insects in the soil may influence nematode communities. Effect of soil microorganisms may be suppressive or stimulatory on the development of nematode populations.

Suppressive Effect

The suppressive effect of microorganisms may occur on nematode population through parasitism or predation.

Parasites

An organism which grows on or inside another organism and obtains its nutrition from the host is classified as a parasite. A parasite is generally able to complete its life cycle and increase the biomass within a single organism. Nematode parasitic fungi and bacteria, present in soil, parasitize nematodes and as a result their population may decrease. Some of the commonly occurring parasites of nematodes are *Lagenidium caudatum*, *Pochonia chlamydosporia*, *Paecilomyces lilacinus* and *Pasteuria penetrans* (detailed in Interaction of Nematodes...).

Predators

An organism which actively seeks out other organisms and consumes them is known as a predator. A single predator consumes large numbers of organisms during its life span. A number of insects, fungi and nematodes are predators of phytonematodes. *Arthrobotrys robusta*, *Dactylaria candida*, *Monacrosporium drechsleri*, *M. bembicoides*, etc., are predacious fungi against a large number of nematodes in natural soils. Predatory insects, *Entomobyroides dissimilis* can consume 1,000 nematode larvae in 24 hours, *Tyrophagus similis* feeds on females of *Heterodera avenae*, *Caloglyphus berlesei* consumes *M. javanica* eggs, and *Hypoaspis aculeifer* feeds on egg masses of *Meloidogyne* spp. *Mononchus aquaticus*, *Panagrellus redivivus*, *Seinura* spp., etc., predate upon a variety of nematodes (detailed in Interaction of Nematodes...).

Organic Matter

Organic matter is an essential component of all soils. The soil rich in organic matter reveals high population of nematodes. The organic matter is not directly responsible for the increase, infact higher concentration of organic matter makes the soil nutritive and suitable for the multiplication and development of microorganisms (fungi, bacteria, etc.), upon which the nematodes feed. *Aphelenchus avenae* feeds on the spores/mycelium of *Rhizoctonia solani*. *Aphelenchoides* spp. feed on the spores of *Fusarium* and *Curvularia* species. *Ditylenchus* species feeds on fungal mycelium. The decomposing organic matter, however, may also adversely affect nematode populations, e.g., plant debris of sweet clover may cause decrease in *H. schachtii* population due to release of toxic gases and organic material during decomposition. Similarly, decomposition of rye residues may decrease the population of *Meloidogyne incognita* and *Pratylenchus penetrans*. Several parasites and predators of nematodes also grow and multiply efficiently on soils rich in organic matter. Raw manures are also known to decrease the population of plant parasitic nematodes (Nahara et al., 2006).

Root Exudates

Plant roots release certain chemicals of complex form in soil which may stimulate or depress nematode populations. Root exudates from potato stimulate hatching of eggs of potato cyst nematode, *Globodera rostochiensis* and out migration of larvae from cyst. The exudates also help the larvae in locating the roots through the chemical gradient. Similarly, exudates from susceptible cultivars of tomato, tobacco, cotton, etc., attract juveniles of *Meloidogyne* for penetration.

Effects of root exudates on nematode populations are not always stimulatory, sometimes they may adversely influence nematode populations. Mustard exudates contain allyl isothiocyanate which can neutralize the stimulatory effect of potato root exudates on *G. rostochiensis*. Similarly, thionemone, nimbidin, azadirachtin, etc., from neem can reduce the population of several nematode genera such as *Meloidogyne, Heterodera, Rotylenchulus, Tylenchulus, Hoplolaimus, Helicotylenchus* and *Tylenchorhynchus*. Root exudates of *Tagetes erecta* and *T. patula* contain α-terthienyl and α-diethienyl, respectively which can reduce the population of *Meloidogyne, Pratylenchus, Paratylenchus, Rotylenchulus, Tylenchorhynchus*, etc. In addition, *in vitro* studies have shown that extract of certain plants (stem + roots) may cause adverse effects on plant nematodes (Choi et al., 2007; Cristobal-Alejo et al., 2006; Oka et al., 2006).

Microbial Metabolites

Metabolites or by-products of certain fungi and bacteria may suppress nematode population and pathogenicity. Metabolites of *Aspergillus niger, A. flavus, Paecilomyces lilacinus, Fusarium* spp., *Penicillium* spp., *Rhizoctonia* spp. and *Trichoderma* spp. may cause inhibitory effects on nematode population. A number of bacteria produce toxins such as bulbiformin (*Bacillus subtilis*), Bt toxin (*B. thuringiensis*), pyoluteorin, phenazin, oomycin A (*Pseudomonas fluorescens*), H_2S (*Desulfovibrio desulfurican*), butyric acid (*Clostridium butyricum*), valinomycin (*Streptomyces anulatus*), and avermectins (*S. avermitilis*), which may inhibit nematode development and reproduction (Khan et al., 2005; Faske and Starr, 2006).

Abiotic Factors

The relationship of nematodes with the abiotic factors, especially soil environment, starts with their total dependence on two factors, i.e., water (for survival) and pore space (for movement). Other factors which affect nematode populations are soil aeration, soil temperature, soil solute, pH, osmotic pressure, rainfall, etc.

Soil Texture and Structure

Texture and structure of soil is determined by the type of soil particles distributed, soil pore and mechanical strength of soil. These three factors can independently influence nematode populations. Primary particles, viz., clay (less than 0.002 mm), silt (0.002-0.02 mm) and sand (0.2-2.0 mm) are bound together by cementing material forming secondary aggregates of various sizes and types such as crumbs, granules, platy and prism. Relative proportions of clay, silt and sand determine the soil texture which are classified as clayey loam, sandy clay loam, silty clay loam, sandy loam, etc. The space between the crumbs occupied by water and air is referred to as pore space. Volumetrically the soil surface of silt loam contains 50% pore space and 50% solid phase. Volume of a pore depends on the size and shape of crumbs and their packing arrangements.

Soil texture and soil pore have profound effect on the movement and distribution of nematodes. In coarse textured soils (sandy or loamy) nematode juveniles move freely without any hindrance. Root knot nematode (*Meloidogyne* spp.), cereal cyst nematode (*Heterodera avenae*), burrowing nematode of citrus and banana (*Radopholus* spp.), wheat gall nematode (*Anguina tritici*), stubby root nematodes (*Trichodorus* spp.), and dagger nematodes (*Xiphinema* spp.) reproduce and multiply efficiently and cause greater damage in light sandy soils than in heavy clayey soils. Proper porosity and aeration are responsible for the enhanced activity of the nematodes mentioned above. In coarse textured soils (sandy or loamy), diffusion of root exudates and carbon dioxide is greater and larval movement is rapid. Optimal soil crumb size for movement of larvae of *G. rostochiensis* is reported as 150-250 microns in sandy loam and 250-400 micron in a peat soil (rich in decomposed plant material). Movement is faster in peat soil because of less friction between larvae and crumbs (particles) compared to clayey or sandy loam where friction is more. Maximum multiplication of the reniform nematode, *Rotylenchulus reniformis* has been recorded in sandy clay loam followed by clayey loam, coarse sandy loam and sandy soils.

Soil Moisture

About 40-50% of the bulk volume of soil body is generally occupied by soil pores which may be completely or partially filled with water. Different textured soils have variable pore areas. Hence the amount of soil moisture also varies with soil type. Sandy soils contain large pores and consequently higher volume of water, but it drains off rapidly compared to clayey soils. In soils rich in clay, the pores are not only narrow and numerous, but also subdivided into micropores and capillaries which retain water for longer duration. In sandy soils, water drains off soon because micropores or capillaries are not present, as a result pore space becomes empty and is occupied by air. In such soils nematodes cannot move or survive. In clayey soils, however, the narrow pores remain filled with water for longer periods leaving little or no space for air. This is also not a favourable condition for nematodes. In sandy loams or loamy sands, the water is retained in the pores around air tunnels, providing ideal condition for nematodes.

The extremes of soil moisture condition are generally harmful to nematodes. Root knot nematodes are killed by flooding or drying of soil. Water flooding creates anaerobic condition which facilitate decomposition of organic matters. This results in the increase of concentration of fatty acids, formic acid, acetic acid, propionic acid, butyric acid, etc., in soil. The higher concentration of the fatty acids is harmful for nematodes. Some bacteria, particularly anaerobic bacteria (*Clostridium* spp.), produce nematicidal toxins in saturated soils. However, certain nematodes, viz., *Hirschmanniella oryzae, Aphelenchoides besseyi* and *Ditylenchus angustus,* have adapted to high moisture conditions and show increased activity in saturated soils, probably because of co-adaptation with their preferred hosts (rice) which grows in standing water. *Rotylenchulus reniformis* shows high activity at a moisture level just below the field capacity (FC). The FC is defined as the amount of water that remains in soil capillaries, once the gravitational water is drained off. *Tylenchorhynchus martini* shows good activity in the soil having moisture level 40-60% of FC. In dried soil having less than 2% gravitational water, 20-80% of nematodes became inactive (Treonis and Wall, 2005).

Soil Aeration

Soil pores provide avenues for gaseous exchange between the soil layers and outer environment. Carbon dioxide is released in soil during respiration by various microflora, fauna and plant roots, and chemical reactions. Although a part of CO_2 is recycled by microbes, excess is harmful which is removed to the outer environment and replaced by O_2 through the soil pores. The ratio of CO_2/O_2 increases with the increase of soil depth and soil moisture, or decrease of soil porosity. Nematodes show variable response to CO_2 and O_2 concentrations. Juveniles of *Meloidogyne* are attracted towards high concentration of CO_2 probably because higher concentrations occur around active roots. *Heterodera schachtii* and *Ditylenchus dipsaci* have affinity towards both CO_2 and O_2.

Soil Solute, pH and Osmotic Pressure

The water present in soil pores is never in a pure form. The impurities in soil water are usually referred to as soil solutes. The solutes are derived from many sources such as distribution of colloidal matter, by-products of chemical and biological activities including root exudates, external application of fertilizers, manures, etc., and other agricultural chemicals. Although nematodes do not draw any nutrition from this solution, they are indirectly or sometime profoundly influenced by the soil solution because it affects pH and osmotic potential of water surrounding the nematode body.

In cultivated lands, soil pH is rarely below 4 and above 8. The hatching of root knot nematodes was maximum at pH 7 compared to pH 3 to 8. Greater root penetration of *Meloidogyne incognita* and disease intensity may occur at pH 5.6 compared to pH 3.2 or 6.8. The soil pH has little effect on *Globodera rostochiensis* and the hatching is more or less similar in acidic or alkaline soil. In many instances it has been found that varied response of nematodes to pH is usually indirect through the host. For example, pH has little effect on potato, consequently *G. rostochiensis* remains unaffected.

Difference in osmotic pressure between nematode body fluid and the soil solution occurs with change in concentration of dissolved salts in soil water. When a difference occurs, water or solutes tend to move through the cuticle to maintain equilibrium. This harms the nematode body. It has been found that the solutions of various salts and dextrose at a molar concentration equivalent to 15 atm prevent emergence of larvae of *Meloidogyne* spp. and *G. rostochiensis*. Eggs of *M. arenaria* did not hatch in 0.3 M sodium chloride solution. Similarly, *Hoplolaimus indicus* exhibits irreversible damage and mortality when exposed to sodium chloride solution of different concentrations.

Soil Temperature

Various factors such as net radiation (difference between heat gain and heat loss by the soil surface), rate of heat transfer from soil surface to subsurface, from subsurface back to soil surface and from soil surface to atmosphere, contribute to energy (temperature) balance of soil. Heat flows downwards in the soil as long as the surface is at a higher temperature than the subsurface layer. This happens in day time. During night when the surface cools rapidly, the heat flows in opposite direction, subsurface to surface, and surface to atmosphere. The major properties of soil that influence energy balance are thermal conductivity and thermal capacity. Thermal conductivity is the ability of soil to conduct heat, and thermal capacity is the quantity of heat required to raise the temperature of soil by a unit degree. Thermal capacity of soil is the sum of the thermal capacities of its components, i.e., solid, liquid and gas. Water has thermal capacity about 6 times greater than the same weight of dry clay or sand. For this reason wet soils require greater calories of heat than dry soil for a unit rise of temperature. The moisture, however, acts as a good conductor of heat and helps in rapid transfer to subsurface layers. The diurnal change of temperature is maximum at the soil surface. During cold months temperature increases with the depth of soil, while in hot-dry season it decreases with increase in soil depth.

Temperature has a major influence on the survival, feeding, reproduction and multiplication of nematodes. Soil temperatures in relation to their effect on nematodes can be categorized into five groups: non-lethal low temperature (5-15°C), optimum temperature (15-30°C), non-lethal high temperature (30-40°C), lethal low temperature (below 5°C) and lethal high temperature (above 40°C).

Temperature requirements of nematodes vary with species and developmental stage. Temperature requirements of *G. rostochiensis* have been found to be 21-25°C for the emergence of larvae from cyst, 15-16°C for invasion of host root, and 18-24°C for development of mature females inside host tissue. The J_2 of root knot nematodes, *M. incognita* and *M. javanica* develop into mature females in 17 days at 27.5-30°C, while in 57 days at 15.4°C. All the stages of *M. incognita* and *M. hapla* can survive at freezing temperatures inside the host root. The second stage juveniles of *M. javanica* and *M. hapla* remain infective in soil at 4-0°C but are injured at subzero temperatures. Eggs and juveniles of root knot nematode are killed at a soil temperature higher than 46°C. Eggs of *M. javanica* are killed when exposed to 46°C for 10 minutes. Endoparasitic nematodes inside host tissue can survive at lethal higher temperatures. However, sometimes these temperatures hinder normal development and induce sex reversal from female into male (McSorley, 2003).

Climate

Rainfall and temperature above the soil level are extremely important for the survival and development of nematodes and their hosts. These two factors, though do not directly influence nematode populations (except those of foliar nematodes), are still considered as key factors. Infact, these factors control the fluctuations of most of the soil factors. For instance, rainfall has direct relationship with soil moisture, soil aeration, soil pH and osmotic potential of soil water. If a graph is plotted between rainfall and soil moisture or aeration, the relationship will appear as an ascending or descending straight line. Similarly, soil temperature changes with change in atmospheric temperature. In addition to indirect effect, rainfall and temperature can directly influence the movement, development and reproduction of foliar nematodes, *Anguina tritici, Aphelenchoides besseyi, A. fragariae, A. ritzemabosi, Ditylenchus dipsaci, D. destructor*, etc.. A thin film of water is essential for the movement of these nematodes on the host surface. This condition prevails under cloudy, cooler and humid climates.

INTERACTION OF NEMATODES WITH OTHER ORGANISMS

Soil is an abode for plants and soil microorganisms. Its porous nature with available moisture, air and nutrients provides an ideal condition for sustenance and multiplication of microorganisms and plant roots. Plant roots also fulfil nutritional requirement of microorganisms. Hence, in root vicinity a number of pathogenic, saprophytic and saprozoic organisms become concentrated and develop an array of relationships. Their relationships may be inhibitory, stimulatory, symbiotic or neutralistic. Plants in nature are rarely infested, affected or infected by a single pathogen. Pathogens seldom occur in pure culture, although population of only a few attain economic injury level. The nature and extent of injury depends on the biological association of the host with each pathogen and mutual influences. Various kinds of associations have been well recognized between nematodes and other organisms which are grouped into six categories, viz., (1) etiological associations often causing disease complexes, (2) competitive associations (nematode - nematode, nematode - mycorrhizal fungi, nematode - rhizobium), (3) nematodes as vector of plant viruses and plant pathogens, (4) other organisms as vector of nematodes, (5) nematodes as parasite and predator, and (6) nematodes as prey or host of other predators and parasites. Etiological associations are extremely important from plant disease point of view.

Disease Etiology

Etiology is the science of cause of a disease, and includes the study of all factors involved directly or indirectly in the initiation and development of the disease. Diseases in terms of etiology may be of two kinds, specific and complex.

Specific etiology

It is a two component system of host and pathogen. A plant infected by a pathogen becomes greatly altered. The effects of pathogen and host response are governed by various factors including all the components of environment. Therefore, plant disease of a specific etiology represents a typical Disease Triangle.

Complex etiology

In a disease of complex etiology the host plant is infected by two or more pathogens, as a result the activity and effect of one pathogen are likely to be influenced by the activity and effect of the other pathogen engaged in the disease complex. The alterations of host plant are of great importance in the disease complex. The nematode and other pathogen become engaged with the host either sequentially or concomitantly, which are referred to as sequential etiology and concomitant etiology.

Sequential and Concomitant Etiology of Disease Complex

In a disease of sequential etiology, one pathogen infects and alters the host in advance, infection of the other pathogen follows. Hence, on the basis of time of invasion, the two pathogens may be recognized as primary pathogen and secondary pathogen. The primary pathogen invades the host prior to the invasion by the secondary pathogen and governs development of the disease complex. The primary pathogen modifies the host plant in such a way that the plant becomes either more suitable or unsuitable for invasion by the secondary pathogen. The time interval in the sequential engagements may vary from a day to several weeks. A primary pathogen possesses specific, inherent and independent capabilities of causing diseases, whereas a secondary pathogen may or may not have this ability. However, under natural conditions, it is difficult to recognize a secondary pathogen. In concomitant etiology, the primary and secondary pathogens invade the host simultaneously. In a disease complex when disease severity increases, it is due to biopredisposition.

Biopredisposition

It is a specific term used to describe predisposition caused by biotic factors. Predisposition is a modification caused by various factors in the expression of host response to a pathogen. It is applicable in sequential etiology. In the disease of complex etiology, nematode infection modifies the host in some way that benefits plant pathogenic fungi and bacteria. The nature and extent of biopredisposition in the disease of sequential etiology is determined by a primary pathogen, the pathogen that invades first. The mechanism of biopredisposition may be mechanical or physiological or both. In biopredisposition, the host plant is altered and becomes more vulnerable to attack by another pathogen (Back et al., 2002). Most of the bacterial pathogens require openings, natural or artificial for entry. The wounding of roots by nematodes generally helps bacterial plant pathogens. Physiological modifications in host plant are important and mainly involved in biopredisposition.

Nematode-Fungus/Bacteria Interaction

Due to coexistence of two or more pathogens in the root zone of a plant, an interaction may occur between the pathogens. The nature and type of interaction depends on the organisms, kind of relationship that develops between the engaged pathogens and the effect of relationship on the host plant. There may be three major types of interactions: synergistic (positive), antagonistic (negative), and additive (neutral) that may develop between nematodes and soil borne pathogens infecting a plant.

Synergistic Interaction

In the synergistic interaction biopredisposition is involved. The primary pathogen makes some mechanical, physical, anatomical, biochemical or physiological modifications in the host that facilitate infection by a secondary pathogen (Khan et al., 2005). As a result, the plant suffers greater pathogenic damage. Hence, the damage caused by the two pathogens together is greater than the sum of the damages caused by both the pathogens separately. The first evidence of interaction reported by Atkinson (1892) was a synergistic interaction between *Meloidogyne* and *Fusarium* species on cotton. A synergistic interaction leads to development of disease complexes such as wilt disease complex (Francl and Wheeler, 1993) and root-rot disease complex (Evans and Haydock, 1993). Various mechanisms are involved in synergistic interactions between nematodes and other plant pathogens, which are as follows:

Nematodes make wounds in healthy roots: This is an original and most simple concept of interaction. Earlier it seemed to be the most appropriate mechanism that during penetration and feeding, nematodes make injuries and wounds in roots which provide avenues for the invasion of fungal/bacterial spores. However, doubts have arisen when artificial wounds have often failed to allow the establishment of fungal pathogens. Occasionally, the wounds do seem to favour fungi and more frequently bacteria. Some bacteria and most fungi are quite capable of invading root tissue in the absence of artificial wounds. Hairy root of rose caused by *Agrobacterium rhizogens* becomes severe in the presence of *Pratylenchus vulnus*. Crown gall of peaches caused by *A. tumifaciens* was severe due to the mechanical damage caused by *M. javanica*. Similarly, severity of bacterial wilt caused by *Pseudomonas solanacearum* greatly increased on tobacco, potato, and tomato in the presence of *Meloidogyne* spp. or *Helicotylenchus nannus*.

Nematodes carry the pathogens to susceptible tissue: Vector role of nematodes in this type of relationship is more specific. Although secondary pathogen is carried externally on the body surface, the inoculum is to be reached specifically to the susceptible tissue which is usually embedded by leaf scales or bud scales. Typical examples are those in which the apical shoot meristem is susceptible to the pathogen. Mostly apical meristem is closely ensheathed by bud scales, leaf sheaths, etc., which make them quite inaccessible to nonmobile fungal spores or bacterial cells. The vector role of the nematode is, however, limited to carrying the pathogens only a few centimeters from soil to plant or from outer plant tissue to the meristem. The classical example is the infection of the developing wheat inflorescence with the bacterium *Clavibacter* (=*Corynebacterium*) *tritici* carried by the second stage juveniles of seed gall nematode, *Anguina tritici,* causing yellow ear-rot disease. A similar situation arises with the same nematode but different pathogen, a fungus, *Dilophospora alopecuri*. The fungus spores carrying bristle-like appendages adhere to the nematode leading to ear twist disease of wheat. *Rhodococcus* (=*Corynebacterium*) *fascians* is unable to reach the meristems of strawberry plants without the aid of *Aphelenchoides ritzemabosi* or *A. fragariae* and causes cauliflower disease of strawberry. Larvae of *Ditylenchus dipsaci* carry the spores of *Clavibacter michiganense* sub sp. *insidiosum* to cause wilt of alfalfa. The same nematode also vectors *Pseudomonas fluorescens* on garlics to cause *Cafe au lait* bacteriosis. Both the bacteria are carried into crown buds and are placed into conducive infection court by the stem and bud nematode, *Ditylenchus* species (Khan and Dasgupta, 1993).

Nematodes make necrotic wound in host tissue: Some pathogens especially facultative parasites, may be unable to establish themselves in healthy tissue even if wounded. The availability of small pockets of necrotic tissue as in root cortex damaged by *Pratylenchus* spp. may, however, provide a micro-food-base on which certain secondary pathogens can become established and from which they may be able to invade healthy roots (Evans and Haydock, 1993). Some of the imperfectly understood root-rots may begin in this way. Some highly specialized fungal pathogens benefit from necrotic lesions caused by nematodes. *Verticil-*

lium dahliae shows a distinctive affinity for lesions caused by *P. penetrans* in eggplant. *Radopholus similis* may also facilitate infection by certain soil-borne fungal pathogens in a similar manner.

Nematodes modify host substrate: Nematode feeding may induce certain anatomical and/or physiological modifications in host tissue. The modifications are mild by ectoparasites, and extensive and complex by sedentary endoparasites. The modifications could help a pathogen in two ways. Firstly, nematode feeding may affect cell permeability leading to leakage of cell sap in soil which may facilitate germination of pathogen propagules. Secondly, by providing a more favourable substrate for the establishment of pathogens. Such effects may be localized or near the site of nematode attack. The nematode injury may also cause systemic changes in the host physiology, as a result the site away from nematode feeding may also become favourable to the pathogen (Khan, 1993).

The localized substrate change may be explained with the example of *Phytophthora parasitica nicotiana*. The fungus was able to penetrate the roots of both susceptible and resistant cultivars of tobacco. The fungal growth was halted in cortex of fungus resistant cultivar. The entry tracts of *M. incognita* larvae had no effect on the resistant reaction but if gall formation had begun, the hyphae adjacent to the gall grew much more luxuriantly destroying their contents. From the galled tissue, hyphae moved rapidly into the vascular tissue in the resistant cultivar. *Fusarium* species grow profusely on giant cells. Strong evidence exists which demonstrates that nematode infection causes certain translocatable/systemic changes in the host substrate that favour secondary pathogens. It has been experimentally shown that split rooted tomatoes with a polygenic resistance to *Fusarium oxysporum* f. sp. *lycopersici*, wilted when the fungus was added to one half of root system and *M. incognita* to the other half. Such plants were as badly affected as those with entire root system inoculated with both the pathogens. Infection by root knot nematode or root lesion nematode caused systemic changes in the host that resulted in increased susceptibility of plants to *Pseudomonas marginata* and *Clavibacter michiganense*.

Nematodes break resistance of cultivars: It has been often observed that a particular cultivar resistant to some bacterial pathogens becomes susceptible in the presence of root knot nematodes. For example, wilt causing bacteria produce some toxins in host tissue which induce the wilt symptoms. In resistant cultivars the mechanism of production of the toxins is inhibited or it is detoxified by the natural defence of the plant, hence, wilting does not take place. However, the nematode infections induce certain changes in host physiology, leading to failure of detoxification mechanism of the resistant cultivars, and the plants become susceptible. An eggplant cultivar, Pusa Purple Cluster, highly resistant to *Pseudomonas solanacearum* became susceptible in the presence of *M. incognita*. Similarly, field resistance in potato to *P. solanacearum* was broken when the plants were infected with *M. incognita*. Wilt fungus resistant cultivars also become susceptible in the presence of root knot nematodes. Wilt resistant cotton cultivars succumbed to *Fusarium* in the presence of *M. incognita*. Tomato cv. Chesapeake resistant to *Fusarium* wilted in the presence of the same nematode species. The wilt severity, however, varied with the nematode species, presence of *M. hapla* caused 60% wilting, whereas 100% wilting occurred with *M. incognita*.

Pathogens benefit from root exudates: Recently attention has been turned to the possibility that physiological changes in host could affect fungal pathogens before entry. In a few instances it has been found that many more propagules of *F. oxysporum* and *Pseudomonas* species occur in the rhizosphere of tomato, potato, etc., infected with *Meloidogyne* spp. It is believed that greater root exudation (resulting due to nematode infection) facilitates germination of dormant fungal spores or bacterial cells.

Antagonistic Interaction

Antagonistic interaction occurs when infection by one pathogen modifies the host or its response in such a manner that it becomes less suitable or unfavourable for infection by the secondary pathogen. Preinfection may result in lesser availability of root mass suitable for feeding or preinduction of resistance in the host against the secondary pathogen, or the engaged pathogens may have mutual detrimental or inhibitory effects. In antagonistic interaction, disease severity is greatly decreased, and such kinds of interactions are agriculturally important, particularly with regard to plant protection. Various mechanisms may be involved in antagonistic interactions depending on the engaged organisms which are summarized as below:

Competition for root space: Since most of the fungi and nematodes involved in the complex disease etiology are root parasites sharing a common habitat, the obvious mechanism by which they would antagonize each other is competition for root space. Hence, time of inoculation, in other words opportunity for root invasion first by one organism, may give it an advantage over the second organism. The prior invasion may result in the reduction of root mass available for colonization by the other organism. A clear evidence on the competition between nematodes and fungi for root space, however, does not exist. The mycorrhizal fungi and nodule forming bacteria can affect nematode pathogenesis by some other mechanisms also. However, on several occasions, it is evident that preinoculation of an organism influences reproduction of the organism inoculated later. Preinoculation of *M. incognita* curtailed development and reproduction of *Pythium graminicola*. In another study, *M. incognita* when inoculated prior to *Rhizoctonia solani* provided protection to bean plants against the fungus. In some cases reproduction of primary pathogen may also be adversely affected, e.g., *Fusarium oxysporum* invades and destructs feeding sites of root knot nematodes, as a result nematode females are starved or unable to lay eggs.

Fungus/bacteria produces metabolites toxic to nematodes: Prior inoculation of a fungus may lead to production of nematicidal metabolites which may adversely affect various stages of nematode infection. Culture filtrates of many fungi have been found to prevent hatching of nematodes and even cause mortality to juveniles. Culture filtrates of eight species of *Aspergillus* decreased hatching and killed the hatched juveniles of *M. incognita*. Similar effects were found with *Sclerotium rolfsii* and *Rhizoctonia solani*. A fungus produces different toxins and their toxicity vary greatly. Out of the nine toxins of *F. oxysporum* tested against *M. incognita*, the toxin T2 was most effective in decreasing the egg hatching. *Aspergillus niger* and *R. solani* culture filtrates suppressed the penetration of *Meloidogyne* juveniles in tomato roots. Culture filtrate of nematophagous fungi (*Arthrobotrys oligospora*, etc.) have also shown to decrease hatching and penetration of *M. incognita* and *Heterodera zeae*. Reproduction of six different plant parasitic nematodes was suppressed to a great extent in the soil treated with culture filtrates of *R. solani*. Among the culture filtrates of 20 isolates of *A. niger* tested, the isolates AnC_2, AnR_3 and AnM_3 greatly decreased hatching of eggs and suppressed the nematode development inside root tissue (Khan and Anwer, 2006, 2008). Culture filtrates of some species of *Fusarium* and *Colletotrichum* cause deterioration of giant cells of *M. javanica* in tomato roots that leads to decline in reproduction of the nematode. Toxin produced by *Bacillus thuringiensis* (δ *endo* toxin), *B. subtilis* (bulbiformin) and *Pseudomonas fluorescens* (phenazin) suppressed the hatching and reproduction of root knot nematodes. Some other bacterial metabolites that have been found nematoxic are butyric acid (*Clostridium butyricum*), hydrogen sulphide (*Desulfovibrio desulfuricans*), and valinomycin (*Streptomyces anulatus*).

Fungi/bacteria infect nematodes: There are numerous fungi which may infect nematodes and kill them or atleast impair their reproduction. This kind of relationship has been exploited for biological control of nematodes (Khan *et al.*, 2005). Several fungi have shown capability to parasitize plant parasitic nematodes. *Fusarium solani* infects *H. zeae* and *M. incognita*. Spores of *Paecilomyces lilacinus* and *Pochonia chlamydosporia* (=*Verticillium chlamydosporium*) infect eggs, larvae and adults of many nematode species. The fungus mycelium invades eggs as well as larvae of *Meloidogyne* spp., *Heterodera* spp., *Pratylenchus penetrans* and *Rotylenchulus reniformis*. The bacterium, *Pasteuria penetrans* possesses great potential to parasitize hundreds of nematode species. The bacterial spores (endospores) infect the reproductive system of nematodes, thereby preventing egg laying.

Nematodes feed on fungi: There are some nematode species which form antagonistic relationship with fungi by feeding upon their spores and sometimes mycelium. *Aphelenchoides hamatum* feeds on plant pathogenic fungi. To control damping-off of cucumber caused by *Rhizoctonia solani* more than 50,000 larvae of *Aphelenchus avenae* per 500 ml of soil are required. *Aphelenchoides composticola* and *Ditylenchus myceliophagus* feed on spores and mycelium of species of *Agaricus, Verticillium, Botrytis*, etc.

Predacious fungi: Numerous fungi, more than 100 species of Hyphomycetes predate upon nematodes; they first trap a wandering nematode larva followed by killing or invasion and consumption of the prey nematode. Such fungi are also known as nematode trapping fungi or nematophagous fungi. These fungi capture a nematode with the help of certain mechanisms or devices developed in the mycelium or spores, and are categorized into two groups on the basis of trap mechanism, i.e., sticky traps and mechanical traps.

Sticky traps

Mycelium and/or spores of certain fungi are provided with sticky materials which help in capturing nematode larvae. Nematodes can be trapped by three types of sticky devices:

Sticky knobs: In certain fungi, single celled, sessile or stalked spores are developed on the mycelium (Fig. 79A). These spores/cells are covered by sticky secretions. When a wandering nematode comes in contact with these cells, it gets stuck to the fungus knobs/mycelium. Sometimes, after a nematode has become attached to an adhesive knob, the violent movement may cause the knob to get detached from the mycelium. Such knobs grow later producing hyphae that penetrate the nematode body and utilize the contents, e.g., *Monacrosporium drechsleri* (Fig. 78A).

Sticky lateral branches: Short lateral branches develop at right angles from the main mycelium. The lateral branches bear swollen cells (single or few) terminally, which are covered by a film of a tenacious adhesive material, e.g., *M. eudermatum* (Fig. 78B). In *M. cionopagum* and *M. gephyropagum*, the lateral branches develop sufficiently close together to form a primitive network.

Sticky nets: It is one of the most common traps and is formed by the anastomosis of lateral branches (fusion of hyphae). A lateral branch origi-

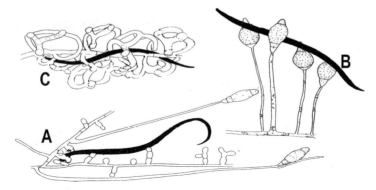

Fig. 78 Predation of nematodes by fungi through sticky traps. A- Knob, *Monacrosporium drechsleri*, B- Lateral branches, *M. eudermatum*, and C- Sticky nets, *Arthrobotrys robusta*.

nating from the main mycelium, curls round to join with the parent hyphae to form a loop. The surface of the loop is provided with some weak adhesive material. Similar loops are formed on or near the first loop, forming a network, e.g., *Arthrobotrys robusta* and *A. oligospora* (Fig. 78C).

Mechanical traps

Some fungi develop certain mechanical devices in the form of a ring which help in nematode trapping. Basically there are two types of mechanical traps:

Simple nonconstricting ring: A branch from mycelium curls round upon itself and forms a three celled ring attached to the parent hypha by a slender stalk also of three cells, e.g., *Dactylaria candida* (Fig. 79A). This ring is commonly called as a simple or nonconstricting ring. A nematode thrusting its body into such loops may become wedged inside and may find it impossible to retract from the ring.

Constricting ring: It is the most dramatic type of trap. The ring is developed in the same manner as the simple or nonconstricting ring. When a nematode enters into a ring, the body touch stimulates swelling of cells, inflation being mainly inwards, while the outer wall of the ring does not change (Fig. 79B). Ring swelling completes within 0.1 second. This inflation tightly occludes the nematode, e.g., *Monacrosporium bembicoides*, *Arthrobotrys dactyloides* and *Dactylaria brochophaga*.

Nematode-Virus Interaction

Hewitt, Raski and Goheen in 1958 were the first who demonstrated that plant parasitic nematodes can transmit plant viruses. When certain nematodes feed on virus infected plants, virus particles are ingested

Fig. 79 Predation of nematodes by fungi through mechanical traps. A- Nonconstricting ring, *Dactylaria candida*, and B- Constricting ring, *Monacrosporium bembicoides*.

along with the cell contents. Out of an estimated total of around 3,000 species of plant-parasitic nematodes only 30 species are virus vectors. These nematodes belong to two families: Longidoridae and Trichodoridae of the order Dorylaimida. The longidorid and trichodorid nematodes are ectoparasites on roots of annual and perennial plants. Within Longidoridae, seven species each of *Xiphinema* and *Longidorus* transmit plant viruses. Similarly, a total of 13 species of *Trichodorus* and *Paratrichodorus* are potential virus vectors. The plant viruses transmitted by nematodes are categorized as nepo and netuviruses (tobraviruses).

Nepoviruses: It is an acronym that stands for "nematode transmitted polyhedral" viruses. The size of virus is 27-30 nm diameter. About 11 serologically different nepoviruses are transmitted by longidorid nematodes, although experimentally 36 nepoviruses have shown transmission through nematodes. The genome of nepoviruses is bipartite with two functional ribonucleic acids (RNA-1 and RNA-2), which are separately encapsidated. The RNA-1 is 8-8.4 kb (kilo base) and RNA-2 of 3.4-7.2 kb and have $5'$ and $3'$ Vpg polyadenylate tail, respectively (Agrios, 2005). The nepoviruses include ring spot viruses and are transmitted primarily by *Xiphinema* and *Longidorus* species and also *Paralongidorus* species (Taylor and Brown, 1997).

- *X. index* transmits arabis mosaic virus, grapevine fanleaf virus.
- *L. elongatus* transmits tomato black ring virus, raspberry ring spot virus.
- *X. diversicaudatum* transmits arabis mosaic virus, strawberry latent ringspot virus.
- *L. macrosom* transmits English strain of raspberry ring spot virus.
- *P. maximus* transmits arabis mosaic virus, strawberry latent ring spot virus.

Netu/Tobra viruses: Netu is an acronym of "nematode-transmitted tubular" viruses. Authentic nomenclature of netuviruses is tobra viruses derived from tobacco rattle virus. The virus particles are rigid and rod-shaped with two different lengths, i.e., 80-110 by 22 nm of 1.8-4.5 kb, containing one gene, and 190 by 22 nm of 6.8 kb and contains four genes. The genomes are bipartite and have two functional RNAs, RNA-1 is larger and RNA-2 is smaller. The short particles are made up of one molecule of RNA-2, whereas the long particle has one molecule of RNA-1 (Agrios, 2005). The smaller particles are non-infective but their presence is essential for the infection. Due to solid stylet and large oral opening, the trichodorid nematodes are able to transmit the long rod-shaped tobra viruses. Three tobra viruses namely, tobacco rattle virus, pea early browning and pepper ring spot virus are transmitted by *Trichodorus* and *Paratrichodorus* species.

T. similis transmits tobacco rattle virus.

T. viruliferus transmits tobacco rattle virus, pea early browning virus.

P. minor transmits tobacco rattle virus, pepper ring spot virus.

Mechanism of Transmission

The mechanism of virus transmission is ingestion-egestion type involving three steps, i.e., acquisition (ingestion), retention, and egestion of virus particles (Weisher, 1993).

Acquisition (ingestion): When a vector nematode feeds on virus infected roots, the virus particles present in the cell sap are ingested by the nematode. The time required for successful acquisition of virus particles is short, e.g., *Xiphinema index* has shown to acquire grapevine fanleaf nepovirus in less than 5 minutes.

Retention: After ingestion, the virus particles are selectively and specifically absorbed at certain retention sites inside the nematode. The viruses not transmitted by the nematodes, if ingested are not retained at the sites. In a rare case if particles of a virus (not to be transmitted) are retained, will not dissociate during the final stage of transmission.

Virus retention sites differ among the nematode genera (Fig. 80). In *Xiphinema*, virus particles are retained at the cuticular lining of the lumen of the 'odontostylet' and of the oesophagus, whereas in *Longidorus* and *Paralongidorus* species, which have very long stylet, the particles are absorbed to the anterior surface of the lumen of 'odontostylet' and with the guiding sheath. In *Trichodorus* and *Paratrichodorus*, they are attached to the lining of oesophagus but not to the onchiostylet. All surfaces serving as retention sites are shed during moulting together with the outer cuticle of the body. Consequently, the virus particles adhering to the retention sites are lost and do not transfer from one developmental stage of nematode to the other. The viruses retained in their vectors do not multiply nor do they get involved in the nematode metabolism.

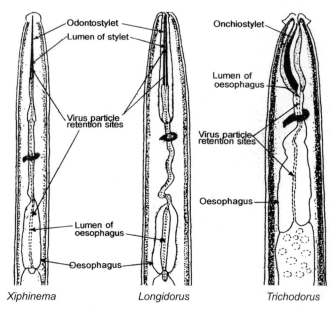

Fig. 80 Virus retention sites in dorylaim nematodes

Egestion: The egestion is the final step of virus transmission. During the feeding of viruliferous nematode, the oesophageal gland secretions pass anteriorly which help in dissociation of virus particles. It is assumed that the saliva modifies the pH within the lumen and alters the surface charge on the virus particles. In most of the cases pH becomes 7.0 (neutral). At this pH, charge on the particles becomes neutral, hence they dissociate from the retention sites. All the virus particles absorbed to retention sites are not detached at one time during a single feeding act. Sometimes the dissociation depends on the virus strains. *Longidorus macrosoma* transmits the English strain of raspberry ring spot virus efficiently, the nematode only rarely transmits Scottish strain of the same virus. Though both the strains are absorbed on the retention sites, the virus particles of the English strain dissociates whereas of Scottish strain do not detach.

Nematode-Rhizobium Interaction

Root nodules are hypertrophic structures of the host plant origin and are formed due to symbiotic association between the bacteria and host roots. The hypertrophic structure is commonly called nodule which is oval or irregular in shape and may be lobed (Subba Rao, 2001). The nodules are formed by a diversified group of bacteria which do not form any evolutionary homogenous clade (Friere and Saccol de Sa, 2006). They belong to the order Rhizobiales of the α-2 class of Proteobacteria (Garrity *et al.*, 2001). The genera which form symbiotic root nodules are *Rhizobium*, *Sinorhizobium* (family Rhizobiaceae), *Allorhizobium*, *Mesorhizobium*

(Phyllobacteriaceae), *Bradyrhizobium* (Bradyrhizobiaceae), *Azorhizobium* (Hypomicrobiaceae) and *Methylobacterium* (Methylobacteriaceae) (Garrity *et al.*, 2001; Sy *et al.*, 2001). These organisms produce nodules on roots but a few species of *Allorhizobium* (*A. caulinodans*) may also form nodules on stem. The bacteria develop symbiotic relationship with a majority of leguminous plants but some nonleguminous plants, e.g., *Parasponia*, a member of the family Ulmaceae, also develop the root nodules (Akkermans *et al.*, 1978). A low inoculum level of rhizobia in the rhizosphere (100 rhizobia) is sufficient to induce nodules, because multiplication of the bacteria is very rapid (Gentili and Jumpponen, 2006). The rhizobia generally enter the root via root hairs. The invaded root hair curls due to binding of the rhizobia to the hair surface followed by the formation of an infection thread in which the rhizobia multiply and are arranged usually in a row. The infection thread extends and releases the rhizobia into the cortex. Eventually many cells become filled with the rhizobia followed by hyperplasia giving rise to a nodule. A nodule has four differentiated tissue zones, viz., nodule cortex, meristematic zone, vascular system, and bacteroid zone. The bacteroid zone is the centre and most active part of the nodule, and consists of bacteroid and nonbacteroid cells. In this zone leghaemoglobin (Lb) is present which helps to create oxygen deficient condition to maintain the activity of bacteroid nitrogenase enzyme. The mature nodules after a certain period of time disintegrate and decay, liberating motile bacteria in the soil which normally serve as a source of inoculum for the succeeding crop of a given species of legume (Freire and Saccol de Sa, 2006). Since plant nematodes and root nodule forming bacteria share a common substrate for colonization and drawing nutrition, definite interaction develops between them that varies with the bacteria, nematode and plant species.

Effect of Nematodes on Root Nodulation

Plant parasitic nematodes may influence the process of biological nitrogen fixation at various stages from infection, bacterial establishment to functioning of the nodule. Survival of root nodule bacteria in the rhizosphere and colonization in the rhizoplane may be influenced by root exudates of nematode infected plants (Huang, 1987). It has been frequently observed that plant nematodes irrespective of their mode of parasitism inhibit nodulation. Sometimes the root system may become devoid of nodules due to severe infection by root knot nematodes (Fig. 81). Nutrient depletion by the nematodes, competition between nematode juveniles and root nodule bacteria, devitalization of root tips and suppression of lateral root formation are possible causes for nodule suppression. The reduced nodulation may also result from interference of

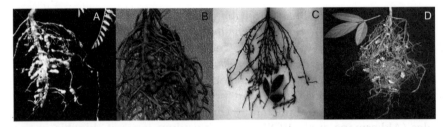

Fig. 81 Chickpea (A) and pigeonpea (C) roots almost devoid of nodules due to severe infection by root-knot nematodes otherwise showing numerous nodules (B, D).

the nematode in lectin metabolism. The nematode infection may reduce the binding of rhizobia to sites on infected roots (Taha, 1993).

The root nodules may also get damaged directly due to the nematode infection. Larvae of *Meloidogyne, Heterodera, Pratylenchus,* etc., invade root nodules on legumes. *Meloidogyne* spp. induce the formation of giant cells in the nodular tissue and other histological changes. Sometimes small galls are formed on the nodule (Khan *et al.,* 2005). The nematode activity may disturb the infection thread and multiplication of rhizobia, as a result development and functioning of the nodule is adversely affected. Moreover, the root system infected by root knot, root lesion, cyst forming nematodes or other nematodes becomes less suitable for symbiosis with the rhizobium. Consequently, lesser number of nodules with smaller size are formed leading to a remarkable decrease in the biological nitrogen fixation. Reduction in size and number of nodules and induction of their premature nonfunctionality on nematode infected plants are considered as two possible reasons for inhibition of nitrogen fixing capability of the roots. The Lb content of nodules may also decrease due to infection by nematodes. Significant reduction in the Lb, bacteroid content and nitrogenase activity in chickpea nodules has been attributed to the infection by root knot nematodes.

Some investigations have indicated that nematode infection may stimulate root nodulation. Huang (1987) observed that infection by *M. incognita, M. hapla, Pratylenchus penetrans* and *Belonolaimus longicaudatus* stimulated nodulation by *Bradyrhizobium japonicum* on soybean. Root nodulation on pea and black beans was also enhanced by *M. incognita.* Some reports, however, show no apparent effect of nematode infection on root nodulation.

Effect of Rhizobium on Nematode Infection

The root nodulation may also influence the parasitism of plant parasitic nematodes. Generally antagonistic effects on the nematode have been recorded. Root knot nematodes cause galls on nodules (Khan *et al.*, 2005). Such galls are shed prematurely, hence a certain degree of nematode population is also expected to be adversely affected (Khan *et al.*, 2002). Stronger host due to great availability of nitrogen makes the plants a bit tolerant to the infection by endoparasites. A discernable effect of rhizobium has not been generally observed on ectoparasites (Taha, 1993). However certain nematodes, e.g., *Tylenchorhynchus*, *Polenchus*, *Cephalenchus*, etc., which specifically feed on root hair, may be suppressed due to lesser availability of their feeding sites in nodulated roots, because nodules are formed on the root hair.

Nematode-Mycorrhizal Fungi Interaction

The feeder roots of most of the flowering plants growing in nature are generally infected by some symbiotic fungi. These fungi do not cause disease but, instead, are beneficial to the host plant. The infected feeder roots are transformed into unique morphological structures called mycorrhizae (meaning fungus roots). Mycorrhizae are symbiotic associations between the hyphae of certain fungi and plant roots. These associations are described as physiologically well balanced reciprocal parasitism. It was once thought that mycorrhizal associations are exceptions in nature and occur only with forest trees especially pines, but now it has been recognized that as a rule of nature the association exists with the roots of both cultivated and wild plants. Mycorrhizal fungi increase the solubility of minerals in the soil, improve the uptake of nutrients to the host plant (Purohit, 2006), produce plant growth hormones (Sharma, 2005) and suppress plant pathogens (Maia *et al.*, 2006). In return, the mycorrhizal fungi obtain carbohydrates from the host plant (Tanu and Adholeya, 2006). There are three types of mycorrhizae, viz., ectomycorrhizae, endomycorrhizae, and ectendomycorrhizae, which are distinguished by the way their hyphae are arranged and associated on the root surface and/or within the cortex.

Ectomycorrhizae: Ectomycorrhizal fungi (EMF) are predominantly higher Basidiomycetes belonging to the order Agaricales. These fungi reproduce by forming fleshy fruiting structures. A few genera of Ascomycetes and Zygomycetes also form similar association. Spores of the ectomycorrhizal fungi are produced aboveground and are disseminated by wind. Important ectomycorrhizal fungi from the order Agaricales are *Suillus, Leccinum, Russula, Amanita,* etc.

Ectomycorrhizal fungi are widespread in nature occurring primarily on forest trees including willow, veech and oak, as well as on all the members of Pinaceae (Dahm, 2006). The hyphae of EMF usually produce a highly interwoven network known as "fungus mantle" around the outer surface of feeder roots. From the mantle the hyphae penetrate the epidermis (epiblema) and form a network lamellae (Hartig net) between root epidermal and cortical cells, sometimes at 2-3 cortical cell depth. The hyphae from the Hartig net extend intercellularly and finally reach the endodermis. Strands extending out of the mantle absorb nutrients from soil and conduct them to plant roots through the net of hyphae extended into the cortex upto the endodermis. The infected roots are morphologically distinct, being shorter, swollen, branched and devoid of root hairs. Ectomycorrhizal roots may appear white, brown, yellow or black depending on the fungal species.

Endomycorrhizae: The fungi forming endomycorrhizae are usually lower fungi and mostly belong to the order Glomales of the class Zygomycetes (Agrios, 2005). Endomycorrhizae are also produced by some Basidiomycetes, i.e., orchid endophytes such as *Rhizoctonia* and *Armillaria mellea*. These fungi form symbiotic relationships with the roots of Angiosperms.

Endomycorrhizal roots externally appear similar to non mycorrhizal roots as these fungi do not produce an extensive mycelium (mantle) on root surface. However, internally the hyphae grow into the cortical cells of the feeder root either by forming specialized feeding hyphae (haustoria) called "arbuscules" or by forming large swollen food storing hyphae called "vesicles". Some endomycorrhizae contain both vesicles and arbuscules and are referred to as a vesicular arbuscular mycorrhizal fungi (VAMF), more frequently as arbuscular mycorrhizal fungi (AMF) because vesicles are not formed by all these fungi. The arbuscules penetrate the cortical cell walls and interfere with the cell plasma membrane much like the haustoria of obligate parasitic fungi. Arbuscules are also regarded as primary avenues of bidirectional transfer of nutrients between root and the fungus. Endomycorrhizae is characterized by the growth of hyphae in an intercellular fashion, penetrating root hairs, epidermal cells and cortical cells. Unlike ectomycorrhizae, the endomycorrhizal roots have root hairs. Endomycorrhizal roots are surrounded by loose mycelial growth on the root surface from which large zygospores (produced sexually) or chlamydospores (asexually) are produced underground. Most of the AM fungi belong to five genera, viz., *Acaulospora, Endogone, Gigaspora, Glomus,* and *Sclerocystis*.

Ectendomycorrhizae: A number of plant genera from the family Ericaceae and some other families in the order Ericales develop symbiotic association with the mycorrhizal fungi, which are somewhat intermediate between ecto and endomycorrhizal fungi and are called ectendomycorrhizae. Roots of the genera *Arbutus* and *Arctostaphylos* (Ericaceae) have shown this kind of mycorrhizal association. Their short roots become swollen and are infested with a hyphal sheath. There is no Hartig net but intercellular coils develop in the outer cortical cells that are eventually lysed. The mycorrhizal fungi from the families, Gentianaceae and Orchidaceae form hyphal coils.

Plant parasitic nematodes and mycorrhizal fungi share plant roots for space and for drawing nutrition. As a result they develop some relationship in which the fungus colonization and/or sporulation are enhanced or depressed. Likewise, the mycorrhizal fungi may have negative or positive effects on nematode penetration, development and reproduction. The nematode reproduction may be suppressed due to competition for availability of root mass for nematode penetration and development (Elsen et al., 2003c), preinduction of resistance (Harrier and Watson, 2004) or production of hormones/enzymes/toxins by the mycorrhizal fungi (Hol and Cook, 2005). Some nematodes may directly inhibit the colonization by mycorrhizal fungi by feeding on the mycelium. The interactive effects of nematode and mycorrhizal fungi on the plant growth may be positive, negative or neutral. The yield, therefore, may depend on the damage caused by nematode to plant and mycorrhizae, fungus colonization on roots, its effects on the nematode pathogenesis and the net benefit the plant derives in the interacting situation. Hence, the interaction between plant nematodes and mycorrhizal fungi is discussed under the following categorization.

Effect of Plant Nematodes on Mycorrhizal Fungi

Considering the feeding habit, in a broader sense, the plant nematodes may be of two types, i.e., fungal feeders or mycophagous nematodes, and plant feeding nematodes or phytonematodes, and their effects on mycorrhizal fungi may vary greatly, hence are treated separately.

Effect of fungal feeding nematodes: The nematodes belonging to the genera *Aphelenchus, Aphelenchoides, Bursaphelenchus, Ditylenchus* and some other nematodes largely from the order Aphelenchida, possess the ancestral habit of feeding on mycelium of fungi including mycorrhizal fungi. The mycophagous activity of the nematodes causes considerable reduction in the fungus colonization. Grazing of the fungal hyphae externally on the root surface may disconnect the mycorrhizal mycelium from the soil (Borowicz, 2001). In this situation additional uptake of nutrients from the soil through the mycorrhizal fungi to host plant decreases or ceases, hence the plant growth and yield are reduced (Hol and Cook, 2005). This kind of adverse effect of nematodes is more pronounced on ectomycorrhizae. Larvae of *Aphelenchus cibolensis* feed and reproduce on numerous species of ectomycorrhizal fungi (Riffle, 1971). It has been observed that *Aphelenchoides bicaudatus* severely decreased the growth of five endomycorrhizal fungi tested, but *Aphelenchus avenae* did not affect the sporulation by *Gigaspora margarita* or *Glomus etunicatum* (Francl, 1993).

Effect of plant feeding nematodes: Phytonematodes can greatly influence the sporulation and/or colonization by mycorrhizal fungi depending on the mode of parasitism (Borowicz, 2001; Annemie *et al.*, 2003). Majority of the nematodes are ectoparasites which feed on root hairs and/or root surface (epidermal cells or subepidermal cells). Ectomycorrhizal fungi are directly affected by the nematode feeding, as a result their colonization on the root surface (fungus mantle) is suppressed (Francl, 1993). Establishment of five ectomycorrhizal fungi, *Scleroderma* spp. was decreased due to the feeding by *Tylenchorhynchus gladiolatus* (Villenave and Duponnois, 2002). It has been observed that ectoparasitic nematodes cause greater damage to ectomycorrhizal fungi (Hol and Cook, 2005). The browsing of ectoparasites may damage extra radical hyphal growth and possibly diminish entry of ectomycorrhizal fungi into the roots (Tuffen *et al.*, 2002). Several species of *Tylenchoryhnchus* are reported to suppress the colonization of *Glomus* spp.

Migratory endoparasitic nematodes such as *Radopholus*, *Pratylenchus* and *Hirschmaniella* spp. damage the cortical tissue of roots where mycorrhizal fungi also colonize. As a result, the sporulation and colonization of mycorrhizal fungi may be influenced. The feeding and migration of nematodes cause necrosis to the cortical cells which also happen to be the site of infection by mycorrhizal fungi. In addition, the nematode activity may also disturb arbuscules of the fungus. Elsen *et al.* (2003a) reported 6-34% suppression in the colonization of *Glomus mosseae* due to infection by *R. similis*. Similarly, *P. coffeae* (Elsen *et al.*, 2003b) and *P. penetrans* (Talavera *et al.*, 2001) adversely affected the colonization of *Glomus* spp. Infection of banana plants by *Radopholus similis* resulted in a decreased colonization by *G. fasciculatum*. The destructed cortical tissue presumably reduced the habitable site or substrate for the AM fungus. It has been frequently noticed that migratory endoparasitic nematodes cause greater suppression in the colonization of endomycorrhizal fungi.

Generally the adverse effects of sedentary endoparasitic nematodes on mycorrhizal fungi are not as intense as those caused by migratory endoparasites. Some researchers have shown that mycorrhizal activity is not affected due to the infection by root knot nematodes (Jothi and Sundarababu, 2002; Rao *et al.*, 2003) or cyst forming nematodes, *Heterodera* and *Globodera* spp. (Ryan *et al.*, 2000; 2003). Other researchers, however, have observed notable suppression of fungal colonization due to root knot nematode infection. The mycorrhizal fungi are sensitive to transformation in roots such as galls, syncytia and nurse cells induced by sedentary endoparasites. The fungus, *Glomus macrocarpum* could not colonize on the galls induced by *M. incognita* and the adjacent areas in roots of soybean, but sporulated normally (Maia *et al.*, 2006). Waceke *et al.* (2001) and Diedhiou *et al.* (2003) have recorded 10-11% and 6-22% suppression of colonization by *Glomus etunicatum* and *G. coronatum* due to infection by *Meloidogyne* spp. on chrysanthemum and tomato, respectively. Todd *et al.* (2001) have reported 27-28% suppression of *G. mosseae* due to infection by *H. glycines* on soybean. *Heterodera cajani* also suppressed the root infection and sporulation of *G. fasciculatum* on cowpea.

Effect of Mycorrhizal Fungi on Plant Nematodes

The mycorrhizal association of plants generally have adverse effects on the pathogenesis of nematodes sharing the root. Biochemical alterations in the root rather than physical competition appear to be mainly involved in the nematode suppression. Phenolic compounds and phosphorus/potassium play an important role in the disease resistance. All these chemicals are found to be in higher amount in the roots colonized by mycorrhizal fungi. Phytoalexin production has been noticed greater in mycorrhizal roots (Edwards *et al.*, 1995), as a result migratory ectoparasitic nematodes did less damage to AMF infected plants (Salonen *et al.*, 2001).

Changes in root exudates that may decrease attraction to nematode larvae may also contribute to a lower infection by plant nematodes (Ryan *et al.*, 2000; Marschner and Baumann, 2003). Reduced penetration of root knot nematodes in the AM fungus colonized roots (Mahanta and Phukan, 2000) may also be due to changes in cell wall composition (Van Buuren *et al.*, 1999). Strong evidence exists for induction of host resistance and/or tolerance by mycorrhizal fungi against plant nematodes (Harrier and Watson, 2004). Mycorrhizal fungi can also directly affect sedentary nematodes by infecting them. Cysts and adults of *Heterodera* species collected from AM fungi infested soils have been found full of the fungus spores (Tribe, 1977). The information available on nematode-mycorrhizal interaction has shown that effects of mycorrhizal fungi on nematodes largely depend on parasitic behaviour of nematodes, which is explained accordingly.

Ectoparasitic nematodes: Researches have revealed that mycorrhizal fungi may cause stimulatory, suppressive or no effect on plant parasitic nematodes. Borowicz (2001) have suggested that ectoparasitic nematodes benefit from the mycorrhizal association. As a result they are more damaging on mycorrhizal roots. In a study on interaction of nematode (*Helicotylenchus digonicus*) and ectomycorrhizal fungus (*Suillus* *luteus*), population of the nematode increased in the pine rhizosphere. However, colonization by *Glomus fasciculatum* suppressed the reproduction of *Tylenchorhynchus vulgaris* by 7-50%. In another study, any effect of *Glomus* spp. was not observed on *Criconemella* spp. and *Tylenchorhynchus* spp. (Hol and Cook, 2005).

Migratory endoparasitic nematodes: It has been observed that colonization of mycorrhizal fungi may suppress the penetration and development of *Hoplolaimus, Pratylenchus, Radopholus*, etc. *Hoplolaimus galeatus* and *Pratylenchus penetrans* penetrated and parasitized normally both mycorrhizal and non-mycorrhizal roots, but the invasion and development of *P. brachyurus* was lower on the cotton roots inoculated with *Gigaspora margarita* than non-mycorrhizal roots. Similarly, penetration of *Radopholus similis* larvae was significantly lower in the banana roots colonized by *Glomus mosseae* (Elsen *et al.*, 2001). A few studies have shown that the colonization by AM fungi (*Glomus intraradices*) may provide protection to plants from the infection of lesion nematode, *Pratylenchus coffeae* (Elsen *et al.*, 2003c).

Sedentary endoparasitic nematodes: Most of the interaction researches on mycorrhizal fungi and plant nematodes have involved sedentary nematodes, especially cyst forming and root knot nematodes. The feeding sites of the fungus and nematode differ greatly, the fungus colonizes in the cortex whereas the nematode feed on nurse cells in stelar tissue. However, penetration and movement of the nematode may be influenced due to intercellular mycelium of mycorrhizal fungi. Population of *M. incognita* in cotton root reduced due to inoculation with *G. fasciculatum*. Galling and reproduction of *M. incognita* was significantly low on black pepper colonized by *G. fasciculatum* or *G. etunicatum*. In another study, penetration of *M. incognita* in tomato and cotton root remained unchanged in the presence of *G. fasciculatum* or its absence. *G. fasciculatum* also suppressed the cyst production and reproduction of *Heterodera cajani* on cowpea. The reduction in cyst number could be either due to reduced penetration of roots by the juveniles, or pathogenic effects of *Glomus* spp. on sedentary females or cysts.

Nematode-Insect Interaction

Considerable information on nematode-insect interaction is available which has revealed that nematodes and insects can interact in different ways and develop a synergistic or antagonistic relationship. Infact, antagonistic effects of nematodes as entomopathogenic nematodes have been known for almost a century and a lot of information is available on this aspect of interaction (Gaugler, 2002). Other kinds of interactions, however, have not been studied adequately, although some information is available which is summarized under:

Synergistic Interaction

In a synergistic relationship, it is the nematode that largely draws benefit from the insect's presence. The insect basically acts as a vector to carry nematodes internally from a diseased host tree to a healthy tree. In the primarily vector relationship, nematode develops two distinct associations with the insect. One type is where the nematode does not draw nutrition from the vector i.e., phoretic relationship. The other type is where the nematode draws nutrition from the vector and the insect larvae feed in the galls induced by the nematodes i.e., mutualistic relationship.

Phoretic relationship: The role of insects under this relationship much resembles the role of nematodes in nematode-virus interactions. There are a few diseases caused by aphelenchid nematodes where the role of insect as a vector is an essential phase of the disease development (Hunt, 1993). The nematode larvae in the dauer state/stage (dispersion adapted state) are carried internally in the trachea or haemocoel and are introduced into the host plant during feeding or egg laying by the vector insect. After introduction the nematode larvae and insect eggs develop and attack the host independently without having any relationship. Red ring of coconut is a classical example to represent this relationship. The disease is caused by the nematode, *Rhadinaphelenchus cocophilus* which is transmitted by the black palm weevil (*Rhynchophorus palmarum*) and sugarcane weevil (*Metamasius* spp.). The adult weevil, while feeding on diseased coconut palm, swallows hundreds of dauers of *R. cocophilus* (J_3) along with the red wood. These larvae reach internally at the site of ovipositor of the weevil and are deposited along with the insect eggs in leaf/bud axil of healthy palms. Another good example is the pine wilt disease caused by *Bursaphelenchus xylophilus*. The dauer nematodes (J_4) are transmitted by beetles (*Monochamus alternatus* and *M. carolinensis*). The dauers enter into the respiratory system via the mouth of the insect larvae during pupation. The emerging adult female beetles have 15,000-20,000 J_4 in the trachea which migrate out via the mouth while the insect feeds on small wounds in susceptible conifers.

Mutualistic relationship: The nematode develops two distinct relationships with the vector insect. In addition of being transmitted, the nematode parasitizes the vector insect without inducing mortality, hence the nematode parasitizes the host plant as well as the vector insect in haemocoel. A typical example to demonstrate the dual parasitism is flower bud gall disease of eucalyptus in Australia and syzygium in India caused by the nematode, *Fergusobia tumifaciens* which is transmitted by *Fergusonina* fly. The adult female fly oviposits into young leaves of *Euca-*

lyptus stuartiana, E. comendulosa, Syzygium cumini or into flower buds of *E. macrorrhyncha* along with nematode juveniles. The juvenile activity causes the formation of galls. The fly also injects some gall-inciting chemical during oviposition. Inside the galls, the nematode juveniles soon develop into parthenogenetic females which feed on the plant tissue and lay eggs. The hatched larvae feed on mesophyll cells and become obese as they do not migrate to initiate a new feeding site. By this time *Fergusonina* larvae emerge from the eggs and enter into the galls, feed on the parenchymatous masses and develop to the third instar. The infective female and male nematodes penetrate the fly larvae and grow rapidly in the fat body of the pupa and become adults. They lay eggs in the haemocoel of the fly which hatch, and the juveniles migrate to the oviduct. Concurrent with this development, pupation completes and the adult female of *Fergusonina* flies to feed on healthy trees.

The other important example of dual parasitism is flower gall disease of figs. The adult female of fig wasp, *Blastophaga psenes* carries 200-400 juveniles and adults of the nematode, *Schistonchus caprifici* in the haemocoel to the inflorescence of caprifig and cultivated fig (Hunt, 1993). Nematode larvae are introduced during oviposition into male and female florets (developing flowers), where they develop and reproduce. Endoparasitic feeding of the nematode causes necrosis and cavities in the cortical parenchyma of florets leading to formation of flower galls. Wasp larvae also develop during this time and penetrate the galls which contain nematodes. The nematode larvae enter into the wasp larvae through mouth, and feed on fat bodies to grow further. The nematode infested wasp pupate and overwinter inside the flower galls. A gall may have upto 20 small gallets.

Antagonistic Interaction

The antagonistic interaction between nematodes and insects is mainly based on predation. However, both nematodes and insects do not predate each other simultaneously, rather some nematodes predate upon insects, whereas other nematodes act as a prey to insects.

Entomopathogenic nematodes: Nematodes from 30 families have the ability to parasitize the insect's body and cause mortality to it (Forst and Clarke, 2002). These nematodes feed internally on the insect tissue and are commonly known as entomopathogenic nematodes (EPN). The nematodes from seven families, viz., Mermithidae, Allantonematidae, Neotylenchidae, Sphaerularidae, Rhabditidae, Steinernematidae, and Heterorhabditidae have great potential in insect pest management (Lacey *et al.*, 2001; Riga *et al.*, 2006; Daniel *et al.*, 2006). Important entomopathogenic nematodes, however, largely belong to Steinernematidae and Heterorhabditidae (Adams and Nguyen, 2002; Shapiro-Ilan et al., 2006). The EPNs from Steinernematidae and Heterorhabditidae are terrestrial and form a symbiotic association with certain gut bacteria from two genera, *Xenorhabdus* and *Photorhabdus*, respectively to kill the insect host by septicemia (Flores-Lara et al., 2007; Goodrich-Blaire and Clarke, 2007). Nematodes of the family Mermithidae are aquatic and kill the host by tearing through the cuticle. The entomopathogenic nematode, *Steinernema carpocapsae* (Steinernematidae) enter the host through natural openings and also by penetrating the host cuticle to reach the haemocoel (Dowds and Peters, 2002). The nematode releases *Xenorhabdus nematophila* (bacterium) into the insect haemocoel. The bacterium has symbiotic association with the nematode in gut region (Wright and Perry, 2002). The bacterium multiply very rapidly and cause septicemia within 24-48 hours, as a result the insect is killed. The bacterium produces some toxins which inhibit putrefaction of the insect, providing an opportunity to the nematode larvae to feed and reproduce, and at the same time remain protected inside the insect cadaver. During feeding the nematode also ingest bacterial cells which remain viable in the gut of nematode. An infected cadaver may contain over 100,000 nematode larvae. In a similar manner *Heterorhabditis bacteriophora* (Heterorhabditidae) in symbiotic association with the bacteria *Photorhabdus luminescens/P. laumondii* predate upon insects (Boemare, 2000). Some important entomopathogenic nematodes along with bacterial symbionts and insect hosts (order) are presented in Table 16.

Nematophagous insects: Numerous insects, especially the micro arthropods, are known to prey on nematodes and may sometimes be involved in the population balance in natural soil (Stirling, 1991). Some insects are fast but nonspecific feeders of nematodes. The collembolla, *Entomobyroides dissimilis* is an indiscriminate feeder and consumes around 1,000 nematode larvae in 24 hours. Mesostigmatic mite, *Lasioseius scapulatus* may feed specifically on *Aphelenchus avenae* and may reduce the nematode population. *Tyrophagus similis* preys on the females of *Heterodera avenae*, whereas the mite, *Hypoaspis aculeifer* feeds on egg masses of *Meloidogyne chitwoodi*, and *Caloglyphus berlesei* quickly consumes egg masses of *M. javanica*.

Nematode-nematode Interaction

Plant pathogens, in nature even within a group, do not occur in pure culture but in cohabitance. However, only a few attain the threshold level and are able to cause a disease, but all the pathogens present in the rhizo-

Table 16 Important entomopathogenic nematodes and their symbiotic bacteria

Entomopathogenic Nematodes	Symbiotic Bacteria	Host Insect (Order)
Steinernema kraussiae	Xenorhabdus bovienii	Lepidoptera, Diptera, Coleoptera
Heterorhabditis bacteriophora	Photorhabdus luminescens	Lepidoptera, Coleoptera
H. bacteriophora subgroup HP 88	Photorhabdus luminescens laumondii	Lepidoptera, Coleoptera
H. bacteriophora subgroup NC	P. temperata	Coleoptera, Lepidoptera
H. indica	P. luminescens akhurstii	Coleoptera
H. megidis Nearctic group	P. temperata	Hemiptera
H. megidis Palaearctic group	Photorhabdus temperata temperata	Coleoptera
H. zealandica	P. temperata	Coleoptera, Lepidoptera
S. abbasi	Xenorhabdus spp.	Lepidoptera
S. affine	X. bovienii	Lepidoptera
S. bicornutum	Xenorhabdus spp.	Lepidoptera
S. carpocapsae	X. nematophila	Lepidoptera, Coleoptera, Diptera
S. feltiae	X. bovienii	Isoptera, Diptera, Fungus gnats, Sciarid flies
S. glaseri Sub group Brecon	X. poinarii luminescens	Coleoptera
S. kushidai	X. japonica	Coleoptera
S. longicaudum	X. beddingii	
S. riobrave	Xenorhabdus spp.	Lepidoptera, Coleoptera
S. scapterisci	Xenorhabdus spp.	Orthoptera

sphere influence the population development of each other. This is generally true for plant parasitic nematodes also. Plant nematodes generally occur in polyspecific communities because of their persistence, polyphagous feeding habits, wide distribution and weak interspecific competition (Landi and Manachini, 2005). Manachini et al. (2005) have recorded nematodes belonging to 14 families and 24 genera from a canola field. Despite weak competition, the nematode community is very dynamic and its members interact continuously with each other. The interaction may be among the genera, species or even races. In interaction, one population of the nematode influences the other, as a result reproduction of a population may be enhanced or inhibited. Entomopathogenic nematodes (EPN) also cohabit in soil but they usually do not interact mutually (Neumann and Shields, 2006; Daniel et al., 2006). The EPNs, however, may affect pathogenesis and reproduction of phytonematodes (Lewis et al., 2001; Crow et al., 2006).

Feeding habit, survival mechanism and ecological requirements of different plant parasitic nematodes vary considerably. Presence of a particular species/race in a location is related to its method of dissemination, host suitability, host range and effects of other factors (Weischer, 1993). The competition is usually strongest among organisms that are most alike with respect to their ecological and physiological demands. A single species may dominate if it is better adapted to the host or to the ecological conditions. Species usually coexist if they utilize different ecological niches or have a similar life cycle.

Species of plant parasitic nematodes in cohabitance may interact ecologically as competitors and etiologically through their effects on plant growth and yield. Both these relationships are interrelated, since the severity of disease often depends on population of the nematode, of course, it is regulated by many other factors such as host, environmental factors, rhizospheric community, etc. Hence, nematodes in the root zone of a plant may be engaged in ecological and/or etiological interactions which are discussed here separately.

Ecological Interactions

Interaction between nematodes may have adverse or favourable effects on one or all the cohabiting partners. Adverse effects may result from physiological destruction of host, spatial occupancy of feeding sites or by changes in host physiology, which alter suitability of the host (Van-der Putten et al., 2006). In a broader sense the competition is related to host suitability and pathogenicity, and persistence of nematode, and both may be density and time dependent.

The nature of parasitism may greatly influence the competition between species. The competition is more severe among species with similar feeding habits and competitive advantage increases as the host-parasite relationship becomes complex or as persistence of the nematode increases. There can be different types of ecological interactions in relation to mode of parasitism, such as among ectoparasites, ecto and migratory endoparasites, ecto and sedentary endoparasites, migratory endoparasites, migratory endo and sedentary endoparasites, sedentary endoparasites, etc., and inter and intraspecific interaction (Eisenback, 1993). Entomopathogenic nematodes also inhabit in soil, hence they may also develop some ecological relationship with plant nematodes.

Ectoparasitic nematodes: The ectoparasites which feed on root hairs and epidermal cells are more antagonistic to each other than deep-feeding nematodes. *Tylenchorhynchus claytoni* was antagonistic to *Paratrichodorus minor* on one cultivar of corn whereas *P. minor* was antagonistic to *T. claytoni* on other three cultivars of the same crop. Relative pathogenicity and competence may affect population development, e.g., highly pathogenic *Belonolaimus longicaudatus* suppressed reproduction of the less pathogenic *Dolichodorus heterocephalus* on corn. *D. heterocephalus* had no effect on the reproduction of *B. longicaudatus*. Sometimes cohabitance of ectoparasites may stimulate the reproduction of the participating species. Reproduction of *B. longicaudatus* was higher in the presence of *Hoplolaimus galeatus* on cotton.

Ecto and migratory endoparasitic nematodes: Migratory endoparasites alter root morphology and physiology as they move through the plant tissue (Morgan et al., 2002) and are, therefore, often antagonistic to ectoparasites. Reproduction of *Tylenchorhynchus martini* was suppressed 75-90% by *Pratylenchus penetrans* on red clover, but *T. claytoni* inhibited the reproduction of *P. penetrans* by 25-90% in tobacco roots. Mutual inhibitory effects of *Paratrichodorus minor* and *Pratylenchus zeae* have also been observed on different cultivars of corn (Landi and Manachini, 2005).

Ecto and sedentary endoparasitic nematodes: These two groups have very different feeding habits and sites. They usually coexist but rarely appear to affect each other. In a few cases, however, antagonism has been observed. Development and reproduction of *M. incognita* was lower in the presence of *T. brassicae* on cauliflower (Khan and Khan, 1986). Sedentary endoparasitic nematodes can also suppress ectoparasites though they occupy different plant tissues probably through some alteration in host physiology. Infection by *M. hapla* has resulted to a decline in the population of *Xiphinema americanum* in alfalfa fields.

Migratory endoparasitic nematodes: These nematodes possess similarity in feeding behaviour, and generally utilize the same feeding site that results in a competitive interaction. Host suitability and climatic conditions may be important factors in the interaction (Eisenback, 1993). In a study, *Radopholus similis* and *P. coffeae* were found mutually suppressive on citrus while in cohabitance, but fine textured soils favoured the reproduction of *P. coffeae*, whereas coarse textured soil was found beneficial for *R. similis*.

Migratory and sedentary endoparasitic nematodes: These two groups of nematodes differ in the feeding sites, but it has been frequently observed that migratory endoparasitic nematodes disrupt the feeding of sedentary endoparasitic nematodes. Development and reproduction of root knot nematode, *Meloidogyne incognita* was greatly inhibited in the presence of *Pratylenchus penetrans* in tobacco roots. Sometimes sedentary endoparasites may also antagonize the migratory endoparasitic nematodes, this usually occurs with the preinoculation with sedentary endoparasites. Prior inoculation with *M. incognita* inhibited penetration of *P. brachyurus* in tomato roots. Synergism may also develop between the migratory endo and sedentary endoparasitic nematodes in cohabitance. Reproduction of *P. brachyurus* was significantly higher in the presence of *M. incognita* on tobacco. Moens *et al.* (2006) have reported mutually inhibitory effects of *Helicotylenchus multicinctus*, *P. coffeae*, *Radopholus similis* and *M. incognita* on banana.

Sedentary endoparasitic nematodes: These nematodes have highly specialized parasitism and develop a long lasting relationship with their host. Competition between the genera/species is generally mutually suppressive because they share the same site for feeding. Root knot nematodes may inhibit, have no effect or stimulate cyst forming nematodes. *Heterodera oryzicola* was inhibited by *M. graminicola* on rice. *M. hapla*, *M. incognita* and *H. cajani* did not influence each other on cowpea. The prior inoculation of *M. hapla* on sugar beet stimulated *H. schachtii*. Similarly, *Heterodera* species may also inhibit or have no effect on root knot nematodes. *H. zeae* inoculated three days prior to *M. incognita* antagonized the latter. Likewise, *M. hapla* was inhibited by *H. schachtii* on tomato.

Interspecific interaction: Information on interaction among species of a genus are more common with root knot and cyst nematodes. A few interactions for *Pratylenchus* species are also known. Concomittant infection by two or more species of *Meloidogyne* is of common occurrence in a field/plant root or even a gall (Khan and Khan, 1997). Factors other than competition may be important in the domination of a species. Among the four common species of *Meloidogyne*, *M. incognita* dominates at higher temperatures, e.g., *M. incognita* dominates over *M. hapla* at or above 20°C, as the latter is a low temperature species. Interaction between *M. javanica* and *M. incognita* have been found mutually inhibitory at 20-25°C (Khan and Haider, 1991). In interspecific interaction between *Globodera* species, one species dominated the other. In New Zealand *G. rostochiensis* antagonized *G. pallida*, whereas in the U.K. *G. pallida* was dominant. Sometimes two species in cohabitance, i.e., *G. rostochiensis* and *G. pallida*, and *H. glycines* and *H. schachtii* produce infertile hybrids due to intense interspecific competition. Interspecific competition between *P. penetrans* and *P. alleni* led to a change in the female to male ratio in *P. penetrans* towards maleness. Similarly, sex reversal towards maleness occurs due to overcrowding of *Meloidogyne* species (Sasser, 1989).

Phyto and entomopathogenic nematodes: Entomopathogenic nematodes (EPN), especially from the families Steinernematidae and Heterorhabditidae, inhabit in soil along with plant nematodes. Hence, they may develop some relationship while in cohabitance, although feeding habits of both groups differ drastically. There are some reports which indicate that cohabitance of EPN and plant nematodes leads to antagonistic interaction. As a result, population of plant parasitic nematodes may be suppressed (Gouge et al., 1994; Grewal et al., 1997). The mechanism involved in the interaction may be competitive in nature; the EPN are bigger in size and move faster than plant nematodes while in search of insect prey. Allelochemicals (repellant, suppressant, toxins, etc.) produced by EPN and the bacterial symbiont may play a vital role in the suppression of plant nematodes (Grewal et al., 1999). Most of the studies conducted to examine the effect of EPN on plant nematodes have used root knot nematodes, and have demonstrated reduction in the galling and soil population of *Meloidogyne* spp. (Smitely et al., 1992; Gouge et al., 1994). Application of *Steinernema feltiae* @ 100 to 5,000 infective juveniles decreased the gall formation, egg mass production and egg hatching of *M. incognita* (Lewis et al., 2001). In most of the interactions so far investigated, specificity with regard to plant parasitic nematode and host plant growth has not been determined. Suppression of plant nematode has been found dependent on EPN population. These interactions, however, may be of great agricultural importance because reduction in the population of plant nematodes will have an additional and off target effect of EPN that may lead to an effortless increase in the crop productivity.

Aetiological Interaction

Studies dealing with the causes of disease have generally considered that a single nematode species may be responsible for development of symptoms of the disease, as well as plant growth reduction. However, nematodes often occur in polyspecific communities, they may influence each other to alter the course of the disease. The final course of the disease and amount of the damage to host are mainly controlled by the type of association or relationship that develops between the nematode species in cohabitance (Eisenback, 1993). Hence, the interactive effects of two or more nematode genera/species may be synergistic, antagonistic or neutralistic.

Synergistic effects on plants: In the synergistic interaction presence of one nematode species modifies the host in such a manner that it favours the invasion and infection by the other nematode species in cohabitance, or there may be unilateral or mutually beneficial effects. As a result greater damage to plants by two or more nematode species generally occurs. The advantage may come through predisposition, involving changes in host physiology which may render the plants easier for penetration and/or suitable for feeding (Khan, 1993). Most of the synergistic interactions so far reported for the mixed populations of nematodes involve a species of *Meloidogyne*. In an experiment, concomitant inoculation of tomato with *H. schachtii* and *M. hapla* caused 61-65% greater reduction in the plant growth compared to the growth of the plants inoculated with either nematode. *M. incognita* and *M. javanica* have also interacted synergistically, producing severe galling on tomato (Khan and Haider, 1991).

Antagonistic effects on plant growth: Sometimes the disease severity decreases when two or more nematode species attack a plant. This happens due to strong competition between the nematode species in cohabitance (Eisenback, 1993). The less pathogenic species reduce the number of infection sites otherwise available to more pathogenic nematode species. Other factors, in particular the morphological/physiological changes in the host, may also be inhibitory for one or both the cohabiting species. For instance, migratory endoparasites disrupt feeding sites of sedentary endoparasites, as a result the female starve and galls are not formed (Khan and Khan, 1997). Root knot nematode, *M. incognita* while acting alone caused more damage to tomato than with *Pratylenchus penetrans* (Estores and Chen, 1970). A similar effect has been observed on black pepper with *M. incognita* and *Rotylenchulus reniformis*.

References

Adams, J.B. and Nguyen, B.K. 2002. Taxonomy and systematics. *In:* Entomopathogenic Nematology. R. Gaugler (ed.), CAB International, Wallingford, Oxon, UK. pp. 1-33.

Agrios, G.N. 2005. Plant Pathology. V Edition, Elsevier Academic Press, California, USA. pp. 922.

Akkermans, A.D.L., Abdulkader, S. and Trinick, M.J. 1978. Nitrogen fixing root nodules in Ulmaceae. Nature 247: 190.

Ali, S.S., Pervez, R., Hussain, M.A. and Ahmad, R. 2005. Artificial media for mass production of entomopathogenic nematodes. Pulses Newsletter, IIPR, Kanpur, India. 16: 3.

Andrassy, I. 1974. A nematodak evolucioja es rendszerezese. MTA Biol Oszt. Koal. 17: 13-58.

Andrassy, I. 1976. Evolution as a Basis for the Systemization of Nematodes. Pitman Publishing, London, UK. pp. 288.

Annemie, E., Stephan, D. and Dirk, D.W. 2003. Use of root organ cultures to investigate the interaction between *Glomus intraradices* and *Pratylenchus coffeae*. Applied and Environmental Microbiology 69: 4308-4311.

Anonymous, 2006. http://www.aber.ac.uk/mp gwww/Edu/Anti DrugTxt.html.

Atkinson, G.F. 1892. Some diseases of cotton. Bulletin of Alabama Agricultural Experimental Station 41: 61-65.

Ayyar, R.K. 1926. http://www.ikisan.com/links/tn turmeric Nematode Management. shtml.

Ayyar, R.K. 1933. Some experiments on the root gall nematode *Heterodera radicicola* Greeff in S. India. Madras Agricultural Journal 21: 97-107.

Back, M.A., Maydock, P.P.J. and Jenkinson, P. 2002. Disease complexes involving plant parasitic nematodes and soil borne pathogens. Journal of Phytopathology 150-469.

Bajaj, H.K. 2002. Diversity of semiendo and endoparasitic nematodes of India. *In*: Nematode diversity. M.S. Jairajpuri (ed.), Silverline Printers, Hyderabad, India. pp. 259-268.

Barber, C.A. 1901. Department of land records and agriculture. Madras Agricultural Branch 2 Bulletin 45: 227-234.

Bastian, H.C. 1865. Monograph on the Anguillulidae or free nematodes, marine, land and fresh water, with description of 100 new species. Transactions of the Linnean Society of London 25: 73-184.

Beaver, J.B., McCoy, C.W. and Kaplan, D.T. 1983. Natural enemies of subterranean *Diaprepes abbreviatus* (Coleoptera: Curculionidae) larvae in Florida. Environmental Entomology 12: 840-843.

Berkeley, M.J. 1855. Vibrios forming cysts on the roots of cucumber (editorial) Gdnr's Chron. No. 14 pp. 220.

Bird, A.F. 1980. The nematode cuticle and its surfaces. *In*: Nematodes as Biological Models. B.M. Zuckerman (ed.), Vol II, Academic Press, London, UK. pp. 213-236.

Bird, D.M. 2004. Signaling between nematodes and plants. Current Opinion Plant Biology 7: 372-376.

Bird, D.M. and Koltai, H. 2000. Plant Parasitic Nematodes: Habitats, Hormones and Horizontally acquired genes. Journal of Plant Growth Regulators 19(2): 183-194.

Blaire, 1964. Red ring disease of coconut palm. Journal of Agriculture Society Trinidad and Tobago 64: 31-49.

Boemare, N. 2000. Biology, taxonomy and systematics of *Photorhabdus* and *Xenorhabdus*. *In*: Entomophathogenic Nematology. R. Gaugler (ed.), CAB International, Wallingford, Oxon, UK. pp. 35-56.

Borowicz, V.A. 2001. Do arbuscular mycorrhizal fungi alter plant pathogen relations? Ecology 82: 3057-3068.

Bridge, J., Coyne, D.L. and Kwoseh, C.K. 2005a. Nematode parasites of tropical roots and tuber crops. *In*: Plant Parasitic Nematodes of Tropical and Sub Tropical Agriculture. M. Luc, R.A. Sikora and J. Bridge (eds.), CAB International, Wallingford, Oxon, UK. pp. 221-258.

Bridge, J., Ploweright, R.A. and Peng, D. 2005b. Nematode parasites of rice. *In*: Plant Parasitic Nematodes of Tropical and Sub Tropical Agriculture. M. Luc, R.A. Sikora and J. Bridge (eds.), CAB International, Wallingford, Oxon, UK. pp. 87-130.

Brody, J.R., Calhoun, E.S., Gallmeier, E., Creavalle, T.D. and Kern, S.E. 2004. Ultrafast high resolution agarose electrophoresis of DNA and RNA using low molarity conductive media. Biotechniques 37: 598-602.

Brown, D.J.F., Dalmasso, A. and Trudgill, D.L. 1993. Nematode pests of soft fruits and vines. In: Plant Parasitic Nematodes in Temperate Agriculture. K. Evans, D.L. Trudgrill and J.M. Webster (eds.), CAB International, Wallingford, Oxon, UK. pp. 427-462.

Buena, A.P., Lopez-Perez, J.A., Bello, A., Diez-Rojo, M.A., Robertson, L., Escuer, M. and Leon, L. de. 2006. Screening of cuarentino pepper (*Capsicus annuum* L.) for resistance to *Meloidogyne incognita*. Nematropica 36: 13-24.

Burton, 1926. MIT biography of Hillier http://web.mit.edu/Invent/iow/hillier.html.

Burton, E.F., Hillier, J. and Prebus, A. 1939. A report on the development of the electron microscope at Toronto. Physical Review 56: 1171-1172.

Butler, E. 1913. An eelworm disease of rice. Agriculture Research Institute, Pusa Bulletin 34.

Campos-Herrera, R., Escuer, M., Robertson, L. and Gutierrez, C. 2006. Morphological and ecological characterization of *Steinernema feltiae* (Rhabditida : Steinernematidae) Rioja strain isolated from *Bibio hortulanus* (Diptera: Bibionidae) in Spain. Journal of Nematology 38: 13-19.

Carne, M.W. 1926. Earcockle (*Tylenchus tritici*) and a bacterial disease (*Pseudomonas tritici*) on wheat. Journal of Departmental Agriculture, West Australia. 31: 508-512.

Carter, W. 1943. A promising new soil amendment and disinfectant. Science 97: 383-384.

Castillo, P., Di Vito, M., Volvas, N. and Jimenez-Diaz, R.M. 2001. Host parasite relationships in root-knot disease of white mulberry. Plant Disease 85: 277-281.

Chitwood, B.G. 1937. A revised classification of the Nematoda. Papers in Helminthology, 30 year Jubileum K.I. Skrjabin, Moscow. pp. 69-80.

Chitwood, B.G. 1949. Root-knot nematodes-Part I. A revision of the genus *Meloidogyne* Goeldi. 1887. Proc. Helminth. Soc. Washington 16: 90-104.

Chitwood, B.G. 1950. General structure of nematodes. In: An Introduction to Nematology. B.G. Chitwood and M.B. Chitwood (eds.), Section 1, Anatomy, Monumental Printing Co., Baltimore, USA. pp. 7-27.

Chitwood, D.J. 1987. Inhibition of steroid or hormone metabolism or action in nematodes. In: Vistas on Nematology. J.A. Veech and D.W. Dickson (eds.), SON Publications, Hyattsville, Maryland, USA. pp. 122-130.

Choi, I.H., Shin, S.C. and Park, I.K. 2007. Nematicidal activity of onion (*Allium cepa*) oil and its components against the pine wood nematode (*Bursaphelenchus xylophilus*). Nematology 9: 231-235.

Christie, J.R. 1941. Life history (zooparasitica). Parasites of invertebrates. In: An Introduction to Nematology. J.R. Christie (ed.), Babylon, NY, USA. pp. 246-266.

Christie, J.R. 1942. A description of *Aphelenchoides besseyi* n. sp., The summer dwarf nematode of strawberries with comments on the identity of *Aphelenchoides subtenius* (Cobb. 1926) and *Aphelenchoides hodsoni* Goodey, 1935, Proceedings of the Helminthological Society of Washington 9: 82-84.

Christie, J.R. 1945. Some preliminary tests to determine the efficacy of certain substances when used as soil fumigants to control the root-knot nematode, *Heterodera marioni* Goodey. Proc. Helminth. Soc. Washington 12: 14-19.

Cobb, N.A. 1893. Nematodes, mostly Australian and Fijian. Macleay Memorial Volume, Linnean Society of New South Wales pp. 252-308.

Cobb, N.A. 1918. Estimating the nema population of soil. Agricultural Technical Circulation USDA. 1: 48.

Cobb, N.A. 1919. The orders and classes of nemas. Contributions to the Science of Nematology VIII, 213-216.

Cristobal-Alejo, J., Tun-Suarez, J.M., Moguel-Catzin, S., Marban-Mendoza, N., Medina-Baizabal, L., Sima-Polanco, P., Peraza-Sanchez, S.R. and Gamboa-Angulo, M.M. 2006. *In vitro* sensitivity of *Meloidogyne incognita* to extracts from native yucatecan plants. Nematropica 36: 89-98.

Crow, W.T., Porazinska, D.L., Giblin-Davis, R.M. and Grewal, P.S. 2006. Entomopathogenic nematodes are not an alternative to fenamiphos for management of plant parasitic nematodes on golf courses in Florida. Journal of Nematology 38: 52-58.

Curtis, R.H.C. 2007a. Do phytohormones influence nematode invasion and feeding site establishment? Nematology 9: 155-160.

Curtis, R.H.C. 2007b. Plant parasitic nematode proteins and the host-parasitic interaction. Briefings in Functional Genomics and Proteomics. doi: 10.1093/bfgp/elm 06.

Dabirea, K.R., Saliou, N., Jean-Luc, C., Sabine, F., Mamodou, T.D. and Thierry, M. 2005. Influence of irrigation on the distribution and control of the nematode *Meloidogyne javanica* by the biocontrol bacterium *Pasteuria penetrans* in the field. Journal Biology and Fertility of Soils 41: 205-211.

Dahm, H. 2006. Role of mycorrhizae in forestry. *In*: Handbook of Microbial Biofertilizers. M.K. Rai (ed.), International Book Distributing Co., Lucknow, India. pp. 241-270.

Daniel, S.G., Karthik, R. and Donald, R.S. 2006. Soil mediates the interaction of coexisting entomopathogenic nematodes with an insect host. Journal of Invertebrate Pathology 94: 12-19.

Dasgupta, M.K. 1998. Phytonematology. Naya Prakash, Kolkata, India. pp. 846.

Dastur, J.P. 1936. A nematode disease of rice in the Central Provinces. Proceedings of Indian Academy of Sciences 4: 108-122.

De Coninck, L., Theodorides, J., Roman, E., Ritter, M. and Chabaud, A.G. 1965. Systmatique des nematodes. *In*: Traite de Zoologie. P.P. Grasse (ed.), Masson, Paris. pp. 586-1200.

de Man, J.G. 1884. Di frei in der reinen Erde und im sussen Wasser lebenden Nematoden der Niederlandischen Fauna. Brill, Leiden. pp. 206.

Desaeger, J.A. and Csinos, A.S. 2006. Root knot nematode management in double-cropped plasticulture vegetables. Journal of Nematology 38: 59-67.

Dhawan, S.C. and Pankaj 1998. Nematode pests of wheat and barley. *In*: Nematode diseases in plants. P.C. Trivedi (ed.), CBS Publishers, New Delhi, India. pp. 13-25.

Di Vito, M., Volvas, N. and Castillo, P. 2004. Host parasite relationships of *Meloidogyne incognita* on spinach. Plant Pathology 253: 508-514.

Diedhiou, P.M., Hallmann, J., Oerke, E.C. and Dehne, H.W. 2003. Effects of arbuscular mycorrhizal fungi and a non pathogenic *Fusarium oxysporum* on *Meloidogyne incognita* infestation on tomato. Mycorrhiza 13: 199-204.

Dowds, A.C.B. and Peters, A. 2002. Virulence mechanisms. *In*: Entomopathogenic Nematology. R. Gaugler (ed.), CAB International, Wallingford, Oxon, UK. pp. 79-114.

Duncan, L.W. 2005. Nematode parasites of citrus. *In*: Plant Parasitic Nematodes of Tropical and Sub tropical Agriculture. M. Luc, R.A. Sikora and J. Bridge (eds.), CAB International, Wallingford, Oxon, UK. pp. 87-130.

Edwards, R., Mizen, T. and Cook, R. 1995. Isoflavonoid conjugate accumulation in the roots of Lucerne (*Medicago sativa*) seedlings following infection by the nematode (*Ditylenchus dipsaci*). Nematologica 41: 51-66.

Eisenback, J.D. 1993. Interactions between nematodes in cohabitance. *In*: Nematode Interaction. M.W. Khan (ed.), Chapman & Hall, London, UK. pp. 134-174.

Elsen, A., Declerck, S. and De Waele, D. 2001. Effects of *Glomus intraradices* on the reproduction of the burrowing nematode (*Radopholus similis*) in dixenic culture. Mycorrhiza 11: 49-51.

Elsen, A., Baimey, H., Swennen, R. and De Waele, D. 2003a. Relative mycorrhizal dependency and mycorrhiza-nematode interaction in banana cultivars (*Musa* spp.) differing in nematode susceptibility. Plant and Soil 256: 303-313.

Elsen, A., Beeterens, R., Swennen, R. and De Waele, D. 2003b. Effects of an arbuscular mycorrhizal fungus and two plant parasitic nematodes on *Musa* genotypes differing in root morphology. Biology and Fertility of Soils 38: 367-376.

Elsen, A., Declerck, S. and De Waele, D. 2003c. Use of root organ cultures to investigate the interaction between *Glomus intraradices* and *Pratylenchus coffeae*. Applied and Environmental Microbiology 69: 4308-4311.

Endo, B.Y. 1986. Histology and ultrastructural modification induced by cyst nematodes. *In*: Cyst Nematodes. F. Lamberti and C.E. Taylor (eds.), NATO ASI Series. Plenum Press, New York, USA. pp. 133-146.

Endo, B.Y. 1987. Histopathology and ultrastructure of crops invaded by certain sedentary endoparasitic nematodes. *In*: Vistas on Nematology. J.A. Veech and D.W. Dickson (eds.), Society of Nematologists Inc. Hyattsville, Maryland, USA. pp. 196-210.

Estores, R.A. and Chen, T.A. 1970. Interaction of *Pratylenchus penetrans* and *Meloidogyne incognita acrita* in cohabitance on tomatoes. Journal of Nematology 4: 170-174.

Evans, K. and Haydock, P.P.J. 1993. Interactions of nematodes with root-rot fungi. *In*: Nematode Interactions. M.W. Khan (ed.), Chapman & Hall, London, UK. pp. 104-133.

Fargette, M., Freitas, H., Hol, W.G.H., Kerry, B.R., Maher, N. and Mateilk, T. 2006. Nematode interactions in nature: models for sustainable control of nematode pests of crop plants. Advances in Agronomy 89: 227-260.

Faske, T.R. and Starr, J.L. 2006. Sensitivity of *Meloidogyne incognita* and *Rotylenchulus reniformis* to abamectin. Journal of Nematology 38: 240-244.

Filipjev, I.N. 1934. The classification of the free living nematodes and their relation to the parasitic nematodes. Smithson. Misc. Cillect. 89: 1-63.

Fischer, M. 1894. Uber eine Clematis-krankheit. Bericht aus dem Physiolischen Laboratorium des Landwirthschaftlichen Instituts der Universitat Halle 11 (3): 1-11.

Flores-Lara, Y., Renneckar, D., Forst, S., Goodrich-Blair, H. and Stock, P. 2007. Influence of nematode age and culture conditions on morphological and physiological parameters in the vesicle of *Steinernema corpocapsae*. Journal of Invertebrate Pathology (In Press).

Forst, S. and Clarke, D. 2002. Bacteria-nematode symbiosis. *In*: Entomopathogenic Nematology. R. Gaugler (ed.), CAB International, Wallingford, Oxon, UK. pp. 57-77.

Fourie, H., Leswifi, C., McDonald, A.H. and De Waele, D. 2007. Host suitability of vetiver grass to *Meloidogyne incognita* and *M. javanica*. Nematology 9: 49-52.

Francl, L.J. 1993. Interactions of nematodes with mycorrhizae and mycorrhizal fungi. *In*: Nematode Interactions. M.W. Khan (ed.), Chapman & Hall, London, UK. pp. 203-216.

Francl, L.J. and Wheeler, T.A. 1993. Interaction of plant-parasitic nematodes with wilt-inducing fungi. *In*: Nematode Interactions. M.W. Khan (ed.), Chapman & Hall, London, UK. pp. 79-103.

Friere, J.R.J. and Saccol de Sa, E.L. 2006. Sustainable Agriculture and the rhizobia/legumes symbiosis. *In*: Handbook of Microbial biofertilizers. M.K. Rai (ed.), International Book Distributing Co., Lucknow, India. pp. 29-49.

Gadea, E. 1973. Sobre la filogenia interna de los nematodes. P. Inst. Biol. Apll. Barcelona. 15: 87-92.

Gentili, F. and Jumpponen, A. 2006. Potential and possible uses of bacterial and fungal biofertilizers. *In*: Handbook of Microbial Biofertilizers. M.K. Rai (ed.). International Book Distributing Co., Lucknow, India. p. 543.

Geraert, E. 2006. Functional and detailed morphology of the Tylenchida (Nematoda). Brill, UK. pp. 200.

Garrity, G.M., Winters, M., Kuo, A.W. and Searles, D.B. 2001. Taxonomic outline of the prokaryotic genera. Bergey's Manual of Systematic Bacteriology, Second Edition, Springer-Verlag, New York. pp. 350.

Gaugler, R. 2002. Entomopathogenic Nematology. CAB International, Wallingford, Oxon, UK.

Gaur, H.S. 2002. Diversity in survival adaptations of nematodes. *In*: Nematode Diversity. M.S. Jairajpuri (ed.), Silverline Printers, Hyderabad, India. pp. 72-95.

Goeldi, E.A. 1887. Relatorio sobre a molestia do cafeeiro na provincia do Rio de Janeiro. Archos. Mus. Nac. Rio de J. 8: 1-121.

Goodey, T. 1934. Plant Parasitic Nematodes and the Diseases they Cause. Dutton and Co. Inc., UK. pp. 306.

Goodey, T. 1951. Soil and Freshwater Nematodes. Methuen and Co. Inc., London, UK. pp. 390.

Goodwin, T.W. and Mercer, E.I. 2001. Introduction to Plant Biochemistry (II edition). CBS Publishers and Distributors, New Delhi, India. pp. 677.

Goodey, J.B. 1963. Soil and Freshwater Nematodes by T. Goodey, rewritten. Methuen and Co. Inc. London, UK. pp. 544.

Goodrich-Blaire, H. and Clarke, D.J. 2007. Mutualism and pathogenesis in *Xenorhabdus* and *Photorhabdus*: two roads for the same destination. Molecular Microbiology 64: 260-268.

Gouge, D.H., Otto, A.A., Schiroki, A. and Hague, N.G.M. 1994. Effects of steinernematids on the root knot nematode, *Meloidogyne javanica*. Annals of Applied Biology 124 (Suppl.): 134-135.

Grewal, P.S., Martin, W.R., Miller, R.W. and Lewis, E.E. 1997. Suppression of plant parasitic nematode populations in turfgrass by application of entomopathogenic nematodes. Biocontr. Sci. Technol. 7: 393-399.

Grewal, P.S., Lewis, E.E. and Venkatachari, S. 1999. Allelopathy: A possible mechanism of suppression of plant parasitic nematodes by entomopathogenic nematodes. Nematology 1: 735-743.

Griffith, R., Gablin-Davis, R.M., Koshy, P.K. and Sossama, V.K. 2005. Nematode parasites of coconut and other palms. *In*: Plant Parasitic Nematodes of Tropical and Sub tropical Agriculture. M. Luc, R.A. Sikora and J. Bridge (eds.), CABI, Wallingford, Oxon, UK. pp. 493-528.

Han, H.R., Jayaprakash, A., Weingartner, D.P. and Dickson, D.W. 2006. Morphological and molecular biological characterization of *Belonolaimus longicaudatus*. Nematropica 36: 37-52.

Harrier, L.A. and Watson, C.A. 2004. The potential role of arbuscular mycorrhizal (AM) fungi in the bioprotection of plants against soil borne pathogens in organic and/or other sustainable farming systems. Pest Management Science 60: 149-157.

Hewitt, W.B., Raski, D.J. and Goheen, A.C. 1958. Nematode vector of soil borne fan leaf virus of grapevine. Phytopathology 48: 568-595.

Hol, W.H.G. and Cook, R. 2005. An overview of arbuscular mycorrhizal fungi-nematode interactions. Basic and Applied Ecology 6: 489-503.

Huang, C.S. 1985. Formation, Anatomy and Physiology of giant cells induced by root knot nematodes. *In*: An Advanced Treatise on *Meloidogyne* Biology and Control. J.N. Sasser and C.C. Carter (eds.), Vol. I. North Carolina State Graphics, N.C., USA. pp. 155-164.

Huang, J.S. 1987. Interactions of nematodes with rhizobia. *In*: Vistas on Nematology. J.A. Veech and D.W. Dickson (eds.), Society of Nematologists, Inc. Hyattsville, Maryland, USA. pp. 301-306.

Hunt, D.J. 1993. Aphelenchida, Longidoridae and Trichodoridae: Their Systematics and Bionomics. CABI, Wallingford, Oxon, UK. pp. 352.

Hunt, D.J., Luc, M. and Manzanilla-Lopzec, R.H. 2005. Identification, morphology and biology of plant parasitic nematodes. *In*: Plant Parasitic Nematodes of Tropical and Sub tropical Agriculture. M. Luc, R.A. Sikora and J. Bridge (eds.), CABI, Wallingford, Oxon, UK. pp. 87-130.

Hussey, R.S. 1987. Secretions of esophageal glands of Tylenchida nematodes. *In*: Vistas on Nematology. J.A. Veech and D.W. Dickson (eds.), Society of Nematologists Inc., Hyattsville, Maryland, USA. pp 221-228.

Hyman, L.H. 1951. The Invertebrates: Acanthocephala, Aschelminthes and Entoprocta. The Pseudocoelomate Bilateria. Vol. 3. McGraw-Hill, New York.

Jagadale, G.B. and Grewal, P.S. 2006. Infection behaviour and overwintering survival of foliar nematodes, *Aphelenchoides fragariae* on host. Journal of Nematology 38: 130-136.

Jairajpuri, M.S. and Ahmad, W. 1992. Dorylaimida. Freeliving, Predaceous and Plant Parasitic Nematodes. Oxford and IBH Pub. Co., New Delhi, India. pp. 458.

Jairajpuri, M.S. 2002. Cryptobiosis- nematodes to man. *In*: Nematode Diversity. M.S. Jairajpuri (ed.). Silverline Printers, Hyderabad, India. pp. 59-71.

Jayaprakash, A., Tigano, M.S., Brito, J., Carneiro, R.M.D.G. and Dickson, D.W. 2006. Differentiation of *Meloidogyne floridensis* from *M. arenaria* using high fidelity PCR amplified mitochondrial AT-Rich sequences. Nematropica 36: 1-12.

Jones, J.R., Lowerence, K.S. and Lowerence, G.W. 2006. Evaluation of winter cover crops in cotton cropping for management of *Rotylenchulus reniformis*. Nematropica 36: 53-66.

Jothi, G. and Sundarababu, R. 2002. Nursery management of *Meloidogyne incognita* by *Glomus mosseae* in eggplant. Nematologia Mediterranea 30: 153-154.

Karanastasi, E., Wyss, U. and Brown, D. 2001. Observations on the feeding behaviour of *Paratrichodorus anemorus* in relation to tobravirus transmission. Meded Rijksunivgent Fak Landbouwkd Toeqep BiolWet. 66: 599-608.

Karssen, G., Van Aelst, A.C., Waeyenberge, L. and Moens, M. 2001. Observations on *Pratylenchus penetrans* Cobb, 1917 parasitizing the coastal dune grass, *Ammophila arenaria* (L.) link in the Netherlands. Journal of Nematological Morphology and Systematics 4: 1-9.

Karssen, G. 2002. The plant parasitic nematode Genus *Meloidogyne* Goeldi, 1892 (Tylenchida) in Europe. Brill Academic Publishers, Boston, UK. p. 157.

Karssen, G. 2006. The Plant Parasitic Nematode genus *Meloidogyne* Goeldi 1882 (Tylenchida) in Europe. Brill. UK. pp. 160.

Kaya, H.K. and Stock, S.P. 1997. Techniques in insect nematology. *In*: Manual of techniques in insect pathology, Lawrence Lacey (eds.), Academic Press NY, USA. pp. 281-324.

Khan, A.A., Raghav, D., Kausar, S. and Khan, M.R. 2005. Root-knot nematode parasites of cucurbitaceous crops. *In*: Plant diseases biocontrol and management. S. Nehra (ed.), Aavishkar Publishers & Distributors, India. pp. 325-339.

Khan, M.R. and Anwer, A. 2006. A characterization of some isolates of *Aspergillus niger* and evaluation of their effects against *Meloidogyne incognita*. Tests of Agrochemicals and Cultivars 27: 33-34.

Khan, M.R. and Anwer, M.A. 2008. DNA and some laboratory tests of nematode suppressing efficient soil isolates of *Aspergillus niger*. Indian Phytopathology 61 (2): 212-225.

Khan, M.R. and Athar, M. 1996. Response of wheat cultivars to different inoculum levels of *Anguina tritici*. Nematologia Mediterranea 24: 269-272.

Khan, R.M. and Khan, M.W. 1986. Antagonistic behaviour of root-knot, reniform and stunt nematodes in mixed infection. International Nematological Newsletter 3: 3-4.

Khan, M.R. and Khan, M.W. 1997. Effects of root knot nematode, *Meloidogyne incognita* on the sensitivity of tomato to sulphur dioxide and ozone. Environmental & Experimental Botany 38: 117-130.

Khan M.R., Kounsar, K. and Hamid, A. 2002. Effect of certain rhizobacteria and antagonistic fungi on root nodulation and root knot nematode disease of green gram. Nematologia Mediterranea 30: 85-89.

Khan, M.R., Mohiddin, F.A. and Khan, S.M. 2005. Effect of seed treatment with certain biopesticides on root knot of chickpea. Nematologia Mediterranea 32: 128-135.

Khan, M.W. 1993. Mechanism of interaction between nematode and other plant pathogens. *In*: Nematode Interactions. M.W. Khan (ed.), Chapman & Hall, London, UK. pp. 55-78.

Khan, M.W. 1997. The four major species of root-knot nematodes-Current status and management approach. Indian Phytopathology 50: 445-457.

Khan, M.W. and Dasgupta, M.K. 1993. The concept of interaction. In: Nematode Interactions. M.W. Khan (ed.), Chapman & Hall, London, UK. pp. 42-54.

Khan, M.W. and Haider, R. 1991. Interaction of *Meloidogyne javanica* with different races of *Meloidogyne incognita*. Journal of Nematology 11: 100-105.

Kuhn, J. 1857. Uber das Vorommen von Anguillulen in erkrankten Bluhtenkopfen von *Dipsacus fullonum* L. Z. Wiss. Zool. 9: 129-137.

Kuhn, J. 1881. Die Ergebnisse der Versuche zur Ermittelung der Ursache der Rubenmudikeit und zur Erforschung der Natur der Nematoden. Berlin Physiological Laboratory Landwirtsch. Institute University Halle 3: 1-153.

Lacey, L.A., Frutos, R., Kaya, H.K. and Vails, P. 2001. Insect pathogens as biological control agents: Do they have a future? Biological Control 21: 230-248.

Landi, S. and Manachini, B. 2005. Nematode population dynamics. Communications in Agriculture and Applied Biological Sciences 70(4): 937-944.

Lewis, E.E., Grewal, P.S. and Sardanelli, S. 2001. Interactions between the *Steinernema feltiae- Xenorhabdus bovienii* insect pathogen complex and the root-knot nematode *Meloidogyne incognita*. Biological Control 21: 55-62.

Linnaeus, C. von. 1758. Systema naturae regna tria naturae, secundum classes, ordines, genera, species cum characteribus differentiis, synonymis, locis. Editio decima reformata 1: 823.

Linnaeus, C. von. 1767. Systema Naturae, Vermes. Holmiae 1: 1327.

Liu, Q.L. and Williamson, V.M. 2006. Host-specific pathogenicity and genome differences between inbred strains of *Meloidogyne hapla*. Journal of Nematology 38: 158-164.

Luc, M., Maggenti, A.R., Fortuner, R., Raski, D.J. and Geraert, E. 1987. Tylenchina (Nemata). 1. For a new approach to the taxonomy of Tylenchina. Revue de Nematol. 10: 127-134.

Luc, M., Bridge, J. and Sikora, R.A. 2005. Reflections on nematology in tropical and subtropical agriculture. In: Plant Parasitic Nematodes of Tropical and Sub Tropical Agriculture. M. Luc, R.A. Sikora and J. Bridge (eds.), CAB International, Wallingford, Oxon, U.K. pp. 1-10.

Maggenti, A.R. 1961. Morphology and biology of the genus *Plectus*. Proc. Helminth. Soc. Washington 28: 118-130.

Maggenti, A.R. 1982. Nemata. *In*: Synopsis and Classification of Living Organisms. S.P. Parker (ed.), McGraw-Hill Book Co., New York. pp. 879-929.

Maggenti, A.R. 1983. Nematode higher classification as influenced by species and family concepts. *In*: Concepts in Nematode Systematics. A.R. Stone, H. Platt and L.F. Khalil (eds.), Academic Press, London, UK and New York. pp. 25-40.

Maggenti, A.R. 1991. Nemata: higher classification. *In*: Manual of Agricultural Nematology. W.R. Nickle (ed.), Marcel Dekker, Inc., New York, pp. 147-187.

Maggenti, A.R., Luc, M., Raski, D.J., Fortuner, R. and Geraert, E. 1988. A reappraisal of Tylenchina (Nemata). 11. List of generic and supra generic taxa, with their junior synonyms. Revue de Nematol. 11: 177-188.

Mahanta, B. and Phukan, P.N. 2000. Effects of *Glomus fasciculatum* on penetration and development of *Meloidogyne incognita* on blackgram. Journal of the Agricultural Society of North East India 13: 215-217.

Maia, L.C., Silveira da, N.S.S. and Cavalcante, U.M.T. 2006. Interaction between arbuscular mycorrhizal fungi and root pathogens. *In*: Handbook of Microbial biofertilizers. M.K. Rai (ed.), International Book Distributing Co., Lucknow, India. pp. 29-49.

Manachini, B., Landi, S. and Tomasini, V. 2005. Biodiversity of nematofauna of oilseed rape (*Brassica napus* L.). Communications in Agriculture and Applied Biological Sciences 70 (4): 927-935.

Marschner, B. and Baumann, K. 2003. Changes in bacterial community structure induced by mycorrhizal colonization in split root maize. Plant and Soil 251: 279-289.

McDonald, A.H. and Nicol, J.N. 2005. Nematode parasites of cereals. *In*: Plant Parasitic Nematodes of Tropical and Sub tropical Agriculture. M. Luc, R.A. Sikora, and J. Bridge, (eds.), CAB International, Wallingford, Oxon, UK. pp. 131-192.

McSorley, R. 2003. Adaptations of nematodes to environmental extremes. Florida Entomologists 86: 138-142.

Mennan, S. and Handoo, Z.A. 2006. Plant parasitic nematodes associated with cabbage (*Brassica oleracea*) in Samsun (middle black sea region), Turkey. Nematropica 36: 99-106.

Moens, T., Araya, M., Swennen, R. and De Waele, D. 2006. Reproduction and pathogenicity of *Helicotylenchus multicinctus*, *Meloidogyne incognita* and *Pratylenchus coffeae* and their interaction with *Radopholus similis* on musa. Nematology 8: 45-58.

Morgan, G.D., MacGuidwin, A.E., Zhu, J. and Binning, L.K. 2002. Population dynamics and distribution of root-lesion nematode (*Pratylenchus penetrans*) over a three year potato crop rotation. Agronomy Journal 94: 1146-1155.

Muschiol, D. and Traunspurger, W. 2007. *Panagrolaimus* sp. and *Poikilolaimus* sp. from chemoautotrophic movile cave, Romania. Nematology 9: 271-284.

Nahara, M.S., Grewal, P.S., Millera, S.A., Stinnerb, D., Stinnerb, B.R., Kleinhenzc, M.D., Wszelakia, A. and Doohanc, D. 2006. Differential effects of raw composted manure on nematode community and its indicative value for soil microbial, physical and chemical properties. Applied Soil Ecology 34: 140-151.

Needham, T. 1744. A letter concerning certain chalky tubulous concentrations called malm; with some microscopical observations on the ferina of the red lily, and of worms discovered in smutty corn. Philosophical Transaction of the Royal Society of London 42: 634-641.

Nelson, D.L. and Cox, M.M. 2007. Lehninger Principles of Biochemistry, IV Edition, W.H. Freeman and Co., UK. pp. 1119.

Neumann, G. and Shields, E.J. 2006. Interspecific interaction among three entomopathogenic nematodes, *Steinernema carpocapsae* Weiser, *S. feltiae* Filipjev and *Heterorhabditis bacteriophora* Poinar, with different foraging strategies for hosts in multipiece sand columns. Environmental Entomology 35: 1578-1583.

Oka, Y., Ben-Daniel, Bat-Hen and Cohen, Y. 2006. Control of *Meloidogyne javanica* by formulations of *Inula viscosa* leaf extracts. Journal of Nematology 38: 46-51.

Orley, L. 1880. Monograph of the anguillulids. Termeszet. Fuzeteb. 4: 16-150.

Otsubo, R., Yoshiga, T., Kondo, E. and Ishibashi, N. 2006. Coiling is not essential to anhydrobiosis by *Aphelenchus avenae* on agar amended with sucrose. Journal of Nematology 38: 41-45.

Patel, D.J. and Patel, R.G. 2002. Occurrence and distribution of root-knot and cyst nematodes in India. *In*: Nematode Diversity. M.S. Jairajpuri (ed.), Silverline Printers, Hyderabad, India. pp. 296-305.

Patrice, C., Vaughan, W.S. and Don, G. McArthur. 2002. Role of plant parasitic nematodes and abiotic soil factors in growth heterogeneity of sugarcane on a sandy soil in South Africa. Plant and Soil 246: 259-271.

Prasad, J.S. 2002. Nematode diversity of rice crops of India. *In*: Nematode Diversity. M.S. Jairajpuri (ed.), Silverline Printers, Hyderabad, India. pp. 494-500.

Purohit, S.S. 2006. Mycorrhiza: for soil health and sustainable ecosystem. *In*: Trends in Organic Farming in India. S.S. Purohit and D. Gehlot (eds.), Agrobios, Jodhpur, India. pp. 183-198.

Raja, P. 2006. Status of endomycorrhizal (AMF) biofertilizer in the global market. *In*: Handbook of Microbial Biofertilizers. M.K. Rai (ed.), International Book Distributing Co., Lucknow, India. pp. 395-416.

Ramclam, W. and Araya, M. 2006. Frequency of occurrence and abundance of root-knot nematodes on banana (Musa AAA) in Belize. International Journal of Pest Management 52: 71-77.

Rao, M.S., Naik, D., Shylaja, M. and Reddy, P.P. 2003. Management of *Meloidogyne incognita* on eggplant by integrating endomycorrhiza, *Glomus fasciculatum* with bioagent *Pochonia chlamydosporia* under field condition. Indian Journal of Nematology 33: 29-32.

Riga, E., Lacey, L.A., Guerra, N. and Headrick, H.L. 2006. Control of the oriental fruit moth, *Grapholita molesta* using entomopathogenic nematodes in laboratory and fruit bin assays. Journal of Nematology 38: 168-171.

Riffle, J.W. 1971. Effect of nematodes on root-inhabiting fungi in mycorrhizae. Proceedings of the First North American Conferences on Mycorrhizae. E. Hacskayato (ed.), Miscellaneous Publication 1189, USDA, Washington, USA. pp. 97-113.

Ritzema Bos, J. 1891. Zwei neue Nematdenkrankheiten der Erdbeerpflanze (Vorlaufige Motteilung). Zeitschrift fur Pflanzenkrankheiten 1: 1-16.

Roch, K.D., Saliou, N., Danamou, M. and Thierry, M. 2007. Relationships between abiotic soil factors and epidemiology of the biocontrol bacterium *Pasteuria penetrans* in a root-knot nematode *Meloidogyne javanica* infested soil. Biological Control 40: 22-29.

Ryan, N.A., Deliopoulus, T., Jones, P. and Haydock, P.P.J. 2003. Effects of a mixed isolate mycorrhizal inoculum on the potato-potato cyst nematode interaction. Annals of Applied Biology 143: 111-119.

Ryan, N.A., Duffy, E.M., Cassells, A.C. and Jones, P.W. 2000. The effect of mycorrhizal fungi on the hatch of potato cyst nematodes. Applied Soil Ecology 15: 233-240.

Salonen, V., Vestberg, M. and Vauhkonen, M. 2001. The effect of host mycorrhizal status on host plant-parasitic interactions. Mycorrhiza 11: 95-100.

Sambrook, J. and Russell, D.W. 2001. Molecular cloning, a laboratory manual. Cold Spring Harbor Laboratory Press, USA. pp. 1-44.

Sasser, J.N. 1989. Plant Parasitic Nematodes: the farmers hidden enemy. North Carolina State University Graphics. Raleigh, USA. pp. 115.

Scopoli, G.A. 1777. Introductio ad historiam naturalem sistens genera lapidum, plantarum et animalium hactenus detecta, caracteribus essentialibus donata, in tribus divisa, subinde ad leges naturae. Prague, Czech. pp. 506.

Scwartz, M. 1911. Die Aphelenchen der Vielchengallen und Blattflecken an Farnen und Chrysanthemum. Arb. kaiserl. biol. Anstalt Land. Forstwirtsch. 8: 303-334.

Shapiro-Ilan, David, I., Stuart, R.J. and McCoy, C.W. 2006. A comparison of entomopathogenic nematode longevity in soil under laboratory conditions. Journal of Nematology 38(1): 119-129.

Sharma, P.D. 2005. Fungi and Allied organisms, I Edition. Narosa Publishing House, Delhi, India. pp. 543.

Siddiqi, M.R. 1980. The origin and phylogeny of the nematode orders Tylenchida Thorne, 1949 and Aphelenchida n. ord. Helminth. Abstr. Ser. B 49: 143-170.

Siddiqi, M.R. 1986. Tylenchida, Parasites of Plants and Insects. CAB International Wallingford, Oxon, UK. pp. 645.

Siddiqi, M.R. 2000. Tylenchida, Parasites of Plants and Insects. Ed II. CABI, Wallingford, Oxon, UK. pp. 833.

Siddiqi, M.R. 2002. Evolution of parasitism in nematodes. *In*: Nematode Diversity. M.S. Jairajpuri (ed.), Silverline Printers, Hyderabad, India. pp. 8-31.

Singh, R.V. and Sharma, H.K. 1998. Nematode problems and their management in ornamental crops. *In*: Nematode Diseases in Plants. P.C. Trivedi (ed.), CBS Publishers, New Delhi, India. pp. 168-176.

Sinha, R.K. 2004. Modern Plant Physiology. Narosa Publishing House, New Delhi, India. pp. 620.

Smitley, D.R., Warner, F.W. and Bird, G.W. 1992. Influence of irrigation and *Heterorhabditis bacteriophora* on plant parasitic nematodes in turf. Journal of Nematology 24: 637-641.

Southey, J.F. 1993. Nematode pests of ornamental and bulb crops. *In*: Plant Parasitic Nematodes of Temperate Crops. K. Evans, D.L. Trudgill and J.M. Webster (eds.), CAB International, Wallingford, Oxon, UK. pp. 463-500.

Sriwati, R., Takemoto, S. and Futai, K. 2007. Cohabitation of the pine wood nematode *Bursaphelenchus xylophilus* and fungal species in pine trees inoculated with *B. xylophilus*. Nematology 9: 77-86.

States, K.M. 2003. Ethidium bromide in the waste paper. The Hazardous Waste Disposal Monthly Update 6(2): 17-19.

Stirling, G.R. 1991. Biological Control of Plant Parasitic Nematodes: Progress, Problems and Prospects. CAB International, Wallingford, Oxon, UK. pp. 282.

Subba Rao, N.S. 2001. Soil Microbiology (Fourth edition of Soil Microorganisms and Plant Growth). Oxford & IBH Pub., New Delhi, India. pp. 327-341.

Suenga, E. and Namura, H. 2005. Prestaining method as a useful tool for the agarose gel electrophoretic detection of polymerase chain reaction products with a fluorescent dye SYBR Gold nucleic acid gel stain. Annals of Science 21: 619-623.

Sy, A., Giraud, E., Jourand, P., Garcia, N., Willems, A., Lajudie, P., Prin, Y., Neyra, M., Gillis, M., Boivin-Masson, C., Dreyfus, B. and de-Lajudie, P. 2001. Methylotrophic *Methylobacterium* bacteria nodulate and fix nitrogen in symbiosis with legumes. Journal of Bacteriology 183: 214-220.

Taha, A.H.Y. 1993. Nematode interactions with root-nodule bacteria. In: Nematode Interactions. M.W. Khan (ed.), Chapman and Hall, London, UK. pp. 175-202.

Takeuchi, Y. and Futai, K. 2007. Asymptomatic carrier trees in pine stands naturally infected with *Bursaphelenchus xylophilus*. Nematology 9: 243-250.

Takeuchi, Y., Kanzaki, N. and Futai, K. 2005. A nested PCR-based method for detecting the pine wood nematode, *Bursaphelenchus xylophilus* from pine wood. Nematology 7: 775-782.

Talavera, M., Itou, K. and Mizukubo, T. 2001. Reduction of nematode damage by root colonization with arbuscular mycorrhiza (*Glomus* spp.) in tomato-*Meloidogyne incognita* (Tylenchida: Meloidogynidae) and carrot-*Pratylenchus penetrans* (Tylenchida: Pratylenchidae) pathosystems. Applied Entomological Zoology 36: 387-392.

Tanu, P.A. and Adholeya, A. 2006. Potential of arbuscular mycorrhizae in organic farming systems. In: Handbook of Microbial Biofertilizers. M.K. Rai (ed.), International Book Distributing Co., India. pp. 223-239.

Taylor, C.E. and Brown, D.J.F. 1997. Nematode Vectors of Plant Viruses. CAB International, Wallingford, Oxon, UK. pp. 286.

Thorne, G. 1941. Some nematodes of the family Tylenchidae which do not possess a valvular median esophageal bulb. Great Basin Naturalist 2: 37-85.

Timper, P., Davis, R.F. and Tillman, P. 2006. Reproduction of *Meloidogyne incognita* on winter cover crops used in cotton production. Journal of Nematology 38: 83-89.

Todd, T.C., Winkler, H.E. and Wilson, G.W.T. 2001. Interaction of *Heterodera glycines* and *Glomus mosseae* on soybean. Supplement to the Journal of Nematology 33: 306-310.

Treonis, A.M. and Wall, D.H. 2005. Soil nematodes and desiccation survival in the extreme arid environment of Antarctic dry valleys. Integrative and Comparative Biology 45: 741-750.

Tribe, H.T. 1977. Pathology of cyst nematodes. Biological Review 52: 477-507.

Tuffen, F., Eason, W.R. and Scullion, J. 2002. The effect of earthworms and arbuscular mycorrhizal fungi on growth of and ^{32}p transfer between *Allium porrum* plants. Soil Biology and Biochemistry 34: 1027-1036.

Tuma, R.S., Beaudet, M.P., Jin, X., Jones, L.J., Cheung, C.Y., Uue, S. and Singer, V.L. 1999. Characterization of SYBR Gold nucleic acid gel stain: a dye optimized for use with 300 nm ultraviolet transilluminators. Annals of Biochemistry 268(2): 278-288.

Van Buuren, M.L., Maldonado-Mendoza, I.E., Trieu, A.T., Blaylock, L.A. and Harrison, M.J. 1999. Novel genes induced during an arbuscular mycorrhizal (AM) symbiosis formed between *Medicago trunculata* and *Glomus versiforme*. Molecular Plant Microbe Interactions 12: 171-181.

Van-Den B.W., Rossing, W.A.H. and Grasman, J. 2006. Contest and scramble competition and the carry-over effect in *Globodera* spp. in potato based crop rotations using an extended Rickev model. Journal of Nematology 38: 210-220.

Van der Putten, W.H., Cook, R., Costa, S., Davies, K.G., Fargette, M., Freitas, H., Hol, W.G.H, Kerry, B.R., Maher, N. and Mateilk, T. 2006. Nematode interactions in nature: models for sustainable control of nematode pests of crop plants. Advances in Agronomy 89: 227-260.

Vanholme, B., De Meutter, J., Tylgat, T., Van Montagu, M., Coomans and Gheysen, G. 2004. Secretions from plant parasitic nematodes: A molecular update. Gene 332: 13-27.

Viketoft, M. 2007. Plant induced spatial distribution of nematodes in a semi natural grassland. Nematology 9: 131-142.

Villenave, C. and Duponnois, R. 2002. Interaction between ectomycorrhizal fungi, plant parasitic and free living nematodes and their effects on seedlings of the hardwood *Afzelia africana* sm. Pedobiologia 26: 176-187.

Vos, P., Hogers, R., Bleeker, M., Reijans, M., vande Lee, T., Hornes, M., Frijters, A., Pot, J., Peleman, J. and Kuiper, M. 1995. AFLP: a new technique for DNA fingerprinting. Nucleic Acids Research 23: 4407-4414.

Vovlas, N., Mifsud, D., Landa, B.B. and Castillo, P. 2005. Pathogenicity of the root knot nematode *Meloidogyne javanica* on potato. Plant Pathology 54: 657-664.

Waceke, J.W., Waudo, S.W. and Sikora, R. 2001. Suppression of *Meloidogyne hapla* by arbuscular mycorrhizal fungi (AMF) on pyrethrum in Kenya. International Journal of Pest Management 47: 135-140.

Weischer, B. 1993. Nematode virus interactions *In*: Nematode Interactions. M.W. Khan (ed.), Chapman & Hall, London, UK. pp. 217-231.

Westermeier, 1997. Electrophoresis in practice: a guide to methods and applications of DNA and protein separation. VCH, Weinheim, Germany.

White, G.F. 1927. A method for obtaining infective nematode larvae from cultures. Science 66: 302-303.

Wilcox Lee, D. and Loria, R. 1987. Effects of nematode parasitism on plant-water relation. *In*: Vistas on Nematology. J.A.Veech and D.W. Dickson (eds.), Society of Nematologists Inc., Maryland, USA. pp. 260-266.

Williamson, V.M., Chen, C., Westerdahl, B.B., Wu, F.F. and Caryl, G. 1997. A PCR assay to identify and distinguish single juveniles of *Meloidogyne hapla* and *M. chitwoodi*. Journal of Nematology 29: 9-15.

Williamson, V.M. and Gleason, C.A. 2003. Plant nematode interactions. Current Opinion Plant Biologists 6: 327-333.

Wilmes, P. and Bond, P.L. 2004. The application of two dimensional polyacrylamide gel electrophoresis and down stream analysis to a mixed community of prokaryotic microorganisms. Environmental Microbiology 6: 911-920.

Woodring, J.L. and Kaya, H.K. 1988. Steinernematid and Heterorhabditid nematodes, a handbook of techniques. Southern Cooperative Series Bulletin 331.

Wright, J.D. and Perry, N.R. 2002. Physiology and Biochemistry. *In*: Entomopathogenic Nematology. R. Gaugler (ed.), CABI Publishing, CAB International, Wallingford, Oxon, UK. pp. 145-168.

Wuyts, N., Elsen, A., Van Damme, E., Peumans, W., De Waele, D., Swennen, R. and Sagi, L. 2003. Effect of plant lectins on the host finding behaviour of *Radopholus similis*. Nematology 5: 205-212.

Wyss, U. 2002. Feeding behaviour of plant parasitic nematodes. *In*: The Biology of Nematodes. D.L. Lee (ed.), Taylor and Francis, USA. pp. 233-260.

Zabeau, M. and Vos, P. 1993. Selected restriction fragment amplification: A general method for DNA fingerprinting. European Patent Application No. 92402629.7. Publication number EP, 0534858, A:1.

Zasada, I.A., Rice, C.P. and Meyer, S.L.F. 2007. Improving the use of rye (*Secale cereale*) for nematode management : potential to select cultivars based on *Meloidogyne incognita* host status and benzoxazinoid content. Nematology 9: 53-60.

Abbreviations, Symbols and Glossary

a. A ratio calculated by dividing the length of the nematode by its maximum width.

Abiotic. A non living entity; An agent of non biological origin.

Acetyl coenzyme A (Acetyl CoA). A coenzyme derivative in the metabolism of glucose and fatty acids that contributes substrates to the Kreb's cycle.

Acetylcholinesterase. An enzyme in the membrane of postsynaptic cells that catalyzes the conversion of acetyl choline into choline and acetic acid during neurotransmission.

Acropetal movement. A pesticide absorbed by the plant roots and circulates through xylem.

Acrylamide gels. A polymer gel used in electrophoresis to measure the size of protein (Dalton), DNA (base pairs).

Adanal. Pertaining to bursa which does not envelop the entire tail.

Adaptability. The ability to change or be changed to fit under changed circumstances.

Additive interaction. It is a type of interaction in which two or more pathogens attacking a common plant (host) act independently and do not develop any relationship. In such situation the host damage caused by the two pathogens together is near to the sum of individual damages.

ADP. An acronym that stands for adenosine diphosphate. It is a ribonucleoside 5'- phosphate serving as phosphate group acceptor in the cell energy cycle.

AFLP. Amplified fragment length polymorphism(s) (AFLP) is a tool that allows differentiation between individuals, genotypes and strains and the assessment of genetic diversity and phylogeny. This technique has the characteristics of an ideal system for detecting genetic variation.

AFLP markers. The AFLP markers are generated from a large number of independent loci from different parts of the genome. The profile generated is highly reproducible.

Agamospecies or morphospecies. Species that reproduce asexually, e.g. by parthenogenesis and are distinguished on morphological and not on gene-flow basis.

Agarose gel. A polysaccharide gel used to measure the size of proteins (Daltons) or nucleic acids (bases or base pairs).

Aggressiveness. Relative ability of a plant pathogen to colonize and cause damage to plants.

Alae. Winglike expansions or projections formed by longitudinal thickening of the cuticle.

Aldolase. A crystalline enzyme that occurs widely in living systems and catalyses reversibly the cleavage of a phosphorylated fructose into triose sugars.

Allele. Any of one or more alternative forms of a gene.

Allelochemicals. The chemicals produced by an organism of one species and affect the growth, health, behavior, population or biology of individuals of another species.

Allelopathy. Ability of one species to inhibit or prevent the growth of another species through the production of toxic substances.

Allomones. A pheromone that induces a behavioral or physiologic change

Allopatry and sympatry. Taxa or populations that occupy different or same geographical areas, respectively.

Allozyme. Any variant of an enzyme coded by different alleles at the same gene locus (refers to motility variants identified by electrophoresis).

Alpha (α), Beta (β) and Gamma (γ) taxonomy. In short, descriptive, synthetic and evolutionary taxonomy, respectively. Alpha taxonomy is the process of naming and describing species, and providing diagnosis, emendation, etc. of taxa. Beta taxonomy includes revising and arranging taxa in higher categories that reflect evolutionary history. Gamma taxonomy deals with tracing intraspecific variation and establishing the evolutionary systematic lines, critical analysis of classification and publication of monographs and books on taxonomy.

Ambifenestrate. The occurrence of two openings in the vulval cone which are separated by the vulval bridge, found in the mature females and cysts of *Heterodera*. Also called semifenestra.

Ambimobile. A systemic pesticide that can be absorbed by the plant root and shoot and circulates through xylem and phloem.

Aminoacids. Aminoacids are the molecules that contain both amino and carboxylic acid functional groups.

Amphidelphic. Having two ovaries, one proceeding anteriorly and one proceeding posteriorly.

Amphids. A pair of chemo-sensory organs located in the lip region on lateral sides of the nematode.

Amphimictic. Reproduction in which sperms and eggs come from separate individuals (cross-fertilization) or capable of interbreeding freely and of producing fertile offspring.

Amphimixis. Copulation of two unrelated cells and nuclei e.g., eggs and sperms; reproduction by a sexual process.

Angstrom (Å). Unit of length (10^{-8}cm) used to indicate molecular or electron microscopic dimensions.

Animate plant pathogen. A living entity that may cause a disease in plants. e.g., fungi, bacteria, viruses, nematodes, etc.

Annulation. A transverse depression, appearing as a line, on the cuticle.

Annule. The space between two adjacent annulations on the cuticle, may also be called interstrial zone.

Antagonist. In plant pathology, an organism which suppresses or contends another organism.

Antagonistic interaction. Antagonistic interaction occurs when the infection by one pathogen modifies host response in such a way that it becomes less susceptible to invasion by secondary pathogen. In antagonistic interaction the interacting organisms antagonize or oppose each other leading to lesser interactive effects compared to the sum of individual effects.

Antibiosis. An association between organism and a metabolic product of another organism, that is harmful to one of them.

Antibiotic. A chemical compound produced by one microorganism (usually bacteria, fungi etc.) that inhibits growth or kills other living organism.

ap., apud. In the work of.

Apex. Tip of root or shoot, containing apical meristem.

Apophysis. A swelling or expansion as found in the organe-Z, in species of *Xiphinema*.

Apoptosis. Programmed cell death in which a cell brings about its own death and lysis, signaled from outside or programmed in its genes, by systematically degrading its own characteristics.

Arch. That portion of the perineal pattern in the genus *Meloidogyne* which is located dorsal and ventral to the lateral field or lines.

Areolation. The condition whereby transverse cuticular striae extend into the lateral field of the nematode.

Art. Article (of the International Code of Zoological/Botanical Nomenclature).

ATP. An acronym that stands for adenosine triphosphate. It is a ribonucleoside 5'-triphosphate functioning as a phosphate group donor in the cell energy cycle; carries chemical energy between metabolic pathways by serving as a shared intermediate coupling endergonic and exergonic reactions.

ATPase. An enzyme that hydrolyses ATP to yield ADP and phosphate. In this process energy is generated which is used in metabolic process such as photosynthesis.

auct. Of author(s).

Axenic. Culture in absence of living bacteria or other organism; pure culture.

b. A ratio calculated by dividing the total body length by the length of the oesophagus measured from the anterior end of the body.

b'. Relative measurement of the body length divided by distance from head end to posterior end of oesophageal gland.

Bacteria (bacterium singular). Any of the unicellular prokaryotic organisms of the class schizomycetes which vary in terms of morphology, oxygen and nutritional requirements and motility, and may be free living, saprophytic or pathogenic in plants and animals.

Basal bulb. Posterior portion of oesophagus which may be a distinct bulb or a lobe overlapping the intestine. Also called glandular oesophagus.

Basement membrane. A layer which separates the innermost layer of the cuticle from the hypodermis.

Basipetal movement. A pesticide absorbed by the foliage and circulates through phloem.

Bilateral symmetry. Referring to organisms whose morphology is such that the body can be divided into mirror-image halves by passing a dorsoventral plane through the body axis.

Binomen, trinomen and scientific names. Binomen is the name of a species with two words, first of a genus name beginning with a capital letter and second of species name which is not capitalized, and trinomen is the subspecies name with three words, the last one of which is subspecific and is not capitalized. Specific epithet includes both species and subspecies names. Citation of subgenus name within brackets is not a part of either a binomen or trinomen, but it does imply that the species name has been combined with the subgeneric name. Binomen and trinomen are called scientific names.

Bioassay. Use of appropriate hosts to determine relative population levels. Bioassays are used primarily for endoparasites such as root knot and cyst nematode, but can be used for any nematode under suitable propagation conditions such as a greenhouse. Bioassay can also be defined as the use of a test organism to measure the relative infectivity of a pathogen or toxicity of a substance.

Biocontrol. Exploitation by humans of the natural competition, parasitism and/or antagonism of organisms for the management of pests and pathogens. It is defined as reduction in the pest/pathogen population through the activity of living organism (other than men, host resistance, etc.) accomplished by manipulating the environment or by introduction of antagonists.

Biodiversity. Diversity in number and kind of taxa and/or their combined genetic variations.

Biological species. A species in which gene flow occurs readily and frequently between its members, and not between its members and those of other species, i.e. such species are reproductively isolated from other species. In other words, a biological species is a group of interbreeding natural populations which have a similar genotype but which are reproductively isolated from other such groups.

Biopesticides. A formulation containing antagonists (bacteria, fungi, viruses, nematodes etc.) or their products such as metabolites, toxins, plant products, etc. which adversely affect pests and pathogens.

Biopredisposition. It is the modification caused by biotic factors in the expression of host response to a pathogen.

Biotechnology. A set of biological techniques developed through basic research and applied to research and product development. It refers to the use of recombinant DNA, cell fusion and new bioprocessing techniques.

Biotroph. An organism that derives nutrients from the living tissues of another organism (its host).

Biotype, pathotypes and host races. A sub species of microorganism morphologically similar but physiologically different from other members of the species. Biological or physiological races of a species which can be differentiated on the basis of host reactions; they have no taxonomic status since ICZN does not recognize any taxa below the subspecific level.

Blastula. An early stage in embryogeny in which the embryo is a fluid filled sphere bounded by a single layer of cells.

Blight. A disease characterized by general and rapid killing of leaves, flowers and stems.

Blind diverticulum. An outpocketing or blind ending of the lumen of the intestine usually forming a side branch of the lumen proper.

Bloat. A disease characterized by swollen and rounded spots or blots on leaves, caused by *Ditylenchus dipsaci*

Buccal aperture. The oral aperture; the anterior opens to the digestive system.

Buccal cavity. The stoma; that connects oral opening with anterior portion of oesophagus. Buccal cavity is equipped with feeding devices. In nematodes feeding device is stylet/spear.

Buffer. A system capable of resisting changes in pH, consisting of a conjugate acid-base pair in which the ratio of proton acceptor to proton donor is near unity.

Bullae. Knob like structures located within the vulval cone of cysts of *Heterodera* near the underbridge or fenestra. The bullae are formed due to fat accumulation.

Bursae or caudal alae. Lateral cuticular extentions present at the posterior end of males especially around cloaca used to clasp or grasp the female during copulation.

c. A ratio calculated by dividing the total length by the length of the tail.

c'. Relative measurement of the tail length divided by body width at anus.

C_3 plants. A C_3 plant is that which assimilate CO_2 via ribulose biphosphate (RUBP) carboxylase enzyme leading to production of 3-carbon compound phosphoglyceric acid. The RUBP carboxylase under higher O_2 concentration acts as RUBP oxygenase and converts ribulose biphosphate into phosphoglyceric acid and phosphoglycolate and finally to serine and CO_2 without the production of ATP and NADPH. This wasteful release of CO_2 is energetically inefficient process and called as photorespiration. For this reason C_3 plants have relatively lower productivity rate than C_4 plants.

C_4 plants. Many plants especially members of Poaceae (sugarcane, maize, etc.) have an additional pathway to overcome photorespiration. In C_4 plants CO_2 is fixed in mesophyll cells and the CO_2 acceptor is phosphoenol pyruvate (3 carbon compound) which forms oxalo acetic acid (3 carbon compound) by the action of enzyme PEP carboxylase.

Capillary water. The water held against gravity in small soil pores or capillaries. It is the principal source of moisture for a plant roots.

Capitulum. Medial ventral sclerotization of the spicular pouch.

Carbohydrates. Any of various chemical compounds composed of carbon, hydrogen and oxygen, such as sugars, starch and cellulose.

Cardia or cardiac valve or oesophago-intestinal valve. Valvular apparatus connecting the oesophagus and intestine, and regulates the flow of food from the former to later.

Caudal glands. Three to five glands located in the tail especially in the males.

Caudal. Pertaining to or located at or on the tail.

Caudalids. The paired commissures (nerve connections) linking the lumbar ganglia and the preanal ganglion, which must circumvent the ventral opening of the anus. They are usually present in the family Hoplolaiminae.

cDNA. Complementary DNA produced by reverse transcription from mRNA (messenger RNA), and it is basically used in DNA cloning.

Cell. The basic structural and functional unit of all organisms.

Cellulase. An enzyme complex that breaks down cellulose to beta glucose. It is produced by symbiotic bacteria in ruminating chamber of herbivore but is absent in most animals including humans.

Cephalic framework. A subcuticular framework (inverted basket like) that supports the lip region and to which are attached the stylet protracter muscles.

Cephalic sensory structures. Sensory structures located on, in or near the head. In some nematode species cephalic sensory complex comprises of 18 sensory structures.

Cephalic. Pertaining to or located near the head.

Cephalids. Two structures (posterior and anterior) situated in the cephalic region and extending in a complete circle around the body; possibly part of nervous system, sometimes called hypodermal commissures.

Cervical papillae. A pair of sensory structures present one each on the lateral sides in the region of nerve ring.

Cervical. Pertaining to or located near the neck region or oesophageal region.

cf. *confer,* compare.

Character and character state. Any structure or behavioural system that is used to characterize a taxon, and any condition that a character can display, respectively.

Cheilorhabdion. Specialized zone of cuticle lining the cheilostome formed by the cells of hyp1, hyp2 and hyp3.

Cheilostome. Anterior part of stoma formed by an invagination of the external cuticle.

Chemotaxis. Movement or growth of an organism in response to changing concentration of a chemical stimulus, often in relation to food or mating.

Chord. A longitudinal internal thickening of the hypodermis usually on dorsal, ventral and the lateral sides.

Chromosome. The self replicating genetic structure of cells containing the cellular DNA.

Circomyarian. Muscle cells in which the sarcoplasmic zone is completely encircled by the fibres. The cytoplasmic muscle belly is confined to a central core of the cell.

Circumfenestrate. The condition in some species of *Heterodera* in which a vulval bridge across the vulval cone is not present producing only a single opening.

Circumoesophageal commissure or nerve ring. The central portion of nematode nervous system to which are connected anteriorly and posteriorly directed nerves and nerve chords. This structure encircles the oesophagus.

Clade and polyclade. Any supposedly monophyletic group in a phylogenetic analysis. Polyclade is used as a multiple entry identification key in a computer or in punched cards.

Cladogram, dendrogram and pholygram. A cladogram is a dendrogram or tree diagram based on synapomorphies depicting a phylogenetic hypothesis. It only depicts the branching pattern of the evolutionary history and, unlike a phylogram, it does not indicate, by means of branch length, the degree of evolutionary change that occurred along each lineage.

Cloaca. Common cavity in males, in which the digestive and reproductive systems terminate.

Clone, cloning. A group of genetically identical organisms, resulting from nonsexual cell division. A number of copies of a fragment of DNA produced by cloning or amplifying the number of copies of DNA.

Cluster analysis, OTUs and HTUs. Procedure that links operational taxonomic units (OTUs) into clusters on the basis of their attributes or overall similarities, as used in phenetics. HTUs are the hypothetical ancestral characters or character states used in a cladogram.

Codon. A coding unit, set of three nucleotides in a nucleic acid that specify a particular amino acid during protein synthesis.

Coelomyarian. A type of muscle cell with its protoplasmic portion extending into the pseudocoel with the fibrillar part extending along the hypodermis.

Coenzyme. An organic cofactor required for the action of certain enzymes; often contains a vitamin as a content.

Cohabitance. Living together in same place sharing food and shelter.

Commissure. Connecting bands of nerve tissue.

Commonality principle, plesiomorphy and apomorphy. The principle that plesiomorphies (primitive or ancestral characters or character states) will be commoner on average than apomorphies (derived character or character states).

Community. A group of plant or animals living and interacting with one another in a specific region at a given time under relatively similar environmental conditions.

Compatibility analysis. Procedure for eliminating characters from a phylogenetic analysis on their not being consistent with other characters.

Concomitant etiology. In a disease complex two or more pathogens invade the host simultaneously and the pathogens cannot be differentiated as primary or secondary pathogen.

Congeneric. Two or more species belonging to the same genus.

Congruence. Degree of similarity between two phylogenetic systems or phylogenetic trees derived from different data sets.

Contact action pesticide. A pesticide that through contact action provides protection to a part of the organism or plant where the pesticide is applied.

Corpus. Anterior elongate cylindrical part of oesophagus. The basal portion of the corpus is swollen to form a bulb, in which case the slender anterior portion is termed a procorpus and the bulb termed the metacorpus.

Cortex. In nematode body wall, the outermost layer of the cuticle. In plant stem or root a large zone of parenchymatous cells between epidermis/epiblema and endodermis.

Cotype. A paratype or syntype, not in usage now as it is not recognized by the ICZN.

Crustaformeria. Glandular region of the distal part of uterus that may play a role in formation of the egg wall, sometimes called tricolumella or quadricolumella for respective cell rows present.

Cultivar. A plant that has been created or selected intentionally and maintained through cultivation.

Cuticle. The non cellular external layer of a nematode secreted and maintained by the hypodermis. Cuticular material occurs as lining of natural openings upto oesophagous, vagina and rectum.

Cyst. The oxidized cuticle of dead adult females of genus *Heterodera* which confers protection to the eggs that may be contained within.

Dauer state/stage. A developmentally arrested dispersal adapted stage of an organism inside a vector. For example, J_3 larvae (dauers) of red ring nematode enter into the vector weevil and survive during the transmission. In entomopathogenic nematodes it is free living stage.

De Man's values. A series of measurements and ratios which express a nematode's length relative to width (a), oesophageal length (b) and tail length (c), the position of vulva, etc. (V, T, b', c', v').

Deacylation. Hydrolysis of ester bond between aminoacid and 2'-3' hydroxyl of the 3'-end of tRNA.

Dealkylation. It is the removal of alkyl group from a compound.

Deirids. Paired, pore like sensory organs located in the lateral fields; believed to be sensory in nature.

Denaturation. Partial or complete unfolding of a specific native conformation of a polypeptide chain, protein or nucleic acid on heating.

Detection level. Minimum population density of a nematode species that can be isolated/detected routinely from a soil or plant tissue sample. This level varies with each sampling and extraction technique.

Dichotomy. In a differential key where two new branches (bifurcation) arise from one stem.

Didelphic. Possessing two ovaries.

Dieback. Progressive death of shoots, branches and roots generally starting at the tip.

Dioecious. Having the male reproductive organs in one individual, the female in another; sexually distinct.

Diorchic. Having two testes.

Diploid and haploid. During meiosis, the chromosome number of a diploid cell (2n) is halved to haploid (n).

Disease. A malfunctioning of host cell or tissues due to continuous irritation by a primary causal agent (animate or inanimate pathogen) resulting in morbific cellular activity and expressed in characteristic pathological responses called symptoms.

Disease etiology. The determination and study of the cause of a disease.

Disease of complex etiology or disease complex. A disease caused by two or more pathogens.

Disease of specific etiology. The disease involving a single pathogen and a host. Here the disease development depends on host susceptibility, virulence of the pathogen and the prevailing environmental conditions.

Distance matrix. A taxon by taxon matrix in which distance or dissimilarities are measured.

Diurnal. Active in relation to day time or progress of time e.g., morning, noon, evening etc.

Diverticulum. Pouch or lobe containing the oesophageal glands.

DN. Dorsal oesophageal gland nucleus.

DNA fingerprinting. Technique based on restriction fragment analysis of DNA that reveals polymorphisms at dispersed loci of tandemly repeated DNA, used for evaluating relatedness between closely related individuals.

DNA hybridization. Process by which single stranded DNA forms pair with other homologous strands.

DNA. Deoxyribonucleic acid. A polynucleotide having a specific sequence of deoxyribonucleotide units covalently joined together through 3′, 5′-phosphodiester bonds; serves as the hereditary material in all organisms except for most of plant viruses.

DO. Dorsal oesophageal gland orifice.

Dorsal gland outlet. The point at which the dorsal gland empties into the lumen of oesophagus.

e.g. An acronym that stands for a Latin word *'exempli gratia'* meaning "for example".

Ecological niche. The status of an organism within its environment and community (affecting its survival as species).

Economic injury level (EIL). The lowest population density of a pest that causes economic damage to the host.

Economic threshold level (ETL). Level of pest population or disease severity at which the value of crop damage is equal to the cost of management. At ETL, preliminary control action should be initiated to prevent further increase in pest population (injury) to reach economic injury level.

Ectomycorrhizae. A symbiotic association where a fungus forms a sheath around the outside of root tip of a plant.

Ectoparasite. A parasite feeding on the host from exterior and does not enter into the host cell/tissue at any stage of its life cycle.

Ectophoretic. Pertaining to external phoretic associations between nematode and its vector.

Eelwool. A mass of desiccated *Ditylenchus dipsaci* (J_4) found in some plant tissues.

Efficient host. The plant species susceptible for infection by a pathogen and support its multiplication so that population of the pathogen builds up several times.

Egg. The first stage of the life cycle containing a zygote or a larva.

Electrophoresis and electromorphs. A process of separating molecules in a supporting medium on the basis of their electric charge in combination with other factors. Electromorphs are variants of a protein characterized by their motility properties during electrophoresis

ELISA. Enzyme-linked immunosorbant assay. A serological test in which one antibody carries with it an enzyme that releases a colored compound.

Emasculation. Removal of male reproductive organs without injuring female organs in an individual or flower.

emend. Emended, emendation.

Endomycorrhizae. The symbiotic association of a fungus and higher plants in which the fungus grows within the cortical cells.

Endoparasite. A parasite which enters into host tissue at any stage of life cycle and feeds from within the host.

Endophoretic. Pertaining to internal phoretic associations between nematode and its vector.

Endotoky or endotokia matricida. The phenomenon in which the egg deposition and hatching occurs inside the body of female and juveniles emerge only after the destruction of the maternal cuticle, as in cyst forming nematodes and also EPNs from the family Steinernematidae and Heterorhabditidae.

Enhancer. An enhancer is a nucleotide sequence to which transcription factor (s) bind, and which increases the transcription of a gene. It may be located a few hundred or thousand base pairs away from the gene.

Entomoparasites and phytoparasites. Parasites of insects (and, for practical purposes, mites) and of plants (and fungi), respectively.

Entomopathogenic. A microorganism or a nematode capable of causing diseases in insects.

Enzyme. A biomolecule either protein or RNA, that catalyzes a specific chemical reaction. It does not affect the equilibrium of the catalyzed reaction; it enhances the rate of a reaction by providing a reaction path with a lower activation energy.

Epiboly. The growth of one structure around another during embryo development.

Epiptygma. A cuticular membranous structure located on vulva or vagina as present in Hoplolaimidae and some other genera.

Epithelium. A sheet-like layer of cells lining the internal or external surface of an organ or tissue.

et al. An acronym that stands for the Latin word '*et alii* or *et alia*' meaning "and others".

Excretory pore. Exterior (external) opening of the excretory system. It is generally located on the ventral side of the body in the region of nerve ring.

Exotoky. The phenomenon in which egg deposition and hatching occurs outside the body of the female nematode.

Extracorporeal. Occurring outside of the body.

Facultative parasite. Having the ability to be a parasite. However, in the absence of susceptible host survives on dead and decaying organic matter.

Fallow. Maintenance of land with no plant growth or keeping a field without vegetation. Weeds are controlled by cultivation or by the application of herbicides and no crop planted.

fam. Family.

Familia dubia. A family with dubicious identity or recognition. A proposed family that has neither been recognized nor rejected.

Family group names. Any recognized taxonomic category name between genus and infraorder (tribe, subfamily and superfamily names). ICZN recognizes categories between superfamily and species only and does not cover taxa above superfamily rank.

Fenestra. A window-like opening in the vulval cone of *Heterodera* spp.

Fertilization. The sexual union of two nuclei resulting in doubling of chromosome number.

Field capacity of soil. Amount of water held in soil after gravitational water has drained away.

Flag leaf. A large flag like leaf at the top just before the inflorescence or the first leaf, immediately before the inflorescence (in members of Poaceae).

Flexure. A turn or fold.

Free living. A microorganism that lives freely, unattached, or a pathogen living in the soil, outside its host.

Free-living nematode. Those nematodes which are not known to be parasitic on animals or higher plants.

Fumigant. A toxic gas or volatile substance that is used to disinfest field or an area.

Fungus (fungi plural). Any of the numerous eukaryotic organisms of the kingdom Fungi which lack chlorophyll and vascular tissue and range in form from a single cell to a body mass of branched filamentous hyphae that often produce specialized fruiting structures.

Gall. An abnormal plant structure formed in response to parasitic attack by certain microorganisms (fungi, bacteria, viruses, nematode or insects). Gall may develop either by localized cell proliferation or increase in size.

Ganglia (ganglion singular). A mass or aggregation of nerve cell bodies, usually located within the central nervous system in invertebrates.

Gastrula. An early embryo composed of a hollow two-layered cup shaped structure derived from the blastula.

gen.nov., gen.n., n.g. *genus novum,* new genus.

Gene. The basic unit of inheritance. A segment of DNA that codes for a single functional polypeptide chain or RNA molecule. Also, a linear portion of the chromosome that determines or conditions one or more hereditary characters; the smallest functional unit of the genetic material.

Genital papillae. Tactile or sensory organs located on the tail.

Genital primordium. The initial cells of the reproductive system in nematodes.

Genome. The full chromosome set containing all the genes of a particular organism/individual.

Gene. The smallest functional unit of the genetic material that determines or conditions one or more hereditary characters and consists of a DNA segment that code for a single functional polypeptide chain or RNA molecule.

Genotype or type species. The type species of a genus. Genotype is now obsolete since it causes confusion with genetic make-up.

Genus-group names. Names of genus and subgenus categories; changing rank of a taxon within a genus group does not affect the authorship or year of publication.

Germinative zone. Apical part of ovary or testis characterized by actively dividing cells.

Giant cell. An enlarged, multinucleate and nutritionally rich cell formed due to repeated endomitosis as the host response to parasitism.

Glottoid apparatus. Sclerotized structure at the base of the stoma.

Gluconeogenesis. The biosynthesis of a carbohydrate from simpler, non carbohydrate precursors such as oxaloacetate or pyruvate.

Glume. One of the two chaffy basal bracts of a grass spikelet.

Glyconeogenesis. The making of the polysaccharide glycogen without using glucose or other carbohydrates and instead using things like fat or proteins.

Gonopore. The exterior (external) opening of gonads; the vulva in females, the cloacal opening in males.

Gradualism vs. punctuated equilibrium. A Darwinian evolution concept in which species are believed to have evolved through gradual change over time, as against punctuated equilibrium which postulates that new species appear by sudden jumps or saltations due to mutations during periods of rapid evolution separated by periods of relative constancy, and which gradually dominate or replace the mother species.

Granek's ratio. The ratio of the distance of the anus from the edge of the vulval fenestra and the diameter of the fenestra; useful in differentiating some species of cyst nematodes.

Gravid female. Female containing an egg or eggs.

Gravitational water. The water held between saturation and field capacity.

gub. Gubernaculum.

Gubernaculum. Spicule guide, sclerotized accessory piece.

Guiding ring. A cuticularized structure which surrounds and guides stylet.

Habitat. Living place of an organism.

Haemocoel. A cavity or spaces between the organs of most arthropods through which the blood circulates.

Hatching factor. A material produced by the roots of certain hosts that increases the hatching percentage of eggs within *Heterodera* cyst.

Head. The portion anterior to the base of stoma or stylet.

Hemizonid. Lens like structure situated between the cuticle and hypodermis on the ventral side of the body just anterior to the excretory pore; generally believed to be associated with the nervous system.

Hemizonion. A companion sensory structure to the hemizonid; it is smaller and located posterior to hemizonid.

Hennigian method. Using cladistic approach for defining taxa (clades) by synapomorphies as proposed by Hennig (1950, 1957, 1966).

Hermaphrodite. An individual bearing both functional male and female reproductive organs.

Heterosexuality. Amphimictic reproduction in which females are mated by different male individuals.

Heuristic. Method of progressive improvement of estimates by trial and error search rather than by following a set method. Also a method of algorithms for finding shortest trees, which are not guaranteed to represent the most parsimonious ones.

Holotype, allotype and paratype. A holotype is a member of a type series. The best example on which the concept of a species or type species is based, and it should be well preserved and preferably illustrated in the description. The remainder of the type series are paratypes; an allotype represents the opposite sex to that of the holotype, not in use since a type series includes only the holotype and paratypes.

Homology and analogy. Similarities in characters or character states in two or more taxa that are the result of their inheritance from a common ancestor; analogy refers to the similarities which are not due to their inheritance from a common ancestor.

Homonymy, primary and secondary homonyms. Either of two or more identical scientific names. When two identical names occur in a genus, the junior homonym has to be given a new name, called the nomen novum. Primary or secondary homonyms are the identical names that were proposed for different taxa in the same genus (primary homonym) or brought to the same genus (secondary homonym); secondary homonymy breaks by the transfer of a homonym to another genus.

Homoplasty. In phylogenetic analysis, when an additional number of character states resulting from reversals and parallel or convergent evolution are also considered, besides the minimum number of changes that could theoretically have taken place.

Host parasite relationship. The modifications (morphological, anatomical and physiological) that occur in the host and pathogen as a result of infection by the latter.

Host. An organism that supports growth and development of another organism (parasite) that has infected it.

Hyaline. Colorless, transparent.

Hygroscopic water. It is defined as the water that is held (absorbed or imbibed) by the soil particles at a suction of more than 31 bars and is not available for utilization by the plants.

Hyperplasia. Excessive cell division or an abnormal increase in the number of cells in a part or tissue that leads to cancerous outgrowth (tumor, gall, knot, cyst etc.).

Hypersensitive. The state of being abnormally sensitive. It often refers to extreme reaction to a pathogen (e.g. formation of local lesion or necrotic response).

Hypersensitivity. The expression of extreme reactivity by a plant in response to a potential parasite or pathogen in order to limit or prevent parasitization.

Hypertonic. When the concentration of solute is higher in the cell surrounding than inside, the cell is said to be in hypertonic solution and leads to cell shrinkage.

Hypertrophy. An abnormal increase in size of cells of a tissue that leads to cancerous outgrowth (tumor, gall, knot, cyst etc.).

Hypodermis. Thin tissue layer beneath the cuticle which thickens to form the dorsal, ventral and lateral chords which extend along the length of the body. It is a living layer.

i.e. An acronym that stands for Latin word *'id est'*, meaning "that is".

ibid. *ibidem* (in the same place as above cited reference).

ICZN. International code of zoological nomenclature. Naming of zoological taxa from subspecies to superfamily is governed by the ICZN, which is a collection of rules and recommendations previously adopted by the International Congress of Zoology.

in litt. In correspondence.

In vitro. Cultivated in an artificial, nonliving environment.

In vivo. Within a living organism, under natural conditions.

Inanimate plant pathogen. A nonliving entity that may cause disease in plants. e.g., toxicants, pollutants etc.

incertae sedis. Wrongly inserted, of uncertain position.

Incisure (Involution). A longitudinal cuticular cleft that divides the lateral fields, sometimes called involution or line.

Incubation period. The period of time between penetration of host by a pathogen and the first appearance of symptom on the host.

Infection. The establishment of a parasite within a host plant.

Infective larvae. The stage in the life cycle of a parasite capable of invading host tissue and establishing infection.

Infested. Presence of greater number of an organism (fungi, bacteria, insects, mites, nematodes etc.) in the tissue, soil, field, area, etc. or on plant surface, without causing apparent damage or symptom.

Ingestion. The act or process of taking food and other substances into the animal body.

Ingroup and outgroup. In phylogenetic analysis, comparison with members of the same monophyletic group or with members of other such groups, respectively.

Injury. Damage of plant by a pest, pathogen, physical or chemical agent.

Inoculate. To introduce a microorganism into an environment suitable for its growth; to bring a parasite into contact with a host.

Inoculum (inocula plural). The pathogen or its part that can cause infection. That portion of individual pathogens that are brought into contact with the host.

Inoculum density. A measure of the number of propagules of a pathogenic organism per unit area or volume.

Insect. Any of class (Insecta) of arthropods (as bugs or bees) with well defined head, thorax and abdomen, only three pairs of legs and typically one or two pairs of wings.

Insemination. A process of fertilization of ova (mature oocyte) with sperms.

Integrated control. An approach that attempts to use all available methods of control of a disease or of all the diseases and pests of a crop plant for best control results and have socio-economic and environment friendly consequences.

Intercellular and intracellular space. Space Between or among the cells and the space within a cell, respectively. An organism moving through a cell is called intercellular movement.

Intersex. An individual possessing both male and female characteristics.

Interspecific and intraspecific competition. The competition for resources between the individuals of different species, and same species respectively.

Invagination. Retraction under force of pressure, of an outer surface toward the inside. In nematodes outer cuticle invaginates through anal openings to form lining upto oesophagus, vagina and rectum.

Invasion. Spread of pathogen onto or into the host tissues.

Isoelectric focusing and isoelectric point. Electrophoretic separation of the protein variants based on their migration within a pH gradient to their isoelectric points, where they accumulate (isoelectric focus point).

Isolate. A single spore or culture and the subcultures derived from it. Also used to indicate collections of a pathogen made at different places and times.

Isolation. The separation of pathogen from its host or a substrate and its culture on a medium.

Isomerases. Enzymes that catalyze the transformation of compounds into their positional isomers.

Isozyme and allozyme. Usually in electrophoresis, a form of an enzyme coded for by a gene at a different locus from that of another form of the same enzyme. Isozyme may arise through gene duplication. An allozyme represents one or more variants of an enzyme coded by different alleles at the same gene locus.

Isthmus. Relatively narrow portion of the oesophagus between the metacorpus and the basal bulb.

ITS (internal transcribed spacers). Regions of DNA sequence that separate genes for certain ribosomal RNAs. The genes appear to be nearly identical among a wide variety of species but the DNA spacers between them are quite variable and may be species-specific. This makes them good potential targets for "primer" for PCR amplification.

Juvenile. The life stage of nematode between embryo and the adult, an immature nematode.

Kairomones. A compound released by one organism which evokes a response beneficial to a member of another species but not to the emitter.

Karyotype. Chromosome complement of a cell or organism.

Kilobase (Kb). One thousand continuous bases (nucleotides) of single stranded RNA or DNA.

Kinases. Enzymes that catalyze the phosphorylation of certain molecules by ATP.

L. Latin or Linnaeus.

L. The overall length of a nematode in millimeters.

Labia (labium singular)/lips. In nematodes six lips (labia), 2 subdorsal, 2 subventral and 2, one each of lateral sides are present around the stomal opening (mouth).

Labial disc. A raised area in the cuticle of the most anterior annule of some species of *Criconemoides*.

Labial. Pertaining to or located on the lips.

lapsus calami. Typists' or printers' error, an error in spelling in nomenclature.

Larva (larvae plural) or juvenile. Any life stage between the embryo and the adult; an immature nematode.

Lateral field. A pair of longitudinal cuticular markings located above the two lateral hypodermal chords.

Leaf sheath. The lower part of the leaf which encircles the stem (members of Poaceae).

Leghaemoglobin. A protein containing heme which functions to bind oxygen to create O_2 deficient condition in the nodules of the leguminous plants.

Lesion. A localized spot of diseased tissue or localized diseased area or wound.

Life-cycle. The stage or successive stages in the growth and development of an organism that occur between the appearance and reappearance of the same stage (e.g. spore) of an organism.

Lipids. Substances whose molecules consist of glycerine and fatty acids and sometimes certain additional types of compounds.

Liquefaction. The state of being liquid; the act or operation of making or becoming liquid especially, the conversion of a solid into liquid by the sole agency of heat.

loc. cit. *locus citatus*, local citation, as cited above.

Lumen. Triradiate canal of duct of the oesophagus.

m. Length of conus as percentage of total stylet length.

MAbs. Monoclonal antibodies.

Matrix. The aqueous contents of a cell or organelle with dissolved solutes or the gelatinous substance into which eggs of root knot nematodes and some other sedentary nematodes are deposited to form an egg mass. The matrix is secreted by different cells viz., rectal cells (*Meloidogyne*), excretory cells (*Tylenchulus*), vulval cells (*Rotylenchulus*) and cells of uterine wall (*Heterodera*). (see rectal glands).

MB. Distance between anterior end of body and centre of median oesophageal bulb as percentage of oesophageal length.

Meiosis. Process of nuclear division in which the number of chromosomes per nucleus is halved i.e., converting the diploid state to the haploid state.

Mesenteron. The intestine, the largest part of alimentary canal that occupies the region between oesophagus and rectum. Major phase of digestion of food and absorption of nutrients occur in the intestine.

Mesorhabdion. Specialized zone of cuticle lining the mesostome or buccal region secreted by the pharyngeal epithelium.

Metabolism. The entire set of enzyme catalyzed transformations of organic molecules in living cells; the sum of anabolism and catabolism.

Metabolite. A chemical intermediate in the enzyme-catalyzed reaction of metabolism.

Metacorpus or median bulb. An enlarged portion of the oesophagus located near its mid point frequently containing a valve.

Metaspecies or metataxon. A species or taxon, respectively, which is only defined by symplesiomorphy and is not known to be paraphyletic.

Metatype. A specimen which has been compared with the holotype and believed to belong to the same species.

Microflora. The combination of all microorganisms in a particular environment.

Micrometer (μm). A unit of length equal to 1/1000 of a millimeter.

Micropyle. Minute opening in the membrane of an egg through which the spermatozoa enter.

Microscopic. Very small, smaller than visible limit of human eye; can be seen only with the aid of a microscope.

Microvilli. Minute processes of protrusion from free surface of a cell, especially cells of the proximal convolution in renal tubules and of intestinal epithelium.

Migratory ectoparasite. A nematode which feeds on the host from the outside while remaining capable of free movement.

Migratory endoparasite. A nematode which enters a host, feeds internally, moves freely through the tissue, and never becomes sedentary.

Mitosis. Nuclear division in which the chromosome number remains the same.

Molecular markers. Samples of DNAs of known size are typically generated by restriction enzyme digestion of a plasmid or bacteriophage DNA of known sequence. Alternatively, they are produced by ligating a monomer DNA fragment of known size into a ladder of polymorphic forms. Mainly it is of two size ranges of standards, high-molecular-weight from 1 kb to >20 kb and a low-molecular-weight range from 100 bp to 1000 bp.

Molt/moult. To shed or cast off the cuticle.

Monoclonal antibody. A pure antibody which is specific to a single antigen determinant and is produced from a specific clonal line of hybridoma cells.

Monoculture. The planting of the same crop species in a field or area in a given period.

Monodelphic. Having one ovary.

Monoecious. See hermaphrodite.

Monogenic. Producing of only one sex.

Monophyletic or holophyletic. Terms used for a taxon or group of taxa whose members are supposed to have descended from a common ancestor.

Monorchic. Having one testis.

Monothetic or polythetic. To differentiate on one or many sets of characters, respectively, e.g. as in a key of identification.

Monotype. Monotypic taxon must be designated as type, e.g. when a species is described on a single specimen, it becomes the holotype.

mRNA (messenger RNA). A form of RNA that carries information to direct the synthesis of protein or a chain of ribonucleotides that code for a specific protein.

mtDNA. Mitochondrial DNA. The DNA present in the mitochondria and is not inherited in the manner as the nuclear DNA.

Mucro. A small pointed projection as occurring on the tail tip of certain nematodes.

Mutualistic relationship. A symbiotic relationship between two different species in which both are jointly benefited.

Mycophagous or Fungivorous. The organisms which feed on fungi (spores or mycelium).

Mycorrhiza. A symbiotic association of a fungus with the roots of a plant or a symbiotic association between a non pathogenic or weakly pathogenic fungus and roots of plants.

Myofibrils. A unit of thick and thin filaments of muscle fibres.

NAD and NADP. These are acronyms that stand for nicotinamide adenine dinucleotide and nicotinamide adenine dinucleotide phosphate, respectively. These nicotinamide containing coenzymes act as carrier of hydrogen atoms and electrons in some oxidation-reduction reactions.

Name-bearing type. In 1985 ICZN introduced this term to indicate either a type genus, a type species or any accepted type specimen which provides an objective standard for the application of a scientific name.

Neck. A portion of the body occupied by the anterior part of oesophagus.

Necrosis. The death of cells or tissues.

Nematicide. A chemical lethal to nematodes.

Nematodes. Triploblastic, bilaterally symmetrical, unsegmented, pesudocoelomate and vermiform animals.

NEPO. An acronym that stands for nematode transmitted polyhedral (spherical) viruses. Longidorid nematodes transmit NEPO viruses. e.g., Arabis mosaic virus.

Nerve ring. See circumoesophageal commissure

Nerve. A bundle of nerve fibres located outside the central nervous system.

NETU or Tobra viruses. Nematode transmitted tubular (rod) viruses. Tobra viruses e.g., Tobacco rattle viruses are transmitted by trichodorid nematodes.

Nitrogenase enzyme. An enzyme of nitrogen fixing bacteria that catalyzes the conversion of nitrogen to ammonia. This enzyme functions in oxygen deficient condition which is maintained by leghaemoglobin.

nm (nanometer). A unit of length equal to 1/1000 of a micrometer.

no., n. Number.

Nodule. Small knot or irregular lump on leguminous plants, structures on roots that contain nitrogen fixing bacteria.

nom. dub. *nomen dubium*, dubious name which cannot be applied with certainty to any taxa, e.g. when the description is inadequate and the type specimens are unavailable.

nom. nov. *nomen novum*, new name, equivalent to new replacement name required for example by homonymy.

nomen conservandum. Conserved name. A name conserved by the use of the plenary powers of the Commission for Zoological Nomenclature, which otherwise would have been invalid using ICZN.

nomen nudum (pl. nomina nuda). Naked name(s), invalid name(s) which is not an available name, but can be made available for the same or a different concept with the authorship and date from that act of establishment.

nomen oblitum. Forgotten name, which remained unused as senior synonym in primary zoological literature for more than 50 years, as per ICZN Article 23(b) of the I and II editions, but later omitted from the III edition.

Non host. A plant resistant to a pathogen that neither allows infection and multiplication nor suppresses the pathogen.

Nucleic acid. An acedic substance containing pentose, phosphorus, and pyrimidine and purine bases. Nucleic acids determine the genetic properties of organism.

Nucleoprotein. Referring to viruses: consisting of nucleic acid and protein.

Nucleoside. The combination of a sugar and a base molecule in a nucleic acid.

Nucleotide. The phosphoric ester of a nucleoside. Nucleotides are the building blocks of DNA and RNA.

Numerical taxonomy. A numerical approach to taxonomy, generally applied to refer to phenetic methods involving clustering and classifying purely on similarity rather than phylogenetic grounds.

O. Distance between stylet base and orifice of dorsal oesophageal gland as percentage of stylet length.

Objective synonyms. Objective synonyms result from two or more taxa having the same type (species or specimen) and over which there can be no difference of opinion or when two or more named genus or species names are based on the same species or type specimens, respectively.

Obligate parasite. An organism that is incapable of living as saprophyte and must live as a parasite on live host.

Obliteration. Destruction of internal body organs due to very rapid and enormous development and production of eggs. Obliteration occurs in *Heterodera*, *Globodera* etc. and is responsible for the death of mature females.

Odontostylet or onchiostylet. Buccal stylet; a large tooth which originates in the oesophageal wall. Generally found among some members of the Adenophorea.

Oesophagostome. Posterior part of the stoma surrounded by anterior oesophagus and lies embedded in the oesophageal tissue.

Oesophagus or pharynx. An anterior portion of the digestive system that connects stoma to the intestine. It primarily acts as a food transporter.

Oocyte. An immature female gamete.

Oogonium. An early stage in the development of an oocyte.

Opisthodelphic. Having a single ovary directed posterior to the vulva.

Oral disc. Anterior most annule on which oral opening is located and is surrounded by lips.

Original designation. Designation or fixation of the type species of a genus in an unambiguous way in the original description.

Osmotic potential. The potential for water to move across a selectively permeable membrane, where the osmotic potential of pure water is zero and any water movement measured is a negative value.

Ovary. Female sexual gland in which the ova or eggs are formed.

Overwinter or oversummer. To survive or persist through the winter or summer period, respectively.

Oviduct. Portion of the female reproductive system between the ovary and uterus. The ripe ova formed in the ovary pass through the oviduct to reach the uterus for fertilization and egg development.

Oviparous. Producing eggs that hatch after being laid outside the body of mother.

Ovoviviparous. Producing eggs that hatch within the uterus of the mother.

p. (plural pp). Page.

PAGE. Polyacrylamide gel electrophoresis.

Papillae. Tactile, sensory organs found on various body regions, e.g., labial or cephalic papillae.

Paraphyletic or paraphyly. A taxon or group of taxa supposed to have a common ancestor and defined by a unique apomorph (derived character state), but not all of whose members are included because some may have undergone one or more reversals of that apomorph.

Parasite. An organism deriving all or a portion of its nutrition from another organism, the host, while permanently or temporarily attached to or within the host.

Parasitoid. An insect that spends a part of its life attached to or within a single insect host and ultimately kills the host or an insect parasitizing another insect.

Parsimony. In cladism, the most likely phylogenetic explanation requiring the least number of evolutionary steps (i.e. the most parsimonious explanation).

part, partim. Part, in part.

Parthenogenesis. Reproduction by the development of an unfertilized egg or reproduction from eggs without fertilization by sperm cells.

Pathogen. An entity that may incite a disease.

Pathogenicity. The capability of a pathogen to cause disease.

Patronym. A scientific name based on a person, in his honour or in recognition of his work.

Penetration. The initial invasion of a host by pathogen.

Perineal pattern. Fingerprint like pattern formed by cuticular striae surrounding the vulva and anus of the mature *Meloidogyne* female.

Peristalsis. Wavelike (peristaltic wave), involuntary muscular contraction that moves the food through the digestive system.

pers. comm. In personal communication.

pH. Negative logarithm of the effective hydrogen ion concentration; a measure of acidity.

Phasmid. Pore like sensory structures located one in each of the lateral fields of the posterior region of a nematode belonging to the class Secernentea.

Phenetics, phenon. Now used mainly to describe numerical taxonomy using clustering of taxa (phenons) based on similarity.

Pheromone. Chemical substance which when secreted by an individual into the environment cause specific reaction in other individual usually of the same species. The substance relate only to multicellular organism and includes kairomones and allomones.

Phoretic. A symbiotic relationship in which one organism associates with another in order to obtain transportation, and causing little or no detectable pathology to the vector.

Photorespiration. Increased respiration that occurs in photosynthetic cells in the light due to ability of RUDP carboxylase to react with O_2 as well as CO_2. The RUDP carboxylase at higher O_2 level acts as RUDP oxygenase and converts RUDP into phosphoglyceric acid (3-carbon compound) and phosphoglycolate (2-carbon compound) and finally to serine and CO_2 without the production of ATP and NADPH. The wasteful release of CO_2 is called photorespiration and reduces the photosynthetic efficiency of C_3 plants.

Photosynthesis. Manufacture of carbohydrate from carbon dioxide and water in the presence of chlorophyll (s) using light energy and releasing oxygen.

Phylogeny. The evolutionary history of a group of taxa.

Phytoalexin. Substances produced in higher plants in response to a number of chemical, physical and biological stimuli that inhibit the growth of certain microorganisms.

Phytopathogenic. Term applicable to a microorganism that can incite disease in plants.

Platymyarian. A type of muscle cell in which the fibres are adjacent and perpendicular to the hypodermis.

Polarity. Evolutionary direction of a character state transition, determining a plesiomorphic (primitive) or apomorphic (advanced, derived) state.

Polymerase chain reaction (PCR). A procedure that results in geometric amplification of a specific DNA sequence through repeated annealing cycles of specific temperatures and durations.

Polymorphism. Occurance of two or more phenotypic states of a character among the members of a taxon.

Polyploidy, aneuploidy. Increasing of chromosome number above the normal diploid number (2n) by an integral number of n, but this might have been modified subsequently to have a different (lower) number by aneuploidy.

Poor host. A plant species which is invaded by a pathogen but does not allow the population build up of the organism.

Population density. Number of individuals of a group, genus or species of an organism per unit volume of soil or weight of plant tissue.

Population. A group of organisms of class, genus, species etc. that interbreed and live in the same place at a given time e.g., nematode population, bacterial population, etc.

Pore space. The space within soil particles (aggregates) that are unoccupied by solid material.

Post-uterine sac. The uterus extending posterior to the vulva which may serve to store the spermatozoa.

Preadult. The larval stage just before adulthood; the fourth-stage larva.

Predator. An organism which seeks another organism to kill/anaesthetize/inactivate and feed on later or an organism which lives by preying upon animals.

Prerectum. Portion of the digestive system between the intestine and the rectum present in *Xiphinema*.

Primary infection. The first infection of a plant by the environment or over summering or over wintering pathogen.

Primary inoculum. The overwintering or oversummering pathogen, or its spores that cause primary infection.

Primary pathogen. The pathogen that invades the host first and modifies the host response in such a way that it either becomes tolerant/susceptible to attack by a following pathogen. The primary pathogen have specific, inherent and independent capability of causing a disease.

Primary soil particles and secondary aggregates. Primary soil particle are clay (<0.002 mm), silt (0.002-0.02 mm) and sand (0.2-2 mm) which are bound together by some cementing material forming secondary aggregates of various shapes and types.

Probe. A labeled fragment of nucleic acid containing a nucleotide sequence complimentary to gene or genomic sequence that one wishes to detect in hybridization experiment.

Procorpus. The anterior subdivision of the corpus of the oesophagus located between the base of the stoma and the metacorpus.

Prodelphic. Ovary located anterior to the vulva.

Propagule. Any part of an organism capable of independent growth.

Protectant. A substance that protects an organism against infection by a pathogen.

Protein. A macromolecule composed of one or more polypeptide chains each with a characteristic sequence of amino acids linked by peptide bonds.

Protoplast. A plant cell from which the cell wall has been removed. The organized living unit of a single cell; the cytoplasmic membrane and the cytoplasm, nucleus and other organelles inside it.

Protractor muscles. A muscle that extends a limb or other part.

Pseudocoelom or body cavity. A false coelom; a body cavity not lined with an epithelium of mesodermal origin containing a liquid in which the various internal organs are suspended. The body fluid apparently functions as a respiratory and circulatory system.

Pylorus. The muscular opening from the vertebrate stomach into the intestine.

Quarantine. Control of import and export of food, fibre and other material to prevent spread of dangerous pathogenic organisms, diseases and pests.

Quiescent. A state of relative biological inactivity induced by physical factors such as low/high temperature, desiccation etc. Also refers to dormant or inactive state.

R. Total number of body annules.

Race. A genetically and often geographically distinct mating group within a species; also a group of pathogens that infect a given set of plant cultivars.

Rachis. The central or axial chord of the ovary of a nematode.

Ran. Number of annules on tail.

RAPD. Random amplification of polymorphic DNA. Random amplification of polymorphic DNA or Randomly amplified polymorphic

DNA. RAPDs are generated by using random sequence of ten bases oligonucleotides (at least 50% G+C and lacking inverted repeats) as primers for PCR amplification of genomic DNAs from different strains/species.

RAPD markers/fragments. RAPD markers are the DNA fragments from PCR amplification of random segments of genomic DNA with single primer of arbitrary nucleotide sequence.

Reciprocal parasitism. It is a type of symbiotic association between two organisms where both partners are benefited by the associations and each of them try to maximize their reproductive output at the expense of other partner e.g., mycorrhizal fungi and higher plants.

Rectal glands. Glands in the rectal region, usually 3 in females and 6 in males. In the genus *Meloidogyne*, 6 rectal glands are present in females which secrete gelatinous material that forms egg mass sac in which eggs are deposited. (see matrix).

Renette. A ventral cell present in the excretory system of some nematodes.

Resistance. The ability of an organism to exclude or overcome, completely or in some degree, the effect of a pathogen(s) or other damaging factor(s).

Respiration. Series of specific chemical reactions that makes energy available through oxidation of carbohydrates and fats.

Restriction enzyme. The enzyme that cleaves double stranded DNA at specific sites.

Restriction fragment length polymorphism (RFLP). The occurrence in a taxon of more than one morph defined by the presence or absence of a particular restriction enzyme recognition (cleaving) site.

Retractor muscle. A muscle, such as flexor, that retracts (brings back) the limb or an organ or a part.

Retrorse. Backward-projecting.

Rex. Number of annules between anterior end of body and excretory pore.

Rhizosphere. The area in the soil immediately surrounding plant roots and influenced by them.

Ribosome. A cellular particle which is involved in translation of mRNAs to make proteins. Ribosomes are a complex consisting of ribosomal RNAs (r RNA) and several proteins.

Ripening zone. Posterior part of ovary or testis where maturation of oocytes or spermatocytes takes place.

RNA. Ribonucleic acid. A polyribonucleotide of a specific sequence linked by successive 3'5'-phosphodiester bonds.

Roes. Number of annules in oesophageal region.

Rostrum. Beak like or snout like projections of the anterior part of the head of certain insects such as weevils and also of spicule.

rRNA (ribosomal RNA). RNA molecules forming part of ribosomal structure.

RV. Number of annules between posterior end of body and vulva.

Rvan. Number of annules between vulva and anus.

s. A ratio calculated by dividing stylet length by body diameter measured at base of stylet of the nematode.

s.l. *sensu lato,* in a broad sense (cf. *sensu stricto*).

s.s., s.str. *sensu stricto,* in the strict sense (cf. *sensu lato*).

Sanitation. The removal and burning of infected plant parts, decontamination of tools, equipments, hands, etc.

Saprophytes. An organism that uses dead organic material as food.

Saprozoic. Feeding on dead or decaying animal matter.

Sclerotized. A term which refers to a hardened condition, generally of cuticle.

Scutellum. An enlarged phasmid found in some species of the subfamily Hoplolaiminae.

SD. Standard deviation.

SE. Standard error.

Secondary pathogen. The pathogen that invade the host after invasion by the primary pathogen. The secondary pathogen may or may not have specific, inherent and independent capability of causing a disease.

Sedentary endoparasite. The nematodes which enter a host, migrate to a feeding site and then become nonmotile.

Seed gall or cockle. Deformed and dark colored grains formed due to infection by *Anguina* spp. which internally contains thousands of second stage juveniles of the nematode in quiescent state.

SEM. Scanning electron microscope/microscopy.

Semifenestrae. See ambifenestate.

Seminal vesicle. A portion of the male reproductive tract that functions as a temporary storage organ for sperms.

Semiochemicals. The substances or chemicals produced by organisms especially insects, participate in to communicate or elicit behavioral responses/in activities such as aggregation of both sexes, sexual stimulation and trail following.

Sensillum (sensilla plural). A sensory organ, generally made up of one or more sensory neurons whose dendritic endings are protected by one or two accessory cells, and a thin wrapping of hypodermis.

sensu. According to.

Sequential etiology. In sequential etiology, the host is invaded by two or more pathogens in sequence. The organism that invades the host first is called a primary pathogen. The primary pathogen modifies host response in such a way that it becomes more suitable to invasion by secondary pathogen thus leading to development of disease complex.

Sessile. Permanently attached; not capable of moving about.

Setae. Elongated cuticular structures usually located around the oral opening, tactile sensory organs.

Sex reversal. Transformation of developing female juveniles into males during adverse environmental condition, scarcity of food etc.

Sexual dimorphism. Species in which the male and female individuals differ morphologically.

Sheath. A covering of a structure; refers to a retained or extra cuticle in some nematode species.

Sibling species. Very closely related species which differ only in minute or cryptic morphological characters but which have different biological characteristics and are reproductively isolated.

sic. Thus (an exact transcription).

Soil capillary. The narrow necks of soil pores connected together are soil capillaries. The capillaries contain water which is called as capillary water and is utilized by plants.

Soil inhabitants. Microorganisms able to survive in the soil indefinitely as saprophytes.

Soil solarization. Attempt to reduce or eliminate pathogen populations in the soil by covering the soil with clear plastic so that the sun rays raise the soil temperature to a level that kills the pathogen.

sp. (spp. plural). Species.

sp. inq. *species inquirenda,* demanding further inquiry or study as the characters are insufficient for recognition.

sp. n., sp. nov., n. sp. *species nova,* new species.

Species group names. Names of species and subspecies; changing the rank of a taxon within a species group does not affect the authorship or year of publication.

Species inquirendae. Species of doubtful status because of inadequate descriptions and lack of preserved material.

Sperm. A mature male reproductive cell or gamete.

Spermatheca. An enlarged portion of the female gonad between the oviduct and uterus that functions as a storage organ for sperms.

Spermatocyte. A cell giving rise to sperm cells or spermatozoa.

Sphincter. A ringlike muscle that surrounds a natural opening in the body and can open or close it by expanding or contracting.

spic. Spicule.

Spicate terminus. Spike shaped tail end.

Spicule. A pair of heavily cuticularized male copulatory organs to transfer sperms into vagina.

st. Stylet.

Sterilization or disinfestation. The elimination of pathogens and other living organisms from soil, containers etc. by means of heat or chemicals.

Stomatostylet. A stylet evolved from a fusion of the walls of the stoma commonly found in nematodes belonging to the order Tylenchida.

Stomodeum. The anterior portion of the digestive system formed by the invagination of the body wall.

Strain. A distinct form of an organism or virus within a species, differing from other forms of the species biologically, physically or chemically.

Stylet or spear. A relatively long, slender, axially located, hollow feeding structure of sclerotized cuticle located in the buccal cavity.

Sub median. At some distance from the middle region.

subfam. Subfamily.

Subjective synonyms. Subjective synonyms are those that result from the opinion of a person, seeing that they are in fact the same taxon described under different names.

subsp., ssp. Subspecies.

Substrate. The material or substance on which a microorganism feeds and develops. Also a substance acted upon by an enzyme.

Suppressive soil. Soils in which certain diseases are suppressed due to the presence of microorganisms antagonistic to the pathogen(s).

Susceptibility. The inability of a plant to resist the infection and effect of a pathogen or other damaging factor.

Susceptible. Lacking the inherent ability to resist the infection, disease or attack by a given pathogen.

Suspended animation. Suspension of reversible life processes as occurs in quiescent state.

SVN. Sub ventral oesophageal gland nucleus.

SVO. Subventral oesophageal gland orifice.

Symbiosis. A mutually beneficial association of two or more different kinds of organisms.

Symplesiomorphy and synapomorphy. Sharing of plesiomorphs (ancestral characters) and apomorphs (derived characters), respectively.

Symptoms. The external and internal reactions or alterations of a plant as a result of infection or a disease.

Syncytia. Multinucleated cells in root tissue formed by dissolution of common cell walls induced by secretions of certain sedentary plant parasitic nematodes.

Synergism. Coparasitism of a host to produce symptoms or other effects of greater magnitude than the sum of the effects caused by both parasites acting alone or the concurrent parasitism of a host by two pathogens, which is beneficial to one or both the pathogens.

Synergistic interaction. This is an interaction in which the interacting organisms are synergized leading to greater interactive effects compared to sum of individual effects. Disease complexes usually develops due to synergistic interaction between two or more pathogens.

Syngonic. Pertaining to hermaphroditic reproduction in which both sperm and eggs are produced by the same gonad.

Synonyms. Synonyms are two or more generic or species names that can be applied to a single taxon.

Syntype, lectotype, topotype and neotype. In a type series, if the holotype is lost, all the paratypes become syntypes, from which any one syntype can be chosen and named the lectotype, after which all the syntypes become paralectotypes (or lectoparatypes). If the type series is lost, specimens of the species collected from its type host and locality are called topotypes, from which a neotype is selected and designated.

Systemic pesticide. A pesticide absorbed by an organism or plant and circulates within its system. The systemic pesticides (chemicals) provide protection to the entire body against the pest(s).

T. Percent of distance from cloacal aperture to anterior end of testis divided by the body length.

Tail. Postanal elongation. The portion of body between the anus and the posterior terminus.

Taq **polymerase.** A DNA polymerase isolated from the bacterium *Thermus aquaticus* and which is very stable at high temperatures. It is used in PCR procedures and high temperature sequencing.

Tautonym. Identical spelling of two generic or species names.

Taxon (taxa plural). A group of organisms or a formal taxonomic unit at any level of hierarchic classification from subspecies to kingdom categories.

Taxonomy and systematics. These terms are largely synonymous. Taxonomy deals with identifying, naming and classifying organisms, whereas systematics has a broader definition and includes aspects of the phylogeny, evolution, biogeography, genetics and physiology of the organisms.

Telamon. Rigid sclerotized portion of the cloacal wall present in stronglylids which apparently guides the spicules from the spicular pouch into the cloaca.

Telegonic. That type of gonad in which germ cells are produced only at the distal end.

Terminator gene and traitor gene. A special sequence of nucleotides in DNA that marks the end of a gene; it signals RNA polymerase to release the newly made RNA molecule, which then departs from the gene. In the seeds having terminator gene, gene sequence in the DNA is induced which is responsible for the execution of a controlled death of the seed at ripening. Usually the terminator gene sequence can be activated through a on or off mechanism. There is another sequence called the traitor Gene (The Lazarus Gene) which means, when the seed is ripe it will die, if not treated with a hormone sold by the company to bring the seed back to life.

Testis. The portion of the male reproductive system that produce sperms.

Thermal capacity. The amount of heat required to produce a unit change of temperature in a unit mass of a substance.

Thermal conductivity. It is the ability to conduct heat.

Tissue. A group of cells of similar structure which performs a special function.

Titillae. Small projections on either side of the distal end of the gubernaculum.

Tolerance. The ability of a plant to sustain the effects of a disease without dying or suffering serious injury or crop loss. Also the amount of toxic residue allowable on edible plant parts under the law.

Transamination. The transfer of an amino group from one molecule to another, usually by the action of transaminase.

Transcription. The process of copying DNA to produce an RNA transcript. It is the first step in gene expression.

Transdeamination. Describes reversal of reactions catalyzed by glutamate dehydrogenase. It is used to produce 2-oxyglutarate from glutamate at times when the former is sparse.

Transformation series. In phylogenetic analysis, a hypothesized set of likely transitions between multiple character states.

Transgenic plants. Plants into which genes from other source(s) have been introduced through genetic engineering techniques, and are expressed, and produce the expected/desired function.

Translation. The assembling of amino acids into a protein using messenger RNA, ribosomes and transfer RNA or the process of decoding a strand of mRNA, thereby producing a protein based on the code.

Translocation. Movement of carbohydrates or other material from the leaves to other parts of plant through phloem.

Transmission. The transfer or spread of virus or other pathogens from one plant to another.

Transovarial transmission. The ability of a virus vector to transmit the virus to its progeny by means of eggs.

Trap crop. A crop plant which is penetrated by a parasite but in which it cannot be or is not allowed to reach maturity.

Triploblastic. Possessing three germ layers: ectoderm, mesoderm and endoderm.

Triradiate. Having three radiating shape, parts or members.

tRNA (Transfer RNA). The RNA that moves amino acid to the ribosome to be placed in the order prescribed by the mRNA.

Tumour. An uncontrolled overgrowth of tissue or tissues formed due to hypertrophy and/or hyperplasia.

Type series. All the specimens of a species used at the time of its description.

Underbridge. A structure extending across the vulval cone of *Heterodera* cyst below and parallel to vulval bridge.

Uterus. The portion of the female reproductive system between the oviduct and the vagina. In the uterus fertilization and egg development occurs.

V. The location of the vulva expressed as a percentage of the total body length as measured from the anterior end.

v'. Percent of the distance from head end to vulva divided by the distance from head end to anus.

Vagina. The cuticle-lined canal leading from the uterus to the vulva (vulval pore).

Variability. The property or ability of an organism to change its characteristics from one generation to other.

Vas deferens. The portion of the male reproductive system from the vas efferentia to the seminal vesicle and the ejaculatory duct.

Vector. An agent of dissemination and/or inoculation or an agent which carries organism(s) or gene to the host, target cell, tissue, etc.

Velum. Wing like membranous longitudinal extension of the spicule blade.

Vermiform. Relatively long and slender; worm shaped or thread like.

Vestibule. An outer cavity with an entrance to a (usually) larger, deeper cavity.

vide. See.

Virulence. State of being pathogenic; the degree of pathogenicity of a given pathogen; capable of causing a severe disease; or strongly pathogenic.

Viruliferous. Said of a vector containing a virus and capable of transmitting it.

Viviparous. Those organisms that give birth to young ones and do not lay eggs.

viz. An acronym that stands for Latin word *'videlicet'* and is translated as "namely".

VL/VB. Distance between vulva and posterior end of body divided by body width at vulva.

Voucher specimens. Specimens used in a study, which are not from a type series, but are deposited in a permanent collection for later use or reference.

Vulva. The female gonopore; external opening of the female reproductive system.

Vulval bridge. A narrow connection across the fenestra of vulval cones of some *Heterodera* cysts that forms two semifenestrae.

Vulval cone. The conspicuous elevation or protuberance on the posterior portion of *Heterodera* cysts.

Weighting. Assigning different importance or value to different character systems in phylogenetic analysis or identification systems (e.g. in keys).

Z-organ. It is a specialized organ located in the region between uterus and oviduct of some *Xiphinema* spp. It is characterized in having strong circular muscles, longitudinally striated (fine striae) internal wall and four well developed and sclerotized refringent apophyses in the lumen. It forms an important taxonomic character among *Xiphinema* species. Its function is to slow down the passage of eggs during eggshell formation and to prevent the reflux of sperms.

Zygote. A diploid cell resulting from the union of two gametes.

Index

A
a 5, 298
Abiotic 298
Abiotic factors 237, 242, 245
Absorption of water 234
Acanthocheilidae 96, 119
Acaulospora 264
Acetic acid 34, 35, 41-43, 61, 246
Acetone 43, 47, 56
Acetyl coenzymes A (Acetyl CoA) 205, 298
Acetylcholinesterase 298
Achlysiella 137
Acid fuchsin 37, 42, 43
Acontylinae 136, 137
Acquisition 259
Acrolein 46
Acropetal movement 298
Acrylamide 52, 53, 62, 63, 65, 66
Acrylamide gels 298
Acrylease 65
Acrylic acid 62, 63
Actinolaimidae 92, 107
Actinolaimoidea 90, 92, 107
Acuariidae 97, 121
Acugutturidae 100, 162, 163,
Acugutturinae 100, 165

Adanal 298
Adaptability 242, 298
Additive interaction 298
Adenophorea 89, 91, 100, 120, 167, 169, 180, 186-189, 194, 198, 199, 216
ADP 298
Adult cuticle 74
Aerolaimus 182
Aetholaimidae 93, 108
Aetiological interaction 276
AFLP 298
AFLP markers 299
Agamospecies or morphospecies 299
Agaricales 263
Agaricus 256
Agarose gel 53, 58-60, 67, 299
Agarose gel electrophoresis 59, 60
Agfidae 95, 115
Agrobacterium rhizogens 252
Agrobacterium tumifaciens 252
Agro-ecosystems 237
Aggressiveness 299
Alae 299
Alcohol 34, 201
Alcohol dehydrogenase 201
Aldolase 299
Alimentary canal 178, 185

Allantonematidae 99, 155, 157, 270
Allantonematinae 152, 158
Allele 299
Allelochemicals 275, 276, 299
Allopatry and sympatry 299
Allorhizobium 260, 261
Allotylenchus 78, 189
Allozymes 299, 315
Allyl isothiocyanate 208, 241, 244
Alpha (α), Beta (β) and Gamma (γ) taxonomy 299
Amanita 263
Ambifenestrate 299
Ambimobile 299
Amidostomatidae 96, 118
Aminoacids 300
Ammonia 36, 189, 207
Ammonium persulfate 63, 65
Amphidelphic 103, 104, 112, 120, 133, 134, 137, 141, 142, 172, 173, 300
Amphidial apertures 106-108, 125-129, 132, 135, 142-144, 152, 160, 179, 180
Amphidial ganglia 190
Amphidial gland 195
Amphids 90, 100, 111-115, 119-121, 123, 124, 127, 130, 137, 143, 154, 179, 180, 190, 195, 196, 213, 300.
Amphimictic 170, 199, 207, 238, 300
Amphimictic species 170, 199, 207
Amphimixis 300
Amplification of DNA 56
Amygdlin 220, 227
Anaesthetizing 33, 34
Anandranema 157
Anatonchidae 92, 104
Ancylostomatidae 96, 118
Ancylostomatoidea 96, 117, 118
Angiostomatidae 95, 115
Angstrom (Å) 300
Anguillicolidae 97, 120

Anguillula marioni 4
Anguillulinoidea 131
Anguina 3, 4, 6, 16, 27, 53, 167, 170, 173, 175, 209, 218, 223, 232, 239, 241, 245, 249, 252
Anguinata 98, 125, 130, 131
Anguinidae 78, 98, 131, 172, 173, 180
Anguininae 131
Anguinoidea 98, 131
Anhydrobiosis 209, 223, 232, 240
Aniline blue 43
Animate plant pathogen 300
Anisakidae 96, 119
Anisocytous 185
Annealing 57, 59
Annulations 35, 78, 80, 111, 112, 128, 212, 300
Annule 80, 141, 300
Anomyctinae 99, 164
Anoxybiosis 209, 210, 240
Antagonist 300
Antagonistic effects 276, 277
Antagonistic interaction 254, 270, 300
Antarctenchinae 144, 145
Anterior nervous system 190, 191
Anterior uterine sac 173
Antibiosis 300
Antibiotic 300
Antistatic gun 48, 51
ap., apud. 300
Apex 300
Aphasmatylenchidae 135
Aphasmatylenchus 132, 134, 135
Aphasmidia 89
Aphelenchida 2, 4, 7, 81, 88-90, 99, 160, 161, 169, 183, 184, 216, 222, 224, 232, 265
Aphelenchinae 88, 99, 162
Aphelenchoidea 99, 161, 162
Aphelenchoides besseyi 16, 27, 208, 209, 233, 240, 246, 249

Aphelenchoides bicaudatus 265
Aphelenchoides composticola 256
Aphelenchoides fragariae 224, 233, 252
Aphelenchoides hamatum 256
Aphelenchoides ritzemabosi 5, 10, 29, 206, 207, 224, 233, 240, 249, 252
Aphelenchoididae 99, 162-164
Aphelenchus 2, 4, 5, 88, 161, 209, 265
Aphelenchus avenae 4, 204, 209, 210, 244, 256, 271
Aphelenchus cibolensis 265
Apoptosis 300
Appligene 59
Aproctidae 98, 123
Arabis mosaic virus 216, 258
Araeolaimida 89, 94, 109, 112
Araeolaimina 94, 113
Araeolaimoidea 94, 113
Arbuscular mycorrhizal fungi 264
Arbuscules 264, 266
Arbutus 264
Arch 300
Arctostaphylos 264
Areolation 81, 301
Armillaria mellea 264
Art 301
Arthrobotrys dactyloides 257
Arthrobotrys oligospora 255, 257
Arthrobotrys robusta 243, 257
Ascaridida 96, 119, 120
Ascarididae 96, 119
Ascaridiidae 95, 116
Ascaridoidea 96, 119
Ascaris 1, 3, 200, 201-207
Ascaris lumbricoides 1, 3, 200
Ascaroside 205
Ascarylose 204, 205
Aschelminthes 89
Ascomycetes 263
Aspergillus niger 244, 255

Asymmetrical 115, 124, 137, 139, 152, 158, 189
Ataloderidae 139, 140
Ataloderinae 140
Atkinson 251
Atlenchoidea 126
ATP 301
ATPase 301
Atractidae 95, 116
Atylenchidae 98, 126, 128, 129
Atylenchinae 129
Atylenchus 128
auct 301
Aulosphora 149
Avermectins 245
Axonchium 177
Axonolaimidae 94, 113
Ayyar 8
Azadirachta indica 241
Azadirachtin 241, 244
Azorhizobium 260

B
b 5, 301
b' 301
Bacillus subtilis 244, 245, 255
Bacillus thuringiensis 244, 245, 255
Bacteria 301
Bacterial wilt 252
Baermann funnel 18, 20, 23, 26-30
Baermann funnel technique 27
Baermann trays 17, 26
Bark beetles 131, 164, 165
Barker 8, 15, 26
Basal bulb 125, 126, 141-147, 149, 151, 152, 154-157, 161, 162, 182-185, 301
Basal lamella 75, 182
Basal layer 74, 75
Basal membrane 75
Basement membrane 301

Basic sensory structure 193
Basidiomycetes 263, 264
Basipetal movement 301
Basirolaimus 132
Bastianidae 94, 113
Bathyodontidae 92, 104
Beetles 131, 163, 164, 269
Belondiridae 92, 107
Belondiroidea 90, 92, 105, 107
Belonolaimidae 99, 141-143
Belonolaimus 3, 10, 71, 81, 84, 140, 141, 218, 219
Belonolaimus longicaudatus 10, 262, 274
Benthimermithidae 93, 109
Bilateral symmetry 301
Bilaterally symmetrical 1, 70
Binomen, trinomen and scientific names 301
Bioassay 302
Bioassay of EPN 32
Biochemical techniques 52
Biocontrol 302
Biodiversity 302
Bioefficacy test 32, 33
Biological control 7, 255
Biological species 302
Biopesticides 302
Biopredisposition 250, 251, 302
Biotechnology 302
Biotic factors 242, 243, 250
Biotroph 302
Biotype, pathotypes and host races 302
Bisacrylamide 62, 63, 65
Bisacrylic acid 62, 63
Bisexual species 174
Black palm weevil 6, 234, 269
Blastophaga psenes 224, 270
Blastula stage 167, 302
Blight 302
Blind diverticulum 303

Blind sac 84, 186
Bloat 303
Blue R dye 38
Body cavity 70, 71, 78, 132, 149, 151, 154, 156, 157, 159, 160, 184, 187, 189, 190
Body shape 73, 87, 133, 136, 151
Body sheath 79, 145, 147, 148
Boleodoridae 127
Boleodorinae 127
Books on plant nematology 6, 7
Boomerangia 150
Botrytis 256
Brachydorinae 142, 143
Bradyrhizobiaceae 261
Bradyrhizobium 261, 262
Bradyrhizobium japonicum 262
Brittonematidae 107
Bromophenol blue 53, 54, 61, 63
Bt toxin (δ endo toxin) 245, 255
Buccal aperture 303
Buccal cavity 100, 180, 303
Buffer 303
Bulbiformin 245, 255
Bullae 303
Bunonematidae 95, 115, 116
Bunonematoidea 95, 115
Bursa/Bursae/caudal alae 84, 108, 114, 117, 118, 120, 303
Bursadera 134, 138, 139
Bursaphelenchinae 100, 165
Bursaphelenchus 2, 16, 69, 233, 234, 265, 269
Bursaphelenchus xylophilus 16, 69, 233, 234, 269
Butler 8
Butyric acid 245, 246, 255

C

c. 5, 303
c´ 303

C_4 plants 303
Cacodylate 47, 48, 50
Cacodylate buffer 47, 50
Cacopaurus 71, 73, 152, 216, 219, 221
Cacopaurus pestis 71, 216, 219, 221
Caenorhabditis 168, 171, 200, 206, 208
Caenorhabditis elegans 168, 171, 200, 205, 206
Caenorhabditis xenoplax 208
Cafe au lait bacteriosis 252
Caloglyphus berlesei 271
Caloosia 148
Caloosiidae 99, 148, 149
Camacolaimidae 94, 113
Camallanidae 97, 120
Campaniform organs 199
Campydoridae 93, 108
Campydorina 90, 93, 108
Canada balsam 39-41
Cap cell 173, 175, 176
Capillaries 246
Capillary 303
Capitulum 126, 133, 134, 141, 176, 303
Carabonematidae 95, 115
Carbohydrate metabolism 201
Carbohydrates 303
Carcharolaimidae 92, 107
Cardia 101-105, 109, 125, 126, 128, 130, 134, 141, 144-146, 152, 154, 184, 304
Casting of gels 53, 61
Caudal 304
Caudal alae 84-86, 100, 107, 114-117, 119, 122, 123, 127, 176
Caudal glands 100-102, 104, 110, 111, 189
Caudalids 84, 193, 198, 304
Cauliflower disease 252
Cavities and tunnels 221, 227
cDNA 304
Ceiling level 240, 241

Cell 304
Cellulase 47, 216, 220, 227, 304
Cellulose acetate 47, 52, 53
Cephalenchus 84, 174, 215, 219, 263
Cephalic 304
Cephalic framework 132, 133, 135, 139, 140, 142, 155, 156, 180, 181, 304
Cephalic papillae 127, 128, 144, 160, 179, 194, 195, 197
Cephalic region 82, 101, 106, 110, 111, 117, 118, 122, 125, 126, 128-133, 135-152, 157, 158, 160-162, 164, 190, 192, 194, 195
Cephalic sensory structures 100, 101, 194, 195, 304
Cephalids 196, 304
Cephalobidae 96, 117
Cephalobina 95, 115, 116
Ceramonematidae 112
Cervical 304
Cervical papillae 190, 198, 304
Cesium 56
cf 304
ch DNA 55
Chambersiellidae 96, 117
Character/character state 304
Characteristic symptoms 10
Cheilorhabdion 304
Cheilostomal 117, 118, 181
Cheilostome 102-104, 106, 111, 116-118, 120, 121, 123, 124, 180, 304
Chemoreception 195
Chemotaxis 304
Chitwood 4, 6, 89, 126, 137, 140, 142, 157, 205
Chloral hydrate 43
Chlorine bleach 42
Choanolaimoidea 93, 110, 111
Chords 73, 76, 77, 80, 188-191, 305
Christie 5, 6
Chromadoria 91, 93, 100, 109, 110

Chromadoroidea 93, 110
Chromic acid 43
Chromosomal DNA 55
Chromosome 305
Chrysoidin 38
Cilia 195
Circomyarian cell 76, 77, 305
Circumfenestrate 305
Circumintestinal 155, 158, 161
Circumoesophageal 161-163, 193, 155
Circumoesophageal commissure 189, 193, 305
Clade and polyclade 305
Cladogram, dendrogram and polygram 305
Classification 5, 87-91, 182, 187
Clavibacter (=*Corynebacterium*) *tritici* 252
Clavibacter michiganense 252, 253
Climate 9, 249
Cloaca 72, 74, 84, 169, 174, 175, 177, 187, 305
Cloacinidae 96, 117, 118
Clone, cloning 305
Clostridium butyricum 245, 255
Clove-oil 41
Cluster analysis OTUs and HTUs 305
Coarse striae 78
Cobb 5, 6, 89
Cobb's decanting and sieving method 17
Cobbonchidae 92, 104
Cockroaches 165
Codon 305
Coelomyarian cell 76, 77, 305
Coenzyme 305
Cohabitance 305
Coleoptera 159, 160, 164, 165, 272
Colletotrichum 255
Comb 61, 62, 65, 66
Comesomatidae 93, 110

Commissure 306
Commonality principle, pleiseomorphy and apomorphy 306
Community 306
Compatibility analysis 306
Complex etiology 250, 251
Composite sample 16
Concomitant etiology 250, 306
Cone column 19
Cone top 41, 82
Coneous 177
Congeneric 306
Congruence 306
Constricting ring 257, 258
Contact action pesticide 306
Continuous head 83
Contortylenchidae 157, 158
Contortylenchus 157
Copolymerization 65
Copper earthing 48
Cores 14-16
Corpus 4, 109, 110, 113, 114, 120, 121, 123-128, 145, 146, 151, 152, 154, 157, 176-178, 306
Cortex 306
Cortical layer 74, 75
Cosmocercidae 95, 116
Cosmocercoidea 95, 115, 116
Cotton 3, 6, 11, 12, 15, 37, 39, 40, 42, 242, 244, 251, 254, 268, 273
Cotton blue 39, 40
Cotype 306
Creagrocericidae 97, 121, 122
Crenate 81, 135, 147
Crescentic thickening 184
Criconema 18, 80, 147, 212
Criconematid type 183-185
Criconematidae 37, 147, 181, 184, 185, 189, 198, 212
Criconematina 99, 124, 144, 145, 150, 173

Criconematinae 146, 147
Criconematoidea 99, 145, 146
Criconemella 268
Critical point dryer 51
Critical point drying 50, 51
Crossophoridae 96, 119
Crown gall 252
Crumb 245
Crura 178
Crustaformeria 125, 129, 141, 145, 153, 154, 174, 306
Cryptaphelenchus 164
Cryptobiosis 210, 224, 233, 240, 242
CsTFA 56
Cultivar 306
Cuticle 2, 35, 40, 46, 47, 73-75, 78, 80-82, 84, 100-102, 104-106, 108, 110-113, 118-120, 122-125, 128-133, 136-140, 142-152, 154, 156, 158, 160, 164, 165, 169, 170, 174, 176, 178, 180-182, 184, 187, 188, 192-194, 198-200, 212, 226, 247, 259, 270, 306
Cyatholaimidae 93, 110
Cylindrocorporidae *incertae sedis* 124
Cylindroid tail 84, 85
Cylindrolaimidae 94, 113
Cyst wall pattern 82
Cysts 17, 23-27, 29, 30, 41, 139, 231, 240, 267, 268, 306
Cytokinins 235, 236

D

Dactylaria brochopaga 257
Dactylaria candida 243, 257, 258
Dark reaction 235
Dastur 8
Dauer state/stage 269, 306
De Man 5
De Man's values 307
Dealkylation 307
Decomposition 28, 244, 246

Deep feeders 219
Deep striae 80
Dehydrated glycerol 37, 38, 40
Deirids 114, 126, 127, 129, 130, 132, 133, 135-137, 140, 142-145, 150, 152, 154, 158, 190, 193, 196, 197, 307
Denaturating gels 65
Denaturation 56, 57, 307
Dendrites 193, 195
Desmidocercidae 98, 123
Desmodorida 94, 109, 111, 112
Desmodoroidea 94, 111
Desmoscolecida 93, 109, 111
Desmoscolecidae 93, 111
Desulfovibrio desulfuricans 245, 255
Detection level 307
Diaphanocephalidae 96, 117
Dichotomy 307
Dictyocaulidae 96, 118
Didelphic 104, 105, 107, 110, 112, 113, 115, 116, 119, 124, 133-135, 137, 139, 141-144, 171-173, 307
Didelphic prodelphic 172, 173
Dieback 307
Digonic hermaphroditism 171
Dilophospora alopecuri 3, 252
Dimethyl sulphoxide 46, 62
Dimethylsulfoxide 62
Dioctophymatidae 97, 120
Dioctophymatoidea 97, 119, 120, 171
Dioecious 307
Diorchic 106, 133, 174, 307
Diphtherophoridae 92, 107
Diphtherophorina 107
Diplogasteria 88, 89, 95, 98, 114, 123, 124
Diplogasteridae 98, 124
Diplogasteroididae 98, 124
Diploid/haploid 307
Diplopeltidae 94, 113

Diplotriaenidae 98, 122, 123
Diplotriaenoidea 98, 121, 122
Diptenchus 130, 131, 173
Diptera 156, 160, 272
Diptheraphorinae 90
Direct examination 27
Direct flame method 34
Disease 307
Disease complex 250, 251, 307
Disease of specific etiology 307
Distance matrix 307
Ditylenchidae 131
Ditylenchus 16, 27, 29, 167, 170, 194, 204, 207, 209, 218, 223, 232, 239, 244, 246, 247, 252, 256, 265
Ditylenchus dipsaci 167, 252
Ditylenchus myceliophagus 256
Ditylenchus tricornis 207
Diverticulum 307
DN 308
DNA 52-65, 67, 69, 308
DNA amplification 56, 58
DNA fingerprinting 308
DNA hybridization 308
DNA restriction 58
DNA sequence 54, 56, 57
DO 308
Dolichodoridae 99, 141, 142, 173, 186
Dolichodorinae 141-143, 198
Dolichodoroidea 98, 133, 140, 141
Dolichodorus 274
Dolichodorus heterocephalus 274
Dolichorhynchus 178
Dormancy 208
Dormancy of eggs 169
Dorsal gland outlet 308
Dorsalla 173
Dorylaimellidae 92, 107
Dorylaimida 90, 92, 103, 104, 184, 216, 225, 258

Dorylaimina 90, 92, 105
Dorylaimoidea 90, 92, 105
DPX mountant 39, 40
Draconematidae 94, 112
Draconematoidea 111, 112
Dracunculidae 97, 122
Dracunculoidea 97, 121, 122
Dracunculus medinensis 3
Drilonematidae 97, 121
Drilonematoidea 121, 122
Dual parasitism 269, 270
Duosulciinae 127
Duosulcius 182
Duration of feeding 215

E
E buffer 60
e.g. 308
Ear cockles 223
Ear twist 252
Ecdysteroids 206
Ecological interactions 273
Ecological niche 308
Ecology 237
Economic injury level (EIL) 308
Economic threshold level (ETL) 308
Ecosystems 237
Ecphyadophoridae 126-128
Ecphyadophoroidinae 128
Ectendomycorrhizae 263, 264
Ectomycorrhizae 263-265, 308
Ectomycorrhizal fungi 266
Ectoparasites 218, 308
Ectoparasitic nematodes 267, 273, 274
Ectoparasitism 219
Ectophoretic 165, 308
EDTA 54, 56, 60, 61
Eelworm wool/Eel wool 232, 308
Efficient host 240, 308
Egestion 259, 260

Egg Development 167
Egg masses 27, 30
Eggs 30, 308
Ejaculated spicule 177
Ektaphelenchidae 99, 163, 164
Elaphonematidae 96, 117
Elastase 56
Electromorph 308
Electron microscope 43-46, 48, 49
Electrophoresis 52-55, 59-67, 308
Electrophoresis buffer 53, 60-63, 67
Electrophoresis tank 61-63, 67
ELISA 309
Elongated tails 84
Elutriation techniques 18-24
Embedding of the Specimens 47
Embryonic development 167, 168
Enchinomermellidae 109
Endogone 264
Endomycorrhizae 263, 264, 309
Endomycorrhizal fungi 264-266
Endoparasites 135, 137, 139, 165, 170, 215, 216, 218, 219, 224, 226, 227, 232, 233, 309
Endophoretic 165, 309
Endophytes 264
Endotokia matricida 175
Endotoky 174, 309
Enface view 40, 83
Enhancer 309
Enoplia 89, 100
Enoplidae 91, 101, 102
Enoploidea 91, 101
Entaphelenchidae 100, 163, 166
Entomobyroides dissimilis 243, 271
Entomoparasites 309
Entomoparasitic 152, 154, 155, 157
Entomopathogenic 309
Entomopathogenic nematodes (EPN) 6, 30, 31, 152, 153, 156, 157, 174, 268, 270-273, 275, 276

Environmental conditions 9, 52, 73, 208, 238, 239, 242
Enzyme 309
Epacris 219, 220
Epiboly 309
Epicharinematinae 129, 130
Epidermis 75, 76, 216, 220, 227
Epiptygma 126, 141, 143, 174, 309
Epithelium 309
EPN 6, 30-33, 270, 272, 276
Epon CY212 resin 47
Epoxy 47, 48
Epoxy acetone 47
Epsilonematidae 94, 112
Ericaceae 264
Ericales 264
Erlenmeyer 20-22, 61, 62
Esterase isozyme 53
Ethanol 35, 37, 41, 50, 56, 65, 200
Ethidium bromide 56, 58, 62, 63
Ethidium bromide solution 62
Eucalyptus comendulosa 270
Eucalyptus macrorrhyncha 269, 270
Euparal 41
Eurystominidae 91, 102
Eutylenchinae 129
Eutylenchus 195
Evolved parasitism 228
Excretory cell 101, 103, 105, 114, 115, 117, 149, 187, 189
Excretory pore 309
Excretory system 71, 100, 104, 114, 117, 118, 119, 124, 145, 187-189
Exotoky 174, 309
Exploration 208, 211, 213
Extension 11, 56, 57, 130, 132, 143, 154, 156, 157, 177, 181, 193
External cuticle 74, 75, 78, 81, 100, 102, 105, 111, 169, 170, 180, 192
Exteroreceptors 192, 193

Extracorporeal 310
Extraction buffer 55, 56
Extraction of Cysts 24, 29
Extraction of vermiform nematodes 17
Extraction solution 53
Eye knives 46

F
Facultative parasite 310
Fallow 310
fam 310
Familia dubia 310
Family group names 310
Fasciculi 143, 186
Fast blue R 54
Fatty acids 204
Fecundity 238, 239
Feeding apparatus 2, 74, 179-181
Female reproductive organs 172
Female reproductive system 171
Fenestra 41, 310
Fenwick Can 24, 25
Fergusobia 152, 156, 224
Fergusobia tumifaciens 269
Fergusobiidae 155
Fergusobiinae 156
Fergusonina 269
Fermentation 201
Fertilization 310
Fibrillar layer 74, 75
Fibrillar zone 77
Fibrils 185
Field capacity 246, 310
Fig nematode 224
Filariidae 98, 123
Filarioidea 98, 121, 123
Filiform tail 84
Filipjev 4, 5, 7, 89, 133, 135, 139
Fine striae 78
Fischer 5

Fixation for TEM 46
Fixation of soil sample 36
Fixatives 34, 35, 41, 46, 47, 49
Fixing of nematodes 34, 49
Flag leaf 310
Flask 20-22, 31, 32, 61, 62
Flemming's solution 43
Flexures 103, 116, 173, 310
Floatation apparatus 19
Flower bud gall disease 269
Fluidizing column 23, 24
Fluorometer 56
Foliar nematodes 222, 223, 230-233, 249
Food transporter 182
Foregut 178
Formal aceto-alcohol 35
Formal-acetic acid 34, 35, 41
Formaldehyde 34-36, 41, 46
Formalin 35, 36, 41
Formal-propionic 35
Free living 310
Free living nematode 310
Fumigant 310
Fungal feeding nematodes 265
Fungiotonchium 153, 159
Fungus 310
Fungus mantle 263, 266
Fusarium 3, 244, 251, 253-255
Fusarium oxysporum 253, 255
Fusarium solani 255

G
Galactose 60
Galleria mellonela 32
Galls 4, 10, 218, 230, 262, 263, 266, 269, 270, 277, 310
Ganglia 311
Gastrula 167, 310
Gel electrophoresis 59, 64
Gel loading buffer 60, 61, 63, 67

Gel tank 63
Gelatin 40, 47, 48, 55
Gelatinous matrix 174, 187, 189, 226
gen.nov., gen.n, n.g. 311
Gene 311
General structures of nematode 72, 76
General symptoms 10, 11
Genital papillae 191, 192, 198, 199, 311
Genital primordium 311
Genome 311
Genomic DNA 55
Genotype/type species 311
Gentianaceae 265
Gentle shaker 54
Genus-group names 311
Germinative zone 173, 175, 311
Giant cells 221, 222, 228, 229, 234, 311
Gibberellins 235, 236
Gigaspora 264
Gigaspora margarita 266, 268
Glacial acetic acid 35, 43
Glandular oesophagus 183, 184
Global loss 12
Globocephalidae 96, 118
Globodera 8, 12, 23, 24, 26, 27, 30, 71, 73, 75, 82, 84, 169, 170, 174, 207, 212, 215, 218, 222, 229, 231, 236, 238-241, 244, 247, 266, 275
Globodera pallida 213, 241, 275
Globodera rostochiensis 8, 212, 213, 215, 241, 244
Globular 71, 102, 103, 117
Glomales 264
Glomus 264, 266
Glomus coronatum 267
Glomus etunicatum 265-268
Glomus fasciculatum 265-268
Glomus intraradices 268
Glomus macrocarpum 266, 267
Glomus mosseae 266, 268

Glottoid apparatus 311
Gluconeogenesis 311
Glucose phosphate isomerase 53
Glucose-6-phosphate 202-204
Glume 233, 311
Glutaraldehyde buffer 45, 46, 49
Glutaraldehyde fixative 47
Glyceel 39, 40
Glycerol 35-42, 53, 61, 204, 209
Glycerol jelly 41
Glycerol methods 37
Glycerol-ethanol method 37
Glycolysis 201, 202
Glyconeogenesis 311
Glycosidic linkage 60
Gnathostomatidae 97, 122
Goeldi 4
Goeziidae 96, 119
Gonopore 311
Goodey 5, 7, 155, 158, 162
Goodeyella 151
Gracilacus 219
Gradualism 311
Granek's ratio 311
Grapevine fanleaf virus 258
Gravid female 312
Gravitational water 312
Greeffiellidae 94, 111
Grooved lateral field 81
Growth 9, 10, 37, 154, 155, 169, 171, 173, 175, 200, 216, 217, 223, 225, 229, 236, 237, 241, 242, 253, 263-266, 273, 276
Growth zone 171, 173, 175
gub see gubarnaculum
Gubernaculum 103-105, 107, 110-112, 115, 116, 119, 121, 123-126, 128-130, 133-137, 139, 141-145, 147, 148, 151, 153-163, 176-178, 312
Guiding ring 312

Gymnotylenchidae 155
Gymnotylenchinae 156
Gymnotylenchus 156

H

H_2O_2 solution 41
H_2S 245
Habitat 312
Haemocoel 312
Hairy root 3, 252
Halenchinae 131
Halenchus 84, 85, 189
Haliplectidae 94, 113
Hartig net 264
Hatching 36, 100, 167, 169, 199, 200, 207, 208, 241, 242, 244, 247, 255, 276
Hatching factor 312
Haustoria 264
Head 78, 82, 83, 178, 194-196, 213, 221, 222, 226, 228, 229, 312
Hedruridae 97, 121
Helicotylenchus 12, 71, 73, 84, 170, 173, 198, 215, 218, 219, 238, 244, 252, 267, 274
Helicotylenchus digonicus 267
Helicotylenchus nannus 215, 252
Helicoverpa armigera 31
Heligmosomatidae 96, 118
Hemicriconemoides 182
Hemicriconemoidinae 146, 147
Hemicycliophora 75, 79, 81, 83, 149, 171, 182, 216, 218, 219
Hemicycliophora arenaria 75, 216, 219
Hemicycliophoridae 99, 148
Hemicycliophorinae 81, 149
Hemicycliophoroidea 99, 145, 146,
Hemizonid 194, 197, 198, 312
Hemizonion 193, 198, 312
Hennigian method 312
Hermaphrodite nematodes 157, 171

Hermaphrodites 157, 312
Heterakidae 95, 116
Heterakoidea 95, 115, 116
Heterocheilidae 97, 119
Heterocytous 185
Heterodera 3, 4, 9, 11, 12, 23, 24, 26, 27, 30, 53, 71, 73, 75, 78, 81, 82, 84, 169-171, 173, 174, 178, 200, 207-210, 215, 216, 218, 221, 222, 227, 229, 231, 234, 236, 238-240, 243, 245, 247, 255, 256, 262, 266-268, 275, 277
Heterodera avenae 9, 11, 210, 240, 243, 245, 271
Heterodera cajani 267, 268, 275
Heterodera glycines 241, 267, 275
Heterodera oryzicola 275
Heterodera radicicola 4
Heterodera schachtii 215, 244, 275, 276
Heterodera trifolii 216
Heterodera zeae 255
Heteroderidae 99, 134, 135, 139, 140, 173, 186
Heteroderinae 140
Heteroderoidea 133
Heteromorphotylenchinae 160
Heteromorphotylenchus 159
Heterorhabditidae 95, 115, 270, 276
Heterorhabditis bacteriophora 271, 272
Heterosexual 153, 155-159
Heterosexuality 312
Heterotylenchinae 160
Heterotylenchus 160
Heuristic 312
Hewitt 6, 257
Hexaradiate 104, 123, 124, 130-132, 138, 145, 158, 179, 180, 194
Hexatylidae 155
Hexatylina 99, 124, 152, 154, 173
Hexatylus 152, 153, 182
Hexose Monophosphate Shunt 201, 203

Hidden damage 10
Hierarchical system 87
Hind III 58, 59
Hindgut 178
Hirschmanniella 137, 170, 220, 226, 227, 266
Hirschmanniellinae 138
Histotylenchus 182
Hollow stoma 180
Hologonic 171
Holomyarian 78
Holotype/allotype/paratype 312
Homocytous 185
Homology/analogy 312
Homonymy 312
Homoplasty 313
Homungellidae 122
Hoplolaimidae 98, 134, 135, 181, 198, 220
Hoplolaimina 132
Hoplolaiminae 136, 199
Hoplolaimini 133
Hoplolaimoida 133
Hoplolaimoidi 133
Hoplolaimus 182, 198, 221, 268
Hoplolaimus galeatus 268, 274
Host 313
Host characteristics 240
Host modification 225
Host parasite relationship 224, 313
Host status 241
Hot fixative method 34
Hot water method 34
Hyaline 313
Hydrogen cyanide 220, 227
Hydrolysis of lipid 205
Hydromermis 177
Hygroscopic water 313
Hyperplasia 218, 225, 229, 261, 313
Hypersensitive 313

Hypersensitivity 313
Hypertonic 313
Hypertrophy 218, 226, 229, 313
Hypoaspis aculeifer 271
Hypodermal cells 76
Hypodermal glands 77
Hypodermis 76, 313
Hypodontolaimidae 110
Hypomicrobiaceae 261

I
IAA 235
ibid 313
ICZN 88, 314
Important enzymes 215
Important nematode genera 12
 in litt 314
 In vitro 314
 In vivo 314
Inanimate plant pathogen 314
incertae sedis 314
Incisure (Involution) 314
Incubation period 314
Infection 314
Infection thread 261
Infective juveniles 31, 32, 314
Infested 314
Ingestion 215, 314
Ingroup and outgroup 314
Initial population 239
Injury 314
Inoculate 314
Inoculum 314
Inoculum density 314
Inositol 209
Insect 314
Insect cadaver 31, 32
Insect trap method 31
Insemination 176, 314

Integrated control 314
Intercellular and intracellular space 314
Interchordal areas 76
Internal cuticle 74
Interoreceptors 192
Interrotylenchidae 135
Intersex 315
Interspecific/intraspecific competition 315
Interspecific interaction 275
Interstrial zone 78, 80
Intestinal 186
Intestinal cells 185
Intestine 185
Invagination 315
Invasion 315
Iodine 46
Iotonchiidae 99, 153, 159
Iotonchioidea 99, 154, 158
Iotonchulidae 92, 104
Ironidae 103
Ironina *inquirenda* 103
Isobutanol 41
Isocytous 185
Isoelectric focusing and isoelectric point 315
Isolaimida *inquirenda* 92, 103
Isolaimiidae 92, 103
Isolate 315
Isolation 315
Isolation of EPN 31
Isolation of genomic DNA 55
Isolation of mitochondrial DNA 55
Isolation of nematodes for SEM 49
Isolation of nematodes for TEM 45
Isomerases 315
Isozyme analysis 53
Isozymes 52, 315
Isthmus 183
ITS (internal transcribed spacers) 315

J
Juvenile 315
Juvenile cuticle 74
Juvenile hormones 206

K
Kairomones 315
Karyotype 315
KCl 55
Kidney shaped 71
Killing and fixing of nematodes for SEM 49
Killing of nematodes 33
Kilobase (Kb) 315
Kinases 315
$KMnO_4$ 35, 38
Kreb's cycle 204
Kurochkinitylenchinae 160
Kurochkinitylenchus 160

L
L 316
Labia (labium singular) 316
Labial 316
Labial disc 316
Labial papilla 190
Labial papillae 160, 194, 195
Lactic acid 36, 40, 201
Lactin dehydrogenase 201
Lactoglycerol 40, 42, 43
Lactophenol 36-43
Lamellae 81
Lamina 177
Lapsus calami 316
Large plot 15
Larva (larvae plural) or juvenile 316
Lasioseius scapulatus 271
Lateral chords 76
Lateral fields 80, 316
Lateral ganglion 190
Lateral lines 79-81

Lateral view of nematodes 83
Lauratonematidae 191, 102
Lead citrate 48
Leaf sheath 316
Leccinum 263
Leghaemoglobin 261, 316
Lemon shaped 71
Leptolaimidae 94, 113
Leptosomatidae 102
Lesion 316
Leucine aminopeptidase 208
Life cycle 316
Ligation 59
Light reaction 235
Linhomoeidae 112
Lip arrangements 83
Lip region 179
Lipase 208
Lipid metabolism 204
Lipids 316
Liquefaction 316
Liquid CO_2 51
Liquid nitrogen 50
Liquified propane 50
Lobed caudal alae 86
loc. cit. 316
Longidoridae 106, 217, 218, 225, 258
Longidorinae 106, 107
Longidorus 216, 218, 226, 259
Longidorus elongatus 216, 258
Longidorus macrosoma 258, 260
Longitudinal lines 79, 81
Longitudinal striations 80
Loofia 148, 149
Lumen 182

M

m 316
Mabs 316
Maceration-filtration 28
Macroposthoniidae 147

Macrotrophurinae 144
Macrotrophurus 132, 144
Madinematidae 147
Malate dehydrogenase 53, 54
Male reproductive organs 176
Malic acid 54
Marcinowski 6
Marenoplica 89, 101
Marimermithidae 109
Matchstick extraction 29
Matrix 316
Maupasinidae 117
MB 317
Mechanical traps 256-258
Mechanoreception 195
Median layer 75
Megalobatrachonematidae 122
Meiodoridae 143
Meiodorinae 144
Meiosis 317
Meloidodera 140, 179
Meloidoderidae 139
Meloidoderinae 140
Meloidoderita 145, 149
Meloidoderitidae 151
Meloidoderitinae 151
Meloidogyne 133, 139, 207, 210, 216, 218, 228, 235, 254-256
Meloidogyne chitwoodi 271
Meloidogyne incognita 210, 213, 235, 254, 255, 262, 268, 274, 275, 277
Meloidogyne graminicola 275
Meloidogyne hapla 235, 262, 274, 275
Meloidogyne javanica 210, 213, 255, 271
Meloidogynidae 134, 135, 138
Meloidogyninae 139
Meloinema 138, 139
Merliniidae 143
Merliniinae 132, 141, 143, 144, 146, 147, 178
Mermithida 89

Mermithidae 109, 270
Mermithoidea 109
Meromyarian type 78
Mesentries 78, 187
Mesentron 178, 317
Mesh size 17
Mesidionematidae 122
Mesorhabdion 317
Mesorhabditis labiata 201
Mesorhizobium 260
Metabolism 201, 204, 317
Metabolite 317
Metacarpus or median bulb 317
Metamasius 234, 269
Metaspecies or metataxon 317
Metastrongylidae 96, 117
Metatype 317
Methyl blue 36
Methylbutyrate 203
Methylobacteriaceae 261
Methylobacterium 261
Meyliidae 112
$MgCl_2$ 55
Microbial feeders 1
Microbial metabolites 244
Microflora 317
Microlaimidae 93, 110
Micropleuridae 98, 122
Micropores 246
Microscopic 317
Microvilli 185, 317
Midgut 178, 185
Migratory ectoparasites 218, 318
Migratory endoparasites 220, 318
Migratory endoparasitic nematodes 226, 268, 274
Mineral oil 55
Miscellaneous nematodes 3
Mistifier extraction 28
Mitochondrial DNA (mtDNA) 55

Mitosis 317
Modified glycolysis 202
Molecular markers 317
Molecular techniques 54
Molt 318
Monacrosporium bembicoides 257, 258
Monacrosporium cionopagum 256
Monacrosporium drechsleri 256, 257
Monacrosporium eudermatum 256, 257
Monacrosporium gephyropagum 256
Monarchic 318
Monhysterida 112
Monhysteridae 112
Monochamus alternatus 269
Monoclonal antibody 318
Monoculture 318
Monodelphic 171, 172, 318
Monoecious 318
Monogenic 318
Monochamus carolinensis 234, 269
Mononchida 103, 184
Mononchidae 92, 104
Mononchulidae 92, 104
Monoopisthodelphic 172
Monophyletic or holophyletic 318
Monoposthia 177
Monoposthiidae 112
Monoprodelphic 172
Monorchic 174
Monothetic or polythetic 318
Monotype 318
Morphological modifications 225
Mortality determination 38
Moulting 169
Mounting of nematodes 38
Mounting on SEM Stubs 51
Mouth cavity 180
mRNA (messenger RNA) 318
MseI 58
mtDNA 55, 318

Mucronate 163
Muscle cells 77
Muspiceidae 97, 121
Muspiceoidea *incertae sedis* 120
Mutualistic relationship 318
Mycophagous 224
Mycophagous or fungivorous 318
Mycorrhiza 318
Mycorrhizal fungi 263-265
Mydonomidae 92, 107
Myenchidae 114
Mylonchulidae 92, 104
Myofibrils 318
Myriocytous 185

N
NaCl 56
Nacobbidae 137
Nacobbinae 138
Nacobboderinae 139
Nacobbus 137, 216, 218, 221, 227
NAD 54
NAD and NADP 318
Name bearing type 319
NaOCl solution 42
Napthyl acetate 54
Neck 319
Necrosis 319
Needham 4
Needle 47
Nema 1, 70
Nemata 89
Nematicide 319
Nematoda 88, 224
Nematode characteristics 238
Nematode definition 1, 270, 319
Nematode tails 85
Nematode trapping fungi 30, 85, 223, 256
Nematode wool 232

Nematode-insect interaction 268
Nematode-Mycorrhizae interaction 263
Nematode-*Rhizobium* interaction 260
Nematodes 30, 223
Nematode-Virus Interaction 257
Nematological societies/journals 8, 9
Nematophagous fungi 256
Nemonchidae 135
Neoditylenchidae 132
Neodolichodorus 141, 142
Neotylenchid type 184
Neotylenchidae 155, 184, 198, 270
Neotylenchinae 156
Neotylenchoidea 154
Nepoviruses 216, 258, 319
Nerve 319
Nerve chords 76
Nerve ring 189
Nervous system 189
Netuviruses 258, 319
Nimbidin 244
Nitrocellulose filter 62
Nitrogenase 261, 319
Nm (nanometer) 319
N-N-methylene-bis-acrylamide 62, 63, 65, 67
No., n 319
Noctuid moths 165
Noctuidonematinae 165
Nodule 319
nom dub 319
nomen nudum (pl. nomina nuda) 319
Non host 320
Nonconstricting ring 258
Nondenaturating gels 65
Nothanguina 130
Nothotylenchidae 131
Nothotylenchoidea 131
Nuclear DNA 55

Nucleic acid 204, 320
Nucleotide 320
Nucleopore envelope 50, 51
Nucleopore filter 50
Nucleoprotein 320
Nucleotide 320
Numerical taxonomy 320
Nygolaimellidae 108
Nygolaimidae 108
Nygolaimina 90, 108

O

O 320
Obesity 71
Objective synonyms 320
Obligate parasite 320
Obliteration 174, 320
Ocelli 199
Odontopharyngidae 124
Odontorhabditidae 115
Odontostylet 180, 320
Oesophageal glands 184, 216
Oesophagi 183
Oesophago-intestinal valve 184
Oesophagostome 180, 320
Oesophagus 182, 320
Ogma 80
Ogmidae 147
Oides 1, 70
Oligocytous 185
Ollulanidae 118
Onchocercidae 123
Oncholaimida 102
Oncholaimidae 102
Oocyte 320
Oocyte movement 173
Oogonium 321
Oomycin A 245
Oostenbrink's Elutriator 19
Opisthodelphic 321

Oral disc 321
Oral opening 181
Orchid 264
Orchidaceae 265
Organic matter 243
Orientation 212
Orientation of the Specimens 47
Orientylus 173
Original designation 321
Osmic acid 43
Osmic acid vapours 43
Osmium tetraoxide 46, 50
Osmobiosis 210, 240
Osmotic potential 321
Oswaldofilariidae 98, 123
Outstreched 173
Ovary 173, 321
Overpopulation 239
Overwinter or oversummer 321
Oviduct 173, 321
Oviparous 174, 321
Ovoviviparous 174, 321
Oxaloacetate 202
Oxydiridae 107
Oxystominoidea 101
Oxyuridae 116
Oxyuroidea 116

P

p321
Paecilomyces lilacinus 244, 255
PAGE 64, 321
Palms 165
Panagrolaimoidea 96, 117
Papillae 321
Papillary ganglia 190
Paraffin wax 39
Paralongidorus 1, 258, 259
Paraoxystominidae 91, 101
Paraphelenchidae 99, 162

Paraphelenchinae 163
Paraphelenchus 161
Paraphylectic or paraphyly 321
Pararotylenchidae 135
Pararotylenchus 134
Parasitaphelenchidae 100, 163-165
Parasitaphelenchinae 100, 165
Parasite 243, 321
Parasites of underground parts 218
Parasitoid 321
Parasitylenchidae 159
Parasitylenchinae 160
Parasitylenchus 160
Parasponia 261
Paratrichodorus 214, 216, 259
Paratrichodorus minor 210, 259, 274
Paratylenchidae 146, 150
Paratylenchinae 152
Paratylenchus 152, 218, 219
Parlodion 48
Parsimony 321
Parthenogenesis 171, 238, 321
Parthenogenetic nematodes 158, 159, 170
Pathogen 321
Pathogenicity 322
Paurodontidae 155, 157
Paurodontinae 157
PCR buffer 55
PCR machine 56, 57
PCR-based tests 69
Pea early browning virus 216, 259
Pear shaped 71
Pectinase 227
Penetration 46, 213, 322
Penetration of fixatives 46
Penicillium 244
PEP carboxykinase 202
Pepper ring spot virus 259
Perennial orchard 16

Perforation 213
Perikaryons 195
Perineal Pattern 40, 82, 322
Peristaltic movement 185, 322
Permanent mounting 38
Phaenopsitylenchidae 155
Phanodermatidae 102
Pharynx 182
Phasmidia 89
Phasmids 198, 322
Phenazin 255
Phenazine methosulphate 54
Phenetic 87
Phenetics, phenon 322
Phenol 36, 40
Pheromones 208, 322
Philometridae 122
Phloxin B 38
Phlyctainophoridae 97, 121
Phoretic 322
Phosphate buffers 48
Phosphogluconate 204
Photorespiration 236, 322
Photorhabdus luminescens 271
Photosynthesis 235, 322
Phyletic 87
Phyllobacteriaceae 261
Phylogeny 322
Physalopteridae 97, 122
Physalopteroidea 97, 121, 122
Physiological modifications 234, 253
Physiology of hatching 208
Physiology of reproduction 207
Phytopathogenic 323
Phyto and entomopathogenic nematode interaction 276
Phytopathogenic 323
Phytoalexin 322
Phytohormones 236
Phytonematodes 3

Phytoparasies 309
Phytophthora parasitica 253
Picking device 33
Picric acid 43
Picrofuchsin 37
Pines 165
Plant feeding nematodes 266
Plant material 16
Plant parasitism 216
Plasma membrane 185
Plasmid DNA 57
Plasmid kit 57
Plasticity 242
Platymyarian 76, 77, 323
Plectidae 94, 113
Plectoidea 113
Pleurotylenchinae 129, 130
Pochonia chlamydosporia 255
Pocket cell 193
Polarity 323
Polonium bars 48
Polyacrylamide gels 64, 66
Polycytous 185
Polycytous intestines 186
Polyethylene tube 47
Polymerase chain reaction (PCR) 323
Polymorphism 323
Polymyarian 78
Polynucleotide kinase 59
Polyploidy, aneuploidy 323
Polypropylene 47
Poor hosts 240, 323
Population 237-239, 323
Population density 323
Population dynamics 237
Pore aperture 17
Pore space 245, 323
Post-anal sac 84
Post-embryonic development 169

Posterior nervous system 191, 192
Post-uterine sac 173, 323
Potassium iodine solution 34
Pratylenchidae 98, 134, 135, 137, 220
Pratylenchinae 138
Pratylenchoides 134, 137
Pratylenchus 137, 210, 213, 218, 220, 226, 266, 268, 277
Pratylenchus alleni 275
Pratylenchus brachyurus 275
Pratylenchus coffeae 266, 268, 274
Pratylenchus laumondii 271
Pratylenchus penetrans 213, 216, 220, 228, 235, 253, 256, 262, 266, 268, 274, 275
Pratylenchus vulnus 252
Pratylenchus zeae 274
Preadult 323
Predaceous nematodes 2
Predacious fungi 256
Predators 243, 323
Prerectum 185, 323
Preselective amplification 59
Preservation of Plant Material 41
Prestoma 195
Primary infection 323
Primary inoculum 323
Primary pathogen 250, 251, 323
Primary soil particles and secondary aggregates 324
Primers 56
Prismatolaimidae 103
Probe 324
Processing of nematodes 36
Prochaetosomatidae 112
Procorpus 324
Proctodeum 178
Prodelphic 324
Pro-diorchic 174

Programmed death 168
Propagule 324
Propionic acid 34, 35
Propylene phenoxetol 34, 46
Protectant 324
Protein 324
Protein electrophoresis 52, 53
Protein metabolism 207
Proteinase K 55
Proteobacteria 260
Protoplast 324
Protractor 181, 324
Protractor gubernaculi 178
Protractor muscles 324
prs. comm. 322
prt, partim 321
Pseudhalenchus 130
Pseudocoelom 78, 187, 324
Pseudocoelomic 70
Pseudocoelomic fluid 78
Pseudocoelomic membranes 78, 187
Pseudocoelomocytes 78, 187
Pseudocollagenase 208
Pseudodiplogasteroididae 115
Pseudomonas 254
Pseudomonas fluorescens 245, 252, 255
Pseudomonas marginata 253
Pseudomonas solanacearum 252, 254
Psilenchidae 132, 141, 142, 144
Psilenchinae 99, 144, 145
Psilenchus 132
Pterotylenchus 130
Pterygorhabditidae 116
Puncturing of cuticle 47
Purpose of sampling 14
Pylorus 187, 324
Pyoluteorin 245
Pyruvate 202, 203, 205
Pythium graminicola 255

Q
Quarantine 324
Quiescence 9, 208, 210, 223, 224, 232, 233, 239
Quiescent 209, 223, 224, 232, 233, 324

R
R 324
Race 324
Rachis 173, 324
Radinaphelenchus 224
Radinaphelenchus cocophilus 234, 269
Radopholidae 137
Radopholinae 138
Radopholus 137, 220, 226, 266, 268, 274
Radopholus similis 167, 216, 220, 227, 228, 253, 266, 268, 274
Rainfall 249
RAPD 57, 324
RAPD markers/ fragments 325
Rapid lactophenol method 36
Raspberry ring spot virus 258
Raw manures 244
Rearing of EPN 31
Reciprocal parasitism 263, 325
Rectal cells 174, 226
Rectal glands 187, 187, 325
Rectum 178
Red-ring nematode 224
Reflection electron microscope (REM) 45
Regulators 229
Remounting 39
Renette 188, 189, 325
Reproduction 170
Reproduction rate 171, 239
Reproductive systems 170
Resistance 325
Respiration 200, 236, 325
Restriction enzyme 325

Restriction endonucleases 58
Restriction fragment length polymorphism (RFLP) 325
Retractor gubernaculi 178
Retractor muscles 181, 325
Retrorse 325
Reverse TCA cycle 204
Rex 325
Rhabdiasidae 95, 115
Rhabditia 89, 114
Rhabditida 115, 184
Rhabditidae 115, 270
Rhabditina 115
Rhabditis cucumeris 201
Rhabditoidea 95, 115
Rhabdito-nematidae 95, 115
Rhabdolaimidae 94, 113
RHEED 45
RHELS 45
Rhigonematidae 95, 116
Rhizobiaceae 260
Rhizobiales 260
Rhizobium 260
Rhizoctonia 244, 264
Rhizoctonia solani 255, 256
Rhizosphere 325
Rhodococcus 252
Rhynchophorus palmarum 234, 269
Ribosome 325
Ribulose biphosphate 236
Ribulose-5-phosphate 204
Ridges 81
Ring spot virus 216, 258, 260
Ripening zone 172, 173, 175, 176, 325
Ritzema Bos 5
RL buffer 59
RNA 325
Robertdollfusidae 97, 121
Robertiidae 117
Roes 325

Root exudates 241, 242, 244
Root incubation 29
Root necrosis 228
Root nodules 260
Roqueidae 107
Rostrum 163, 326
Rotylenchoides 173
Rotylenchoidinae 136
Rotylenchulidae 98, 134-136
Rotylenchulinae 137
Rotylenchulus reniformis 222, 226, 227, 256, 277
Rotylenchus 218
Roveaphelenchus 166
Rowed vegetation 15
rRNA 326
Rubzovinematinae 156
Running of electrophoresis 53, 63
Russula 263
RV 325
Rvan 325

S
s. 326
s.l. 326
s.s., s.str. 326
Saliva 216
Salivation 214
Sample area 14
Sample preparation for Isozyme Study 53
Sample splitter 26
Sampling pattern 14
Sanitation 326
Saprophytes 326
Saprozoic 326
Sarcoplasm 76, 77
Sarcoplasmic zone 77
Sarkosyl 56
Sauertylenchus 143

Scanning electron microscope 44
Scanning transmission electron microscope 45
Scaptrellidae 94, 112
Scatonema 157
Schacht 4
Schistonchus 224
Schistonchus caprifici 270
Schmidt 4
Schwartz 5
Sclerocystis 264
Scleroderma 266
Sclerotium rolfsii 255
Sclerotized 326
Scolecophilidae 97, 122
Scolytidae 164, 165
Scopoli 4
Scutellonema 198
Scutellum 326
Scutylenchus 144, 198
SD 326
SE 326
Secernentea 89, 95, 114, 187, 188
Secondary hypertrophy 218, 226
Secondary pathogen 250, 326
Secretory cells 174
Sectioning of the Specimens for TEM 48
Sedentary 71
Sedentary ectoparasites 219
Sedentary endoparasites 221, 326
Sedentary endoparasitic nematodes 228, 268, 275
Sedentary parasitism 221
Seductor gubernaculi 178
Seed galls 223, 326
Seinhorst's Elutriator 20, 21
Seinura celeris 239
Seinuridae 100, 163, 164
Seinurinae 164

Selective amplification 59
SEM 44, 45, 326
SEM stubs 51
Semi-Automatic Elutriator 26
Semiendoparasitic nematodes 222, 226
Semiendoparasitism 222
Semifenestrae 326
Seminal vesicle 175, 326
Semiochemicals 207, 326
Sensilla 192, 327
Sensillar pouch 193, 195
Sensory structures 192, 194, 197
sensu 327
Separation of DNA fragments 64
Sequential etiology 250, 327
Serpentine canals 186
Sessile 327
Setae 327
Setariidae 98, 123
Set-off lips 84
Seuratoidea 97, 119
Sex pheromones 207, 208
Sex reversal 171, 239, 327
Sexual dimorphism 71, 135, 138, 139, 145, 146, 150, 158, 327
Shaft 177, 178, 181
Shapes of amphid 196
Sheath 75, 327
Shoot incubation technique 29
Sibling species 327
sic 327
Sieve shaker 26
Sieves 17, 19, 20, 23-30
Sigmacote 65
Siliconizing fluid 65
Silver paint 51
Simple sample 16
Single plant plot 15, 16
Sinorhizobium 260
Siphanoptera 156, 160

Siphonolaimidae 94, 112
Size and thickness of spear 182
Size of nematodes 70
Slow glycerol method 37
Soboliphymidae 97, 120
Socket cell 193
Sodium azide 34, 46
Sodium bicarbonate 54
Sodium cacodylate 50
Sodium dodecyl sulphate 56
Sodium hypochloride 42
Sodium phosphate 49, 54, 56
Soil aeration 246
Soil capillary 327
Soil collection 14
Soil inhabitants 327
Soil moisture 246
Soil organisms 243
Soil particles 245
Soil pH 247
Soil samples 16
Soil samples for EPN 30
Soil sampling 14
Soil solarization 327
Soil solutes 247
Soil temperature 248
Soil texture 245
Somatic musculature 77
Sonication 50
sp. (spp plural) 327
sp. n., sp. nov., n. sp. 327
sp.inq. 327
Spacers 65
Spear muscles 181
Species group names 327
Species inquirendae 327
Specific etiology 250
Sperm 328
Spermatheca 174, 328
Spermatocytes 175, 328

Sphaerolaimidae 112
Sphaeronema 151
Sphaeronematidae 99, 146, 150, 151
Sphaeronematinae 151
Sphaerotylenchus 226
Sphaerularia 153
Sphaerulariaceae 156
Sphaerularidae 270
Sphaerulariidae 155, 156
Sphaerulariina 89, 152
Sphaerularioidea 154
Spherocrystals 185
Sphincters 173, 328
spic 328
Spicate terminus 328
Spicular pouches 176
Spicules 176, 328
Spilotylenchinae 160
Spilotylenchus 159
Spirinidae 110
Spiruria 89, 118
Spirurida 121
Spiruridae 122
Spiruroidea 121
Sputtering machine 51
Squalene 206
st 328
Staining for TEM 48
Staining of gels 54
Starch 53
Starvation of nematodes 204
Steiner 5
Steinernema carpocapsae 271
Steinernema feltiae 276
Steinernematidae 115, 270, 276
STEM 45
Steps of PCR 57
Sterilization or disinfestation 328
Stichocytes 109
Stichosomida 89, 108

Sticky traps 256, 257
Stoma 180
Stomatal stylet 180
Stomatostylet 328
Stomodeum 178, 182, 328
Storage box 51, 52
Strain 328
Strawberry latent ring spot virus 258
Streptomyces annulatus 245, 255
Streptomyces avermitilis 245
Striae 78
Strongylacanthidae 118
Strongylida 117
Strongylidae 118
Strongyloidea 117
Strongyloididae 115
Stubby root nematodes 225
Stylet and spicule protrusion 36
Stylet cone 181
Stylet guide 181
Stylet knobs 182
Stylet muscles 181
Stylet or spear 328
Sub median 328
Subanguina radicicola 130, 131
Subcrystalline layer 82
subfam. 328
Subjective synonyms 328
Sublimation 50
subsp., ssp 328
Substrate 328
Subsurface feeders 219
Succinate 203
Sucrose 53, 55
Sugarcane weevil 234
Suillus 263, 267
Superimposed moulting 170
Supplementary organs 199
Suppressive soil 328
Surface feed 218, 219

Susceptibility 328
Susceptible 328
Suspended animation 209, 328
SVN 328
SYBR Gold 62, 63
Sychnotylenchidae 131, 132
Sychnotylenchinae 132
Symbiosis 329
Symplesiomorphy and synapomorphy 329
Symplocostomatidae 102
Symptoms 10, 329
Syncytia 232, 329
Syncytium 231
Synergism 329
Synergistic effects on plants 276
Synergistic interaction 251, 329
Syngamidae 118
Syngonic 329
Syngonic hermaphroditism 171
Synonyms 329
Synthesis of sterols 206
Syntype, lectotype, topotype and neotype 329
Syrphonematidae 115
Systema Naturae 88
Systematics 87
Systemic changes 253
Systemic pesticide 329
Syzygium cumini 270

T

T 329
T4 DNA ligase 59
TA cloning kit 57
Tagetes erecta 241
Tail 329
Tanzaniinae 127
Taq polymerase 55, 329
Tautonym 329
Taxon (taxa plural) 330

Taxonomy and systematics 330
TBE 60
TCA cycle 201, 204, 205
TE buffer 59
Telamon 330
Telescopic movement 212
Telogonic 171, 175, 330
Telotylenchidae 141-143
Telotylenchinae 141, 144
TEM 44, 45
TEMED 63, 66
Temporary mounts 38
Terminal galls 218, 225, 226
Terminator gene and traitor gene 330
Terrenoplica 89, 103
Testis 175, 330
Tetradonematidae 109
Tetrameridae 122
Tetramethylethylene diamine 65
Tetrazolium 54
Thadinae 127
The New Systematics 89
Thelastomatidae 116
Thelaziidae 122
Thermal 248
Thermal capacity 248, 330
Thermal conductivity 248, 330
Thermobiosis 210
Thermocyclers 56, 57
Thermolabile membrane 75
Thionemone 244
Thoracostomopsidae 102
Thread-like 70
Timber nematode 234
Tissue 330
Titillae 178, 330
Tobacco rattle virus 216, 259
Tobra viruses 216, 258
Tolerance 330
Tomato black ring virus 216, 258

Toxocaridae 119
TPE 60
Trachypleurosidae 107
Transamination 330
Transcription 330
Transdeamination 330
Transgenic plants 331
Transilluminator 58
Translation 331
Translocation 331
Transmission 331
Transmission electron microscope (TEM) 44-46
Transovarial transmission 331
Transpiration 235
Transverse lines 79
Transverse striations 78
Trap crop 331
Trehalose 201, 209
Tricarboxylic acid 204
Trichinellidae 109
Trichocephalida 89
Trichocephaloidea 109
Trichoderma 244
Trichodoridae 107, 108, 216, 217, 225, 258
Trichodorus 214-216, 225, 258, 259
Trichodorus patula 241
Trichodorus similis 259
Trichodorus viruliferus 259
Trichostrongylidae 96, 118
Trichostrongyloidea 96, 118
Trichosyringida 89
Trichosyringidae 109
Trichotylenchus 182
Trichuridae 93, 109
Triethanol amine 35
Trifluoro acetate 56
Triglycerides 205
Trilobed alae 86

359

Tripius 153, 157
Triploblastic 70, 331
Tripylida 91, 102
Tripylidae 92, 103
Tripylina 102, 113
Tripyloididae 94, 113
Triradiate 182, 331
Tris 53, 55, 56
Tris HCl 55
Tris-acetate 60
Tris-borate 60
Tris-phosphate 60
Triton X-100 53, 55
tRNA (transfer RNA) 331
Trophosome 109
Trophotylenchus 226
Trophurus 141, 172
Tubular type 188
Tumiota 145, 149
Tumour 331
Tungsten needle 47
Tylenchata 125, 126
Tylenchid oesophagus 183, 184
Tylenchida 124, 184, 216, 218, 223, 232
Tylenchidae 126, 127, 181, 198
Tylenchina 89, 125
Tylenchinae 127
Tylenchocriconema 146, 149, 152
Tylenchocriconematinae 152
Tylenchocriconematoidea 149
Tylenchoidea 98, 126
Tylenchorhynchidae 143
Tylenchorhynchinae 186, 197
Tylenchorhynchus 178, 215, 218, 268
Tylenchorhynchus brassicae 241
Tylenchorhynchus claytoni 274
Tylenchorhynchus gladiolatus 266
Tylenchorhynchus martini 246, 274
Tylenchorhynchus vulgaris 268
Tylenchulidae 99, 146, 150

Tylenchulidoidea 99, 149
Tylenchulinae 151
Tylenchuloidea 99, 145, 146, 149
Tylenchulus 145, 146, 149, 210
Tylenchulus semipenetrans 222, 226, 227
Tylenchus 198, 213, 215, 218
Tylodoridae 98, 126, 129
Tylodorinae 129
Type series 331
Types of agaroses 60
Tyrophagus similis 271

U
Uncinariidae 96, 118
Underbridge 331
Underpopulation 239
Undulatory movement 211, 212
Ungellidae 97, 122
Unsegmented 1, 70
Uranyl acetate 48
Uterus 172-174, 331

V
V 331
v´ 331
Vagina 174, 331
Valinomycin 245, 255
Valvular apparatus 183, 184
Variability 331
Variation in stylet 182
Varotylus 173
Vas deferens 175, 321
Vector 332
Velum 177, 178, 331
Ventricular region 185
Vermiform 332
Verticillium 252, 256
Verticillium chlamydosporium 255
Verticillium dahliae 253
Verutinae 137

Vesicles 175, 264
Vestibule 195, 332
Vide 332
Vinyl polymerization 65
Virulence 332
Viruliferous 332
Viviparous 174, 332
viz 332
VL/VB 332
Voucher specimens 332
Vulva 332
Vulval bridge 332
Vulval cone 41, 322

W
Wax-ring method 39
Weighting 332
Well 62, 63
White bowl 24, 25
White tip 233
White trap technique 31
Wilt of alfalfa 252

X
Xenorhabdus 271, 272
Xenorhabdus nematophila 271, 272
Xiphinema 216, 218, 225, 258, 259
Xiphinema americanum 274
Xiphinema diversicaudatum 218, 258
Xiphinema index 216, 218, 250, 258, 259
Xiphinematinae 106, 107
Xyalinidae 112
Xylene cyanol 63, 67

Y
Yellow ear-rot 252
Yolk 173

Z
Z-organ 332
Zygomycetes 263, 264
Zygote 332

α- Keto-glutarate dehydrogenase 204
α-diethienyl 244
α-terthienyl 244
β-glucosidase 215, 220, 226, 227

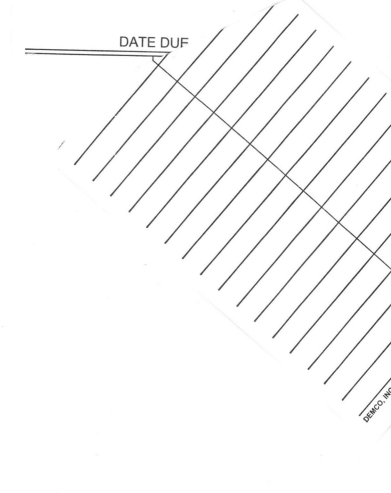